NANOTECHNOLOGY-DRIVEN ENGINEERED MATERIALS

New Insights

NANOTECHNOLOGY-DRIVEN ENGINEERED MATERIALS

New Insights

Edited by

Sabu Thomas, PhD
Yves Grohens, PhD
Nandakumar Kalarikkal, PhD
Oluwatobi Samuel Oluwafemi, PhD
Praveen K. M.

AAP | APPLE ACADEMIC PRESS

Apple Academic Press Inc.
3333 Mistwell Crescent
Oakville, ON L6L 0A2 Canada

Apple Academic Press Inc.
9 Spinnaker Way
Waretown, NJ 08758 USA

Exclusive worldwide distribution by CRC Press, a member of Taylor & Francis Group
No claim to original U.S. Government works
Printed in the United States of America on acid-free paper

International Standard Book Number-13: 978-1-77188-634-5 (Hardcover)
International Standard Book Number-13: 978-1-315-10260-3 (eBook)

Library and Archives Canada Cataloguing in Publication

Nanotechnology-driven engineered materials / edited by Sabu Thomas, PhD, Yves Grohens, PhD, Nandakumar Kalarikkal, PhD, Oluwatobi Samuel Oluwafemi, PhD, Praveen K.M.

Includes bibliographical references and index.
Issued in print and electronic formats.
ISBN 978-1-77188-634-5 (hardcover).--ISBN 978-1-315-10260-3 (PDF)

1. Nanostructured materials. 2. Nanoparticles. 3. Nanocomposites (Materials). 4. Biomedical materials. 5. Nanotechnology. I. Thomas, Sabu, editor II. Grohens, Yves, editor III. Oluwafemi, Oluwatobi Samuel, editor IV. Kalarikkal, Nandakumar, editor V. K. M., Praveen, editor

TA418.9.N35N42 2018 620.1'15 C2018-902030-X C2018-902031-8

Library of Congress Cataloging-in-Publication Data

Names: Thomas, Sabu, editor.

Title: Nanotechnology-driven engineered materials : new insights / Sabu Thomas, PhD [and four others].

Description: First edition. | Ontario ; New Jersey : Apple Academic Press, 2018. | Includes bibliographical references and index.

Identifiers: LCCN 2018018031 (print) | LCCN 2018019345 (ebook) | ISBN 9781315102603 (ebook) | ISBN 9781771886345 (hardcover : alk. paper)

Subjects: LCSH: Nanostructured materials.

Classification: LCC TA418.9.N35 (ebook) | LCC TA418.9.N35 N377 2018 (print) | DDC 620.1/15--dc23

LC record available at https://lccn.loc.gov/2018018031

Apple Academic Press also publishes its books in a variety of electronic formats. Some content that appears in print may not be available in electronic format. For information about Apple Academic Press products, visit our website at **www.appleacademicpress.com** and the CRC Press website at **www.crcpress.com**

ABOUT THE EDITORS

Sabu Thomas, DSc, PhD, CChem, FRSC

Sabu Thomas, PhD, is the Pro-Vice Chancellor of Mahatma Gandhi University, Kottayam, Kerala, India. He is also Founding Director of the International and Inter University Centre for Nanoscience and Nanotechnology and Professor of Polymer Science and Engineering at the School of Chemical Sciences at the same university. Professor Thomas has received a number of national and international awards, including a Fellowship of the Royal Society of Chemistry; Distinguished Professorship from the Josef Stefan Institute, Slovenia; an MRSI medal; a CRSI medal; and the Sukumar Maithy award. He is on the list of most productive researchers in India, holding the fifth position. University of Lorraine, France and Université Bretagne Sud, France have awarded honoris causa (especially of a degree awarded without examination as a mark of esteem) to him. Professor Thomas has published over 600 peer-reviewed research papers, reviews, and book chapters. He has delivered over 300 plenary, inaugural and invited lectures at national/international meetings across 30 countries. He has established a state-of-the-art laboratory at Mahatma Gandhi University in the area of polymer science and engineering and nanoscience and nanotechnology through external funding from various organizations and institutes, including DST, CSIR, UGC, DBT, DRDO, AICTE, ISRO, TWAS, KSCSTE, BRNS, UGC-DAE, Du Pont, USA, General Cables, USA, Surface Treat, Czech Republic, and Apollo Tyres. The *h*-index of Professor Thomas is 87, and his work has been cited more than 23,850 times. He also has four patents to his credit.

Yves Grohens, PhD

Yves Grohens, PhD, is Director of the ComposiTIC Technical Platform and Head of the composites research group of the IRDL-CNRS laboratory (Material Engineering) at the Université de Bretagne Sud, France. His master's and PhD degrees were from Besançon University, France. After finishing his studies, he worked as Assistant Professor and later as Professor in various reputed universities in France. He has been an invited professor to many universities in different parts of the world as well. His areas of interest include physicochemical studies of polymer surfaces and interfaces, phase transitions in thin films confinement, nano- and bio composites design and characterization, and biodegradation of polymers and biomaterials. He has written several book chapters, monographs, and scientific reviews and has published 230 international papers. He is chairman and member of advisory committees of many international conferences.

Nandakumar Kalarikkal, PhD

Nandakumar Kalarikkal, PhD, is Associate Professor of Physics at the School of Pure and Applied Physics, as well as Joint Director of the Centre for Nanoscience and Nanotechnology, Mahatma Gandhi University, India. He has published more than 80 research articles in peer-reviewed journals and has co-edited several books. Dr. Kalarikkal's research group focuses on the specialized areas of nanomultiferroics, nanosemiconductors, nanophosphors, nanocomposites, nanoferroelectrics, nanoferrites, nanomedicine, nanosensors, ion beam radiation effects and phase transitions. The research group has extensive exchange programs with different industries and various research and academic institutions all over the world and is performing world-class collaborative research in various fields. The Dr. Nandakumar Kalarikkal Centre is equipped with various sophisticated instruments and has established state-of-art experimental facilities that cater to the needs of researchers within the country and abroad.

Oluwatobi Samuel Oluwafemi, PhD

Oluwatobi Samuel Oluwafemi, PhD is Professor and a National Research Foundation (NRF), South Africa-rated researcher at the Department of Applied Chemistry, University of Johannesburg, South Africa. His research is in the broad area of nanotechnology and includes green synthesis of semiconductor and metal nanomaterials for different applications which include, but are not limited to, biological (imaging, labeling, and therapeutic), optical, environmental, and water treatment. He has authored and co-authored many journal publications, book chapters, and books. He is a reviewer for many international journals in the field of nanotechnology and has won many accolades both local and international.

Praveen K. M.

Praveen K. M. is Assistant Professor of Mechanical Engineering at Saintgits College of Engineering, India. He is currently pursuing a PhD in Engineering Sciences at the Université de Bretagne Sud – Laboratory IRDL PTR1, Research Center "Christiaan Huygens," Lorient, France, in the area of coir-based polypropylene micro composites and nanocomposites. He has published an international article in Applied Surface Science and has presented posters and papers at national and international conferences. He also has worked with the Jozef Stefan Institute, Ljubljana, Slovenia; Mahatma Gandhi University, India; and the Technical University, Liberec, Czech Republic. His current research interests include plasma modification of polymers, polymer composites for neutron-shielding applications, and nanocellulose.

CONTENTS

LIST OF CONTRIBUTORS

K. Anver Basha
P.G. & Research Department of Chemistry, C. Abdul Hakeem College, Melvisharam 632509, Vellore District, Tamil Nadu, India. E-mail: kanverbasha@gmail.com

Ayşe Çelik Bedeloğlu
Department of Fiber and Polymer Engineering, Bursa Technical University, Bursa 16310, Turkey. E-mail: ayse.bedeloglu@btu.edu.tr

Rajashekhar Bhajantri
Department of Physics, Karnatak University, Dharwad, Karnataka, India

Sukhwinder K. Bhullar
Department of Mechanical Engineering, Bursa Technical University, Bursa, Turkey
Department of Mechanical Engineering, University of Victoria, Victoria, BC, Canada. E-mail: sbhullar@uvic.ca; morhan@uludag.edu.tr

P. K. Brahmankar
Department of Mechanical Engineering, Dr. Babasaheb Ambedkar Technological University, Lonere 402103, Raigad, Maharashtra, India

A. S. Ruby Celsia
Department of Biotechnology, Mepco Schlenk Engineering College, Sivakasi 626005, Tamil Nadu, India

R. Narayana Charyulu
Department of Pharmaceutics, NGSMIPS, Mangalore, Karnataka, India

Rinul M. Dhajekar
Department of Mechanical Engineering, Dr. Babasaheb Ambedkar Technological University, Lonere 402103, Raigad, Maharashtra, India

Krishna Dutt
Bhaskaracharya College of Applied Sciences, University of Delhi, Delhi 110075, India

Nishant Jain
Bhaskaracharya College of Applied Sciences, University of Delhi, Delhi 110075, India

Sanjay Jain
Department of Pharmacognosy, Indore Institute of Pharmacy, Indore, India

Mahantappa S. Jogad
Department of Physics, School of Physical Sciences, Central University, Karnataka, Kadaganchi, Aland Road, Kalaburagi 585367, India
Department of Physics, Karnataka State Women's University, Vijayapura, India. E-mail: jogad1952@rediffmail.com

Rashmi M. Jogad
Department of Physics, Karnataka State Women's University, Vijayapur, India

Bhagwan F. Jogi
Department of Mechanical Engineering, Dr. Babasaheb Ambedkar Technological University, Lonere 402103, Raigad, Maharashtra, India.
E-mail: bfjogi@dbatu.ac.in; bfjogi@gmail.com

M. B. G. Jun
Department of Mechanical Engineering, University of Victoria, Victoria, BC, Canada

Ayan Khan
Department of Physics, School of Engineering and Applied Sciences, Bennett University, Greater Noida 201310, India. E-mail: ayan.khan@bennett.edu.in

Poonam Khullar
Department of Chemistry, BBK DAV College for Women, Amritsar 143005, Punjab, India

G. P. Kothiyal
Glass and Advanced Ceramics Division, Bhabha Atomic Research Center, Mumbai 400085 (Superannuated), India

P. S. R. Krishna
Solid State Physics Division, Bhabha Atomic Research Center, Mumbai 400085, India

Madan Kulkarni
Department of Mechanical Engineering, Dr. Babasaheb Ambedkar Technological University, Lonere 402103, Raigad, Maharashtra, India

V. Madhurima
Department of Physics, School of Basic and Applied Sciences, Central University of Tamil Nadu, Thiruvarur 610005, India

R. Mala
Department of Biotechnology, Mepco Schlenk Engineering College, Sivakasi 626005, Tamil Nadu, India

Ishwar Naik
Government Arts and Science College, Karwar, Karnataka, India. E-mail: iknaik@rediffmail.com

Jagadish Naik
Department of Physics, Mangalore University, Mangalore, Karnataka, India

Ratyakshi Nain
Bhaskaracharya College of Applied Sciences, University of Delhi, Delhi 110075, India

K. Nilavarasi
Department of Physics, School of Basic and Applied Sciences, Central University of Tamil Nadu, Thiruvarur 610005, India

Mehmet Orhan
Department of Textile Engineering, Uludag University, Bursa 16059, Turkey

Balaram Pani
Bhaskaracharya College of Applied Sciences, University of Delhi, Delhi 110075, India

Kamlesh Panwar
Department of Textile Technology, Indian Institute of Technology, Hauz Khas, New Delhi 110016, India

D. Ratna
Naval Materials Research Laboratory (DRDO), Ambernath, Mumbai, Maharashtra, India

S. Mohammed Safiullah
P.G. & Research Department of Chemistry, C. Abdul Hakeem College, Melvisharam 632509, Vellore District, Tamil Nadu, India

Gökçenur Sağlam
Department of Fiber and Polymer Engineering, Bursa Technical University, Bursa 16310, Turkey

Gajendra Saini
Advanced Instrumentation Research Facility, Jawaharlal Nehru University, New Delhi 110067, Delhi, India

Ravinder Singh
Bhaskaracharya College of Applied Sciences, University of Delhi, Delhi 110075, India

Sidhharth Sirohi
Bhaskaracharya College of Applied Sciences, University of Delhi, Delhi 110075, India.
E-mail: siddharth.sirohi@)bcas.du.ac.in

K. Srinivas
Department of Physics, Gitam University, Bengaluru, Karnataka, India. E-mail: Srinkura@gmail.com

Sudhakar C. K.
Department of Pharmaceutics, School of Pharmaceutical Sciences,
Lovely Professional University, Jalandhar-Delhi G. T. Road, Phagwara 144411, Punjab, India
"Smriti College of Pharmaceutical Education, Indore, India"

B. Tanatar
Department of Physics, Bilkent University, Ankara 60032, Turkey

Lavnaya Tandon
Department of Chemistry, BBK DAV College for Women, Amritsar 143005, Punjab, India

Nitish Upadhyay
Smriti College of Pharmaceutical Education, Indore, India

Piyush Wadhwa
Bhaskaracharya College of Applied Sciences, University of Delhi, Delhi 110075, India

K. Abdul Wasi
P.G. & Research Department of Chemistry, C. Abdul Hakeem College, Melvisharam 632509, Vellore District, Tamil Nadu, India

LIST OF ABBREVIATIONS

AFM	atomic force microscopy
AIDS	acquired immunodeficiency syndrome
APETS	amino propyltrimethoxysilane
APU	auxetic polyurethane
BARC	Bhabha Atomic Research Center
BHL	butanoyl homoserine lactone
BPO	benzoyl peroxide
CAUTI	catheter-associated urinary tract infection
CB	chlorobenzene
CCA	constant contact angle
CCNR	current controlled negative resistance
CCR	constant contact radius
CCT	calcium copper titanate
CNTs	carbon nanotubes
Cp	cloud point
CTAB	cetyltrimethyl ammonium bromide
CTAC	cetyl trimethyl ammonium chloride
CTBN	carboxyl group terminated butadiene nitrile
CVD	chemical vapor deposition
DLS	dynamic light scattering method
DMA	dynamic mechanical analysis
DMF	N-dimethylformamide
DMTA	dynamic mechanical thermal analysis
DNA	deoxyribonucleic acid
DOPE	di-oleoyl phosphatidyl ethanolamine
DRP	dense random packed
DSC	differential scanning calorimetry
DSHP	disodium hydrogen phosphate
DTA	differential thermal analyzer
ECP	electron channeling pattern
ED	electron diffraction
EDS	energy dispersion analysis
EG	ethylene glycol

ESFs	embryonic skin fibroblasts
FESEM	field emission scanning electron microscope
FFLO	Fulde–Ferrell–Larkin–Ovchinniov
FTIR	Fourier-transform infrared
GFA	glass formation ability
GMA	glycidyl methacrylate
HNAA	half neutralized adipic acid
HOMO	highest occupied molecular orbit
HPMs	hybrid polymeric micelles
IAEC	Institutional Animal Ethics Committee
INS	inelastic neutron scattering
IRO	intermediate range order
IXS	inelastic X-ray scattering
LMW	low molecular weight
LRO	long-range order
LUMO	lowest unoccupied molecular orbit
MCGR	magnetically controlled growing rod
Mw	molecular weight
MWNT	multiwalled carbon nanotubes
Na-AHA	sodium salt of hexanoic acid
ND	neutron diffraction
NIR	near infrared
NMR	nuclear magnetic resonance
NPs	nanoparticles
NRTI	nucleoside analog reverse transcriptase inhibitor
PBS	phosphate buffer sulfate
PCBM	butyric acid methyl ester
PCM	phase-change memories
PDMS	polydimethyl siloxane
PEO	polyethylene oxide
PET	poly(ethylene terephthalate)
PGMA	poly(glycidyl methacrylate)
Ph	potential of hydrogen
PLA	poly(l-lactic acid)
PNC	polymer nanocomposites
PPO	poly propylene oxide
PU	polyurethane
QA	quaternary ammonium

RCP	random close packing
RH	relative humidity
ROS	reactive oxygen species
SACP	selected area channeling pattern
SANS	small-angle neutron scattering
SBS	specific bond strength
SC	stratum corneum
SDS	sodium dodecyl sulfate
SEM	scanning electron microscopy
SERS	surface-enhanced Raman spectroscopy
SPR	surface plasmon resonance
SRO	short-range order
SS	stainless steel
TBP	triblock polymers
TEM	transmission electron microscopy
TGA	thermogravimetric analysis
TMA	thermomechanical analyzer
TMAH	tetramethylammonium hydroxide
TSB	triptychs soy broth
UTI	urinary tract infection
UTM	universal testing machine
UV–Vis	ultraviolet–visible
WAXD	wide-angle X-ray diffraction
XRD	X-ray diffraction
ZEH	zinc 2-ethylhexanoate
ZnO	zinc oxide

PREFACE

This book is an outcome of contributions from the scientific fraternity in India and abroad. Nanostructured materials are emerging as a new class of materials that exhibit unique microstructures and enhanced mechanical performance under the light of nanoscience and nanotechnology. As an outcome of this, these materials have attracted considerable attention from the scientific community all over the world. Nanoscience and nanotechnology have now emerged as a specific stream taught in universities and practiced in industries for the attainment of new products with specific features and to facilitate product development, thereby improving the product quality and reliability. We hope that our new book, *Nanotechnology-Driven Engineered Materials: New Insights*, will surely be an asset for scientists, engineers, and budding researchers working in the area of nanoscience and nanotechnology.

This book contains 13 chapters. The 13 chapters are categorized under three major streams, which are "Nanoparticles Assembly and Nanostructured Materials," "Nanocomposites Properties," and "Nanostructured Materials for Biomedical Applications." Under the stream "Nanoparticles Assembly and Nanostructured Materials," six chapters are included.

Chapter 1 presents an overview of the various processes involved in the self-assembly of liquid droplets on substrates, with special reference to constrained surfaces. The self-assembling process leads to interesting physics such as depinning of the three-phase contact line. The patterns thus formed deviate from an ideal honeycomb structure and yield interesting patterns. Use of polymeric substrates enhances the pattern formation and also keeps the experimental simplicity of self-assembly. The applications of self-assembled structures are discussed toward the end of this review.

Chapter 2 presents the synthesis techniques and affecting parameters in silver nanowires synthesization. In addition, biomedical applications of silver nanowires are highlighted. Nanostructures such as rods, wires, belts, ribbons, and tubes have been intensively investigated for years due to their unique features, including small sizes and high surface-to-volume ratios. Among these, nanowires, in particular, have been the focus of many recent studies in a variety of different disciplines, such as biotechnology, nanobiotechnology, and bioengineering. Since bulk silver has the highest electrical and thermal conductivity and low-cost among all noble metals, therefore, by combining optical transparency and flexibility, silver nanowire (AgNWs) networks are used in the development of electronic and optoelectronic devices.

Chapter 3 deals with fabrication of metal nanoparticles on poly(glycidyl methacrylate). It also includes discussions on the synthesis, characterization, and catalytic applications. To produce polyglycidyl methacrylate (PGMA) core/metal nanoparticle (NP) shell nanocomposite, the authors have adopted a deposition technique that involves two simple steps: (1) The synthesis of PGMA beads by suspension polymerization followed by (2) direct deposition of metal NPs on activated PGMA beads. The PGMA beads were used as a soft template to host NPs without surface modification. Two different metal NPs were employed to know the viability and variability of the designed method.

Chapter 4 discusses the structural (neutron diffraction) and physical property studies of sb2se3-(cui) chalcohalide glasses and nanocrystals. Halide and chalcogenide glasses have received a great deal of interest as potential candidates for infrared transmitting materials. This study shows its potential as glass for infrared application. It may be mentioned that glasses containing copper (Cu) have better resistance to cracks propagation and a higher hardness but are less stable against crystallization.

Chapter 5 demonstrates the promising applications in materials chemistry in terms of synthesis, characterization, and applications of nanomaterials. Its micellar form is highly effective in the synthesis of gold (Au) NPs by using the surface cavities in the form of nanoreactors. The core–shell configuration of such micelles helps in loading and unloading of the NPs simply by altering the solvents' nature. A change in the hydrophilicity and hydrophobicity parameters induce change in the shape and size of the micelles and, bold hence, control the overall shape and size of the NPs. Both template and seed growth methods have been employed. pH plays an important role in the synthesis of gold nanoparticles. Thus, applications of block polymers in materials chemistry open up a new direction for materials synthesis in different directions. The block copolymer micelles are used as drug delivery carriers for the various hydrophobic anticancer drugs like paclitaxel (PTX), doxorubicin, etc. This chapter proposed that an appropriate choice of block copolymer can achieve a desired synthesis of the nanomaterials and, hence, explores a new direction in materials synthesis, characterization, and applications.

Chapter 6 presents electron–hole bilayer systems in semicondutors from a theoretical perspective. It is well known that the importance of excitonic research lies in its multifaceted possible applicability. The current interests of different scientific and engineering endeavors are mostly getting converged to the field of energy and communication. Already there are different techniques of solar cells and microchips that have made substantial enrichments in these fields, but it is still far beyond the goal. In the domain of information technology and computation, the idea of quantum computers and simulators are already in place. From a theoretical point of view, quantum information processing can be considered as a well-established field by now, but the key issue of the design and realization of concrete solid-state implementation protocols are subject of intense investigation at the moment.

Four chapters are included in the section on "Nanocomposites Properties." Chapter 7 demonstrates polymer nanocomposites (PNC), presenting a case study of rubber toughened epoxy/CTBN matrix in the presence of clay and carbon nanotubes as nanofillers. This research study concludes that the filler, multiwalled carbon nanotubes (MWCNT), has a better effect on the mechanical properties of the matrix materials. However, nanofillers are found to be best suitable if they are uniformly dispersed in the matrix material. According to the application point of view, the clay is a better filler for improvement of mechanical properties, as it is cheaper and more effective. However, MWCNT is also suitable for improving mechanical properties in the manufacturing of multifunctional PNCs. According to the desired applications, the suitable filler may be selected for PNC.

Chapter 8 focuses on the feasibility of the development of a lead-free nanowire-reinforced polymer matrix capacitor for energy storage applications. There is an increasing demand to improve the energy density of dielectric capacitors for satisfying the next-generation material systems. One effective approach is to embed high dielectric constant inclusions such as lead zirconia titanate in polymer matrix. However, with the increasing concerns on environmental safety and biocompatibility, the need to expel lead (Pb) from modern electronics has been receiving more attention. Using high aspect ratio dielectric inclusions such as nanowires could lead to further enhancement of energy density. This chapter also emphasizes the critical role of the interfacial region and presents hypotheses for multi-scale phenomena operating in PNCs dielectrics.

Chapter 9 reports on the antibacterial activity of smart (auxetic) polyurethane foams. Polymeric biomaterials in their woven and non-woven form at macro to nano scale offer huge potentials for external and internal use in biomedical and other fields. Applications of polymers, particularly in the form of foam such as polyurethane (PU) foams are versatile. They are used in a variety of biomedical and other applications including wound care, tissue engineering, implants, prostheses, surgical masks, hospital bedding, packaging, sound insulation, air filtration, shock absorption, and as sponge materials. In this study, smart PU foams, called auxetic PU (APU) foams, are fabricated and characterized for antibacterial activities. APU foam samples are developed based on compression, heating, cooling, and relaxation method from the open literature. Antimicrobial agent's chitosan and silver are incorporated in bulk in fabricated foam samples to test antibacterial activities against Gram-negative and Gram-positive bacteria qualitatively (agar diffusion test) and quantitatively (suspension test). It is mentioned in the concluding remarks that they may offer more potential over the conventional PU foam as smart bandages impregnated with medicine for wound care, tissue engineering, smart filters, and as smart sponge materials in many other applications.

Chapter 10 discusses optically tuned MDMO-PPV/PCBM blend for plastic solar cell. It is known that in the constructional hierarchy, single layer, bilayer, bulk heterojunction, tandem cells, and plasmon enhanced cells follow in steps in an attempt to achieve more efficiency. The main difficulty with plastic solar cells is the electron–hole recombination before the exciton migration to the P–N junction interface. The problem is overcome by introducing the concept of bulk heterojunction blend of donor–acceptor pair. This work is focused to optimize the photoactive blend of poly[2-methoxy-5-(3′,7′-dimethyloctyloxy)-1,4-phenylenevinylene](MDMO-PPV) and [6,6]-phenyl C61 butyric acid methyl ester (PCBM) for maximum absorption of the solar energy.

The section "Nanostructured Materials for Biomedical Applications" consists of three chapters. Chapter 11 reports, the synthesis of well-dispersed ZnO NPs for nanofinishing textiles and biomedical applications. ZnO NPs find tremendous applications in the area of electronics, optoelectronics, bioengineering, catalysis, biosensors, and nanofinishing textiles. Various methods of synthesis are described in the literature, such as hydrothermal, chemical, solvothermal, sol–gel, and electrochemical synthesis. The aggregation tendency of NPs restricts the complete use of their potential. Earlier, the use of surfactants and capping agents has been reported for improving the dispersion behavior of NPs. However, the results are not promising and need further improvement. In this work, the synthesis of ZnO NPs is carried out and the effect of surfactant has been investigated. These well-dispersed NPs may be used as a potential antimicrobial agent and nanofinishing textiles.

Chapter 12 presents crystalline biofilm formation on Foley catheter—a macro problem with microbes and the nano weapons to control blockage. Catheter-associated urinary tract infections are a major hospital-acquired infection. Millions of people suffer from this disease annually. *Proteus mirabilis* is a dimorphic organism that is predominantly associated with crystalline biofilm formation and encrustation in urinary catheters. This review addresses the orchestral performance of *Proteus mirabilis* with various virulence factors in the process of encrustation. The influence of pH on the shape and planes of struvite crystals were elaborated. It explores how silver and other nano weapons are exploited to win over the micro enemy by targeting the different stages of crystalline biofilm formation. The mechanism of silver nanomaterial in reducing the biofilm and its limitations are also discussed. The advantages of a functionalizing Foley catheter with an array of bioactive compounds layer-by-layer assembly were emphasized in a nutshell.

Chapter 13 discusses nanostructure deformable elastic vesicles as nanocarriers for the delivery of drugs into the skin. Dressing of active pharmaceutical ingredient (API) should be in such a manner that it looks simple and in a more beneficial way. Phospholipids (phosphatidylcholine) with accessories like ethanol and water are dressing materials for many drugs that overcome many problems related to poor bioavailability, poor solubility, etc. A plethora of nanomedicine has been highlighted for various purposes. Human immunodeficiency virus (HIV) is not curable, but it can be controlled by nanomedicine or drug delivery systems, which boost the immune system of the body. Ethosomes (deformable vesicles) are the perfect dress for antiretroviral drugs and it is nano drug delivery system, which will enhance the permeation of drugs through the skin and is able to heighten the immune system of body during HIV infection. Different formulations of ethosomes are prepared and characterized for size, zeta potential, and entrapment efficacy, in vitro and in vivo studies.

Result reveals that ratio of phospholipids and ethanol should be optimal to achieve superior entrapment, permeation of model drug into skin, and to have sustained release pattern.

Nanoscience and nanotechnology has emerged as a hot topic from the postgraduate level to the levels of graduation and diversified industrial courses in polytechnics and university colleges. Keeping the young researchers in mind, this book is edited to highlight the latest innovations and principles behind these findings, specific to nanostructured polymeric materials and polymer nanocomposites. This book is devoted to novel architectures at the nanolevel with an emphasis on new synthesis and characterization methods. Special emphasis is given to new applications of nanostructures and nanocomposites in various fields such as nanoelectronics, energy conversion, catalysis, drug delivery, and nanomedicine. Chapter-wise bibliographies have been introduced for further research in these topics.

As authors, we express our sincere gratitude to the creative environments of the International and Inter University Centre for Nanoscience and Nanotechnology (IIUCNN), Mahatma Gandhi University, Kerala, India; University of South Brittany (Université de Bretagne Sud), Lorient, France; and University of Johannesburg, South Africa, and to our colleagues and family for their encouragement and support.

PART I

Nanoparticles Assembly and Nanostructured Materials

CHAPTER 1

SELF-ASSEMBLY OF DROPLETS ON 1D AND 2D PATTERNED SURFACES: A REVIEW

V. MADHURIMA* and K. NILAVARASI

Department of Physics, School of Basic and Applied Sciences, Central University of Tamil Nadu, Thiruvarur 610101, India

**Corresponding author. E-mail: madhurima@cutn.ac.in*

CONTENTS

ABSTRACT

This chapter presents an overview of the various process involved in the self-assembly of liquid droplets on substrates, with special reference to constrained surfaces. The self-assembling process leads to interesting physics, such as depinning of the three-phase contact line. The patterns thus formed deviate from an ideal honeycomb structure and yield interesting patterns. Use of polymeric substrates enhances the pattern formation as well as keeps the experimental simplicity of self-assembly. The applications of self-assembled structures are discussed toward the end of this chapter.

1.1 INTRODUCTION

Controlling wetting of solid surfaces has a wide variety of applications, such as microfluidic biomedical devices, non-stick coatings, self-cleaning surfaces, microelectronics, information processing and storage, nano/microfluidic devices, bio-detection, etc.[1,2] Wetting of a solid by a liquid depends on the one hand on the chemical composition of the solid and liquid and on the other hand, on the roughness of the solid surface. Surfaces are patterned to control their roughness, and hence the wetting. Patterning of surfaces falls broadly into two categories—the top-down patterning and the bottom-up patterning.[3,4]

Top-down approach involves the direct writing of the pattern onto the substrate by applying suitable etching and deposition process;[5] the methods usually used include optical lithography,[6] electron beam lithography, etc. Top-down approaches to patterning suffer limitations including the inefficiency of patterning surfaces with very small structural sizes and aspect ratios.[7]

These limitations to the top-down approach are overcome by the concepts of self-assembly and self-organization which provide a way to pattern very small structures over large areas.[8] Such an approach where layers of the patterning materials are deposited on a surface is called the bottom-up approach.

The process of self-assembly is based on the local interactions of molecules or particles leading to them organizing into structured patterns, starting from a disorganized initial state. It refers to an autonomous organization of molecules into structures under defined boundary conditions.[9]

Liquid drops condensing on a cold surface also self-assemble into hexagonal patterns, called breath figures.[10] Evaporative cooling of the liquid (solvent) can be used to cool the system, instead of using a cold solid surface. The second method is useful in creating self-assembled droplets on polymeric surfaces. This method is well reviewed in Refs. [11–20]. A typical experiment to create breath figures on a polymeric surface will begin with coating a hard solid surface with a polymeric layer, usually through spin-coating. In a closed chamber, containing the substrate coated with polymer solution and flow of moist air, the evaporation of solvent leads to the cooling of the substrate. Water in the chamber starts to condense onto the substrate under conditions of high humidity. With the ongoing evaporation of the solvent, these water droplets can grow until the solution on the substrate becomes too viscous.[21] Influenced by surface convection and capillary attractive force, the droplets of condensed water self-assemble into a hexagonally ordered array at the air/solution interface.[22] The highly volatile solvent evaporates along with the water droplets, thus leaving behind a well-ordered honeycomb pattern of pits on the polymer surface.

Since evaporating liquid droplets are used to template the polymeric surface, there is no need to remove residual processing liquids and hence, this is a useful technique. This simple method also has the advantage of being low-cost while patterning large surface areas.[23]

Pitois and Francois discovered the formation of a highly ordered honeycomb film when a drop of star-shaped polystyrene/carbon disulfide solution was exposed to the flow of moist air.[18] This discovery, along with the work of Rayleigh,[24] demonstrated that these honeycomb films were the result of self-assembly of liquid drops on a solution surface. Following this, there were numerous studies on patterning surfaces by self-assembled condensing liquid drops. Various solid surfaces, such as microscopic glass slides,[25,26] mica,[27] and silicon wafers[28] were also used as substrates. Steyer et al. investigated the growth of breath figures on liquids.[16]

For many years the focus of self-assembled droplets based patterning was on changing polymers and solvents to obtain an ordered array of pores. However, in the recent years, the focus has shifted to obtaining hierarchical patterned structures and building ordered films with different properties, such as hydrophobicity, optoelectronic conversion, anti-reflection, high mechanical strength, etc.[11,14,23] Connal et al. were the pioneers of forming breath figures on non-planar substrates.[29]

The process of formation of drops on a solid surface is governed by conditions of energy equilibrium between the solid, liquid, and the ambient gas phases. The contact between the liquid drop and the solid surface is defined by the three-phase contact line, which is a circle in 2D under ideal conditions.[30] The self-assembled liquid droplets form ideal hexagonal patterns under ideal conditions. If the underlying surface is non-planar (or constrained in any other way) the contact line will undergo pinning, depinning, and movement.[30,31] The underlying constraint in the case of polymer patterning comes from the surface over which the polymer layer is coated.[32]

This review focuses on the formation of breath figure arrays on non-planar substrates, mainly the effect of boundary conditions on the self-assembly of droplets on the surface. Various applications of self-assembled droplet-based patterning are also reviewed. This review is structured into five sections: The first section gives the theoretical background of the formation of breath figures starting from nucleation of droplets and its self-assembly and evaporation. The second section explains briefly the effect of constraints on the formation of a droplet on the surface. The third section discusses various fabrication methods of breath figure formation. The development of breath figure arrays on non-planar substrates, their effect on pore size, arrangement, etc., are discussed in the fourth section and in the fifth section, various applications of self-assembled liquid droplet based patterned films (breath-figure films) is discussed.

1.2 THEORY

1.2.1 SELF-ASSEMBLED DROPLET PATTERN

Energetics demand that the structure of the self-assembled condensed droplets on any ideal surface be a honeycomb.[33,34] The energetics are determined by the interfaces between the solution and the substrate, the solution and water droplets, and the film surface and air thus playing an important role in determining structures of the patterns formed.[35] The condensation of any liquid (say water) on a surface can be caused by the difference in temperature between the atmosphere and the surface, the latter being cooler. Once the solvent evaporates completely, the temperature of the surface rises again to ambient temperature. At this stage, the water droplets evaporate, leaving behind an array of pores.

There are two main pathways for the formation of the self-assembled droplet patterns. One method involves the condensation of water vapor on a cold surface[36] due to nucleation and growth of droplets under partial wetting conditions.[37] This typically results in a poly-dispersion of droplet sizes.

Another method of formation of breath figures is by the evaporative cooling of the polymer solution. In this method evaporation of solvent lowers the temperature of the substrate, thus removing the need to cool the substrate separately. This process usually results in monodispersed distribution of droplet sizes.[38]

Although the general process of droplet condensation is understood completely, the mechanism of formation of self-assembled arrays of droplets is not fully understood. One mechanism is given.[37–40] According to them, evaporative cooling of the solvent reduces the substrate temperature by 0–6°C. The nucleated droplets self-assemble into a hexagonally ordered array without coalescing but the drops start to sink into the polymer solution. A second layer of droplets may be deposited on the top the first layer. After complete evaporation of the solvent and water, polymer is left with 2D/3D array of droplet imprints. According to the authors, non-coalescence is the main factor for the formation of honeycomb structure. Thermocapillary effect (temperature gradient induced motion of droplets of one liquid in another liquid, two liquids being immiscible) and the Marangoni convection (two drops of the same liquid do no coalesce if a sufficiently large temperature gradient exists[41]) are the main contributors for the non-coalescence of droplets until the droplets begin to sink into the polymer solution. Until then the drops levitate on the surface due to the above said two effects. According to Pitois and Francois,[18] there will not be any levitation of droplets on the surface instead they touch the surface and the intervening polymer layer keeps the droplets from coalescing. According to the mechanism proposed by Nishikawa et al.,[42] the droplets neither levitate nor touch the surface. They just submerge into the polymer solution due to combined effects of Marangoni convection and thermocapillary effects.

The formation of self-assembled droplet structures occurs in four stages. The initial stage is nucleation of droplets, the second stage is the growth and self-assembly of droplets, the third stage is the coalescence of droplets, and the final stage is the renucleation of droplets.

1.2.2 NUCLEATION OF DROPLETS

The change in Gibb's free energy during nucleation as a function of cluster size is shown in Figure 1.1. It is seen in the figure that the change in Gibb's free energy increases with increase in the size of the cluster and reaches a maximum at a size r_c. The size of the cluster corresponding to the maximum change in Gibb's free energy is called critical size and this maximum energy is called nucleation energy.[43] Nucleation is the process of formation of super-critical clusters with sizes $r > r_c$ capable of further growth.[44] The change in Gibb's free energy can be written as

$$\Delta G = G_2 - G_1 = -\frac{4}{3}r^3\Delta G_v + 4\pi r^2\gamma_{sl} \tag{1.1}$$

where ΔG_v is the volume free energy difference (in joules) and the second term refers to surface energy (mJ/m²). There are two different modes of formation of nuclei. They are heterogeneous nucleation and homogeneous nucleation.[45] Heterogeneous nucleation requires a preferential nucleation site which includes defects, impurities, or grain boundaries. The droplets are nucleated either on impurity sites or on the substrate directly.[46] Homogeneous nucleation is one which occurs spontaneously and randomly without any explicit nucleation site. The nucleated droplets are found above the solid surface in the chamber. The change is Gibb's free energy is different for homogeneous and heterogeneous nucleation.[46]

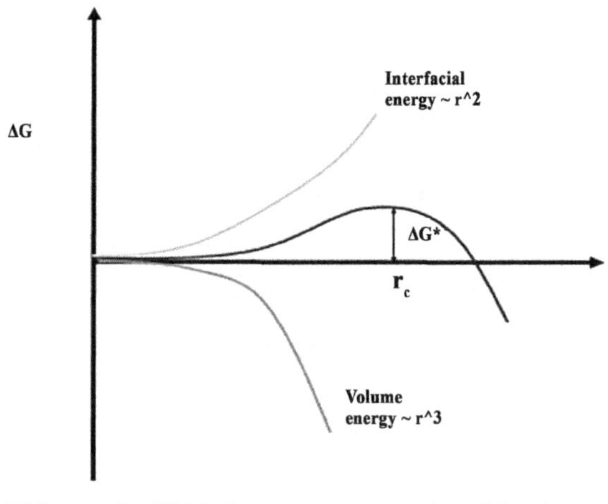

FIGURE 1.1 Difference in Gibb's free energy versus size of the cluster.

1.2.2.1 HOMOGENEOUS NUCLEATION

The change in Gibb's free energy for a homogeneous nucleation process is given by

$$\Delta G = G_2 - G_1 = -V_s\left(G_V^L - G_V^S\right) + A^{SL}\gamma^{SL} \tag{1.2}$$

where Vs is the volume of the cluster, A^{SL} is the solid–liquid interfacial area (mJ/m^2), γ^{SL} is the solid–liquid interfacial energy (mJ/m^2), and $\Delta_{GV} = G_V^L - G_V^S$ is the volume free energy difference (Joules). For spherical cluster of radius "r" the volume of the cluster and the solid–liquid interfacial area is given as

$$V_s = \frac{4}{3}\pi r^3 \text{ and } A^{SL} = 4\pi r^2, \tag{1.3}$$

$$\Delta G = -\frac{4}{3}r^3\Delta G_v + 4\pi r^2\gamma_{sl}. \tag{1.4}$$

1.2.2.2 HETEROGENEOUS NUCLEATION

For the heterogeneous nucleation, the change in Gibb's free energy is given as[46]

$$\Delta G_{hete} = -V_s\left(G_V^L - G_V^S\right) + A^{SL}\gamma^{SL} + A^{SP}\gamma^{SP} - A^{LP}\gamma^{LP}, \tag{1.5}$$

$$= \left(-\frac{4}{3}r^3\Delta G_v + 4\pi r^2\gamma_{sl}\right)S(\theta), \tag{1.6}$$

$$\Delta G_{hete} = \Delta G_{homo}\left(S(\theta)\right). \tag{1.7}$$

1.2.2.3 RATE OF NUCLEATION

In both the cases, Gibb's free energy increases for $r < r_c$ and decreases for $r > r_c$. The radius r_c is the critical nuclear size and this cluster is called critical nuclei. The energy at this radius is called the nucleation energy. Figure 1.2 shows comparison of the variation of difference in Gibb's free energy with the radius of the cluster for the two types of nucleation.[47] The rate of nucleation is given as

$$\frac{dN}{dt} = v_d \exp\left(\frac{\Delta G^*}{kT}\right) \text{nuclei} / \text{m}^3 / \text{s}, \tag{1.8}$$

where v_d is the frequency at which the atoms from vapor attach to the liquid nucleus and this rearrangement needed for forming a stable nuclei also follow the same temperature dependence. The rate of nucleation with respect to temperature for both types of nucleation is shown in Figure 1.3.

$$v_d = \exp\left(-\frac{Q_d}{kT}\right), \tag{1.9}$$

where Q_d is the diffusion co-efficient. Therefore, the rate of nucleation is given as

$$\frac{dN}{dt} \approx \exp\left(-\frac{Q_d}{kT}\right) \exp\left(\frac{\Delta G^*}{kT}\right). \tag{1.10}$$

Special cases: when $\Delta G^* > Q_d$, that is, if ΔG^* is very high, there is no possibility of nucleation to takes place.

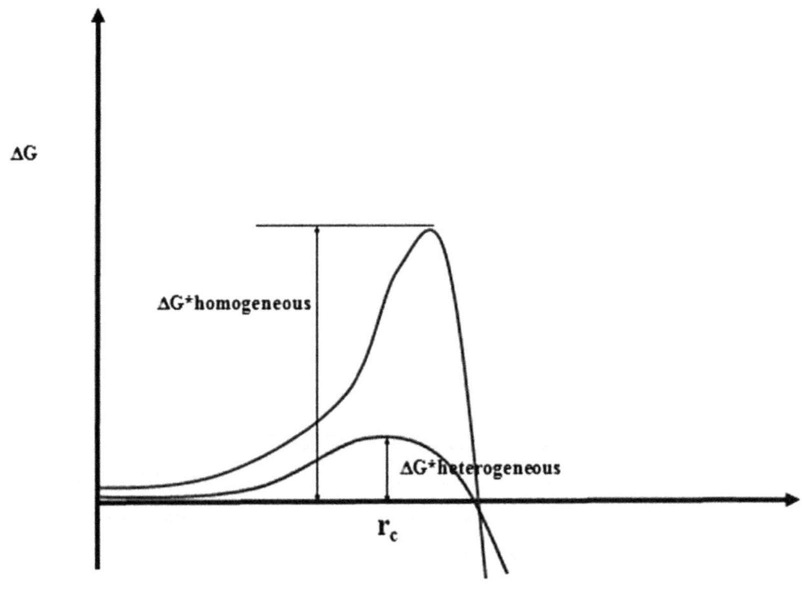

FIGURE 1.2 Comparison of difference in Gibb's free energy for homogeneous and heterogeneous nucleation.

If $\Delta G^* < Q_d$, there is a sharp rise in nucleation rate.

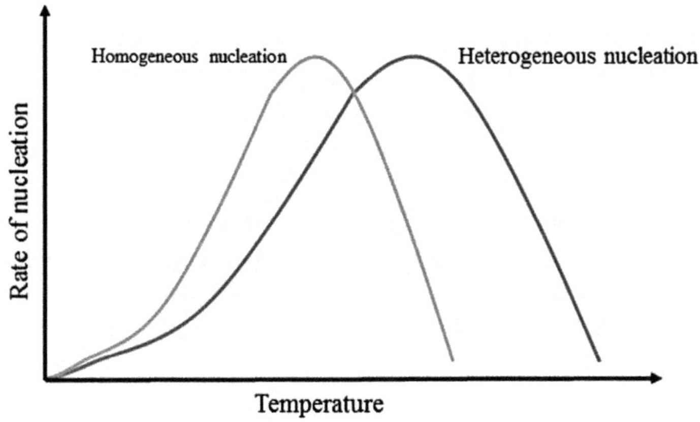

FIGURE 1.3 Rate of nucleation with respect to temperature.

$\Delta G^* = Q_d$, then the nucleation rate is very small. Figure 1.4 shows the variation of nucleation rate with respect to temperature for all the special cases. The rate of nucleation of the liquid phase and thus the pattern generation is governed by the wetting properties of the substrate.[48]

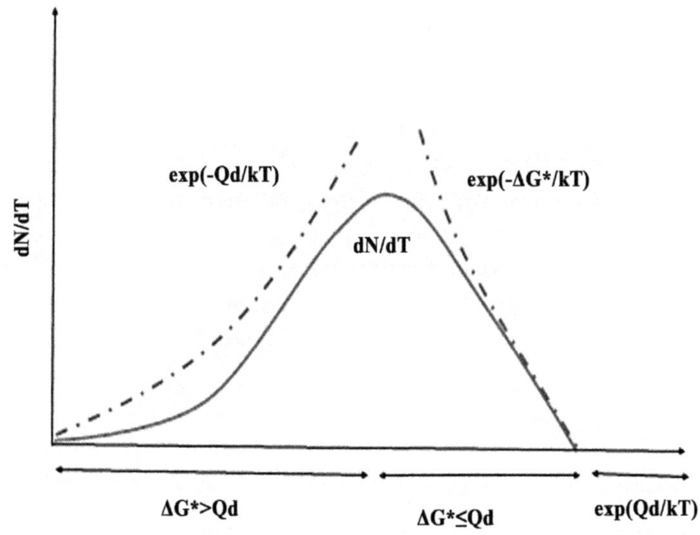

FIGURE 1.4 Nucleation rate versus temperature.

1.2.3 CONTACT ANGLE, CONTACT LINE, AND YOUNG'S EQUATION

Wettability refers to the extent of wetting when a liquid drop interacts with a solid surface and it is characterized by contact angle. Contact angle (θ) is the angle formed between the solid–liquid interface and the liquid–vapor interface as shown in Figure 1.5.[49,50] The interface where the solid, liquid, and vapor co-exists is called three-phase contact line. The three-phase contact line is shown in Figure 1.6.[51,30] Low contact angle indicates high degree of wetting and high contact angle indicates the droplets bead over the solid surface. More specifically if the contact angle is less than 30°, then the surface is called super hydrophilic and if it is greater than 145° it is called superhydrophobic surface. The surface with contact angle $30° \leq \theta$ 90° is called hydrophilic and with $90° \leq \theta \leq 145°$ is called hydrophobic surfaces, respectively.[52] The interfacial tensions of the phases play an important role in determining the contact angle.[53,54] Under equilibrium conditions, the relation connecting the interfacial tensions and the contact angle is given by Young's equation:[55]

$$\cos\theta_y = \frac{\gamma_{sv} - \gamma_{sl}}{\gamma_{lv}}, \tag{1.11}$$

where γ_{sv}, γ_{sl}, and γ_{lv} are the solid–vapor, solid–liquid, and liquid–vapor interfacial tensions, respectively, and θ_y is the Young's contact angle which is the intrinsic contact angle of the surface at the three phase contact line. Young's equation is valid only for smooth surfaces. Real surfaces are rarely smooth and the contact angles observed on such surfaces deviates from the Young's contact angle.[56–58] These contact angles are called apparent contact angles. Wetting of liquid on such rough surfaces can be described by two distinct modes. In one mode the liquid makes a complete contact with the solid surface forming a continuous solid–liquid interface and is shown in Figure 1.7. The apparent contact angle for this mode is given by Wenzel's equation developed by Wenzel:[59]

$$\cos\theta^* = r\cos\theta_y, \tag{1.12}$$

where r is the roughness factor defined as the ratio of actual surface area and the projected surface area. From eq (1.12) it is clear that the roughness amplifies wettability of a solid, that is, if the surface is intrinsically hydrophobic, the roughness will enhance its hydrophobicity and if the surface is intrinsically hydrophilic, then the roughness will enhance its hydrophilicity will increase.

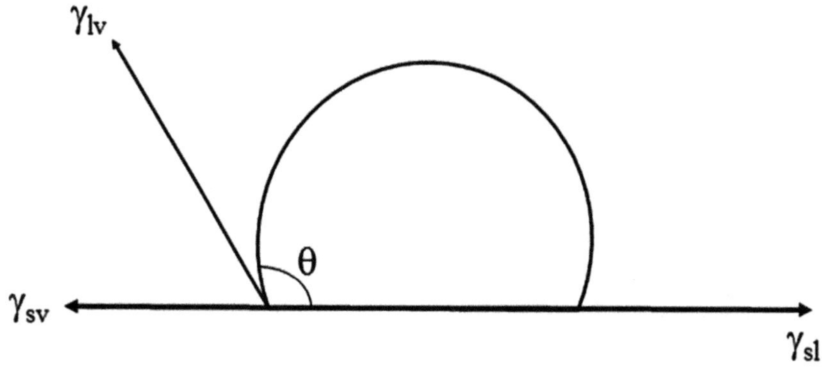

FIGURE 1.5 Contact angle (θ) formed between the solid–liquid interface and the liquid–vapor interface.

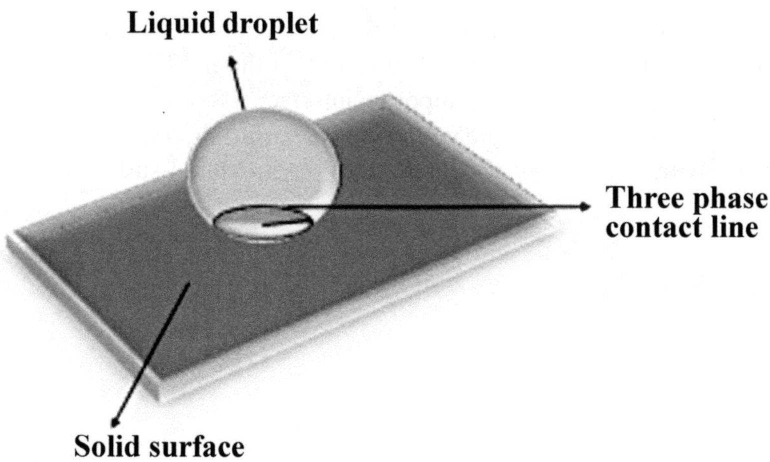

FIGURE 1.6 Figure showing three phase contact line. It is the three-dimensional image of a droplet on the surface, as seen from the top view. Whenever a drop is placed on the solid surface, a contact can occur, as it would between any two surfaces that are brought close together. A contact line is defined where three phases (solid, droplet forming liquid, and ambient vapor) can co-exist. The contour formed by the three-phase contact line is shown in the figure.

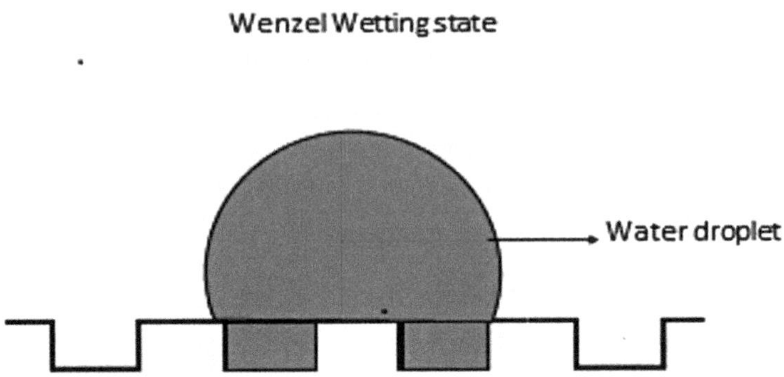

FIGURE 1.7 Wenzel state of wetting.

In the second mode, the liquid will not be in conformal contact with the solid surface; instead, the liquid droplet is suspended on a mixed interface composed of solid protrusions and air packets trapped between them.[60,61] Figure 1.8 shows Cassie's state of wetting. The apparent contact angle in this mode was described by Cassie and Baxter as

$$\cos\theta_{CB} = rf\cos\theta_y + f - 1, \tag{1.13}$$

where f is the fraction of area enclosed by the composite interface. From the Cassie equation it is observed that to obtain the perfect non-wetting situation, the fraction of the composite interface should be more. The interaction of liquid contact line with the solid surface becomes important in dealing with the motion of contact lines on structured surfaces/rough surfaces.

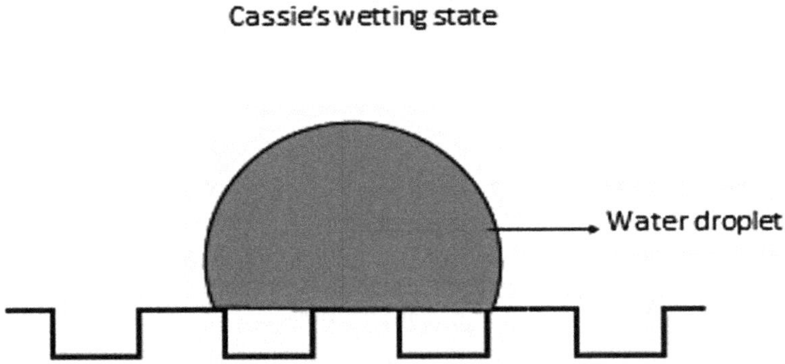

FIGURE 1.8 Cassie's state of wetting.

1.2.4 CONTACT ANGLE HYSTERESIS AND CONTACT LINE MOVEMENT

The contact line on a flat surface will not recede (advance) until the contact angle reaches the critical receding (advancing) contact angle. They are represented by θadv and θrec. The liquid droplet on a rough surface (having asperities) exhibits a range of contact angles bounded by two extreme values due to pinning effect of droplets.[57,62–66] The upper limit is called local advancing contact angle (θadv) and the lower limit is called local receding contact angle (θrec). The difference between these two extremes gives the contact angle hysteresis which characterizes the strength of pinning that inhibits droplet motion relative to the surface.[67–71] Figure 1.9 shows an illustration of advancing and receding contact angle.

$$\text{CAH} = \theta_{\text{adv}} = \theta_{\text{rec}}, \tag{1.14}$$

$$\theta^*_{\text{adv}} = \theta_{\text{adv}} + \varphi, \tag{1.15}$$

$$\theta^*_{\text{rec}} = \theta_{\text{rec}} - \varphi; \theta^*_{\text{rec}} \geq 0, \tag{1.16}$$

where θ_{adv} and θ_{rec} are critical advancing and receding contact angles of at surface, respectively θ^*_{adv} and θ^*_{rec} corresponds to local advancing and receding contact angles of a surface having asperities.

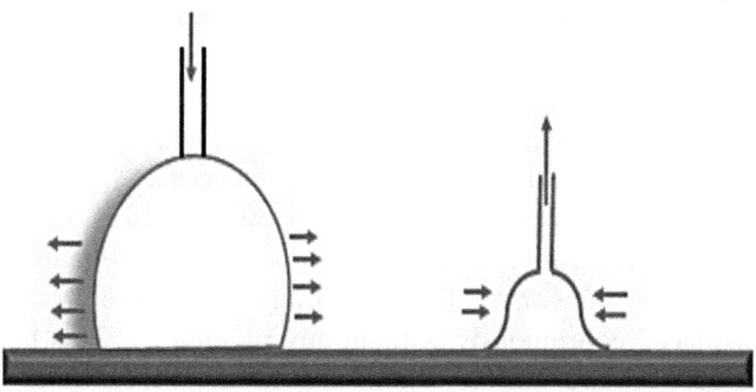

FIGURE 1.9 Depiction of advancing and receding contact angles while adding and removing liquid.

The Gibb's inequalities are given by Kalinin et al.[66] and Gibbs[67]

$$\theta^*_{\text{rec}} = \theta_{\text{rec}} - \varphi \leq \theta \leq \theta^*_{\text{adv}} = \theta_{\text{adv}} + \varphi . \tag{1.17}$$

The presence of contact angle hysteresis gives rise to a force called surface retention force which resists the motion of liquid droplet of characteristic length L on the surface.[72] This surface retention force is given by

$$F_{\text{r}} = \gamma_{\text{LV}} L(\cos\theta_{\text{rec}} - \cos\theta_{\text{adv}}) . \tag{1.18}$$

From eq (1.18) it is clear that minimizing hysteresis leads to decreased resistance to motion of the liquid droplet which results in high mobility of droplets on the surface. There are three main causes for hysteresis. They are roughness, chemical contamination, and solutes in liquids.[30,73]

The presence of solutes in the liquids form a layer over the solid surface, its presence or absence may lead to contact angle hysteresis.[30] If the influence of heterogeneities is strong, the contact line can be pinned or move intermittently.[74] If the scale of roughness is small, the apparent contact angle is sensitive to the average surface plane rather than the true contact angle. If the scale of roughness is too small of the order of 2 nm, the limits of Gibb's inequalities observed with macroscopic asperities may not be reached. Pinning depends on the dimensions of the asperities.[66,70,71] If the height of the asperities is smaller than 30–50 nm, there is no effect of roughness in pinning the contact line. If the height of the asperities is around 2–10 nm, receding line gets pinned.[75–77] If the surface has periodic roughness, for example, parallel grooves, the pinning or depinning depends on the orientation of the contact line with respect to the grooves.[30,56,78,79] For a surface with parallel grooves, there are two possible situations—either the contact line is parallel to the grooves and it is pinned[30,79] or the contact line lies at an angle to the grooves and it can be displaced continuously without any pinning.[30,80] This holds only when the grooves are infinitely long and close to each other like that of number grooves in the case of capillaries normal to the axis.[81] When the contact line moves over the solid surface there are often points that remain pinned, prompting the contact line to suddenly jump to a new position. The pinning is a consequence of opposition that a surface presents to the movement of liquid front.[30,82,83]

1.2.5 SPREADING COEFFICIENT

In the case of partial wetting, it is possible to find the spreading coefficient when the liquid is immiscible with the polymer layer.[84,85] The spreading coefficient S is given by

$$S = \gamma_S - (\gamma_1 - \gamma_{1S}), \tag{1.19}$$

where γ_S, γ_1, and γ_{1S} are the surface energy of the solid, surface tension of the liquid (N/m^2), and the liquid–solid interfacial energy (mJ/m^2), respectively. If $S > 0$ there are spreading behavior and $S < 0$ then there is a non-spreading behavior.

1.2.6 GROWTH OF DROPLETS

The rate of droplet growth is proportional to the concentration gradient at the boundary layer, mass diffusivity, and the saturation pressure difference.[86] This is given as

$$\frac{dV}{dt} = D\Delta p C_0 \delta^{-1}, \tag{1.20}$$

where D is the diffusion coefficient of liquid molecules in the carrier gas, C_0 is the concentration of liquid in the carrier gas, δ^{-1} is the length scale of condensation gradient, and Δp is the saturation pressure difference of water between the environment and the substrate surface. Initially there are isolated droplets nucleated on the surface.

These droplets do not have a strong interaction with each other and the diameter of the droplets grows according to a power law with time.[14,48] Droplet diameter grows with time as[18]

$$d \propto \left(\frac{dV}{dt}t\right)^{1/3}, \tag{1.21}$$

$$d \propto \left(D\Delta p C_0 \delta^{-1} t\right)^{1/3}, \tag{1.22}$$

$$d \propto kt^{1/3}, \tag{1.23}$$

where k is the function of air flow velocity and temperature. In the first stage, the surface coverage of droplets is very low.

In the second stage, called cross-over stage, the droplets grow until they touch each other.[5,48] In this stage the surface coverage by droplets is maximum and the droplets interact closely. The entropy of the system decreases by the rearrangement of droplets in a close-packed honeycomb structure. Coalescence of droplets does not yield enough space for new nucleation, thus leading to a mono-dispersed droplet size distribution. Coalescence of droplets continues with time yielding enough spaces for nucleation of new droplets.[87–91] Hence, in this phase, the diameter of the droplets grows as $d \alpha t$.[92]

1.2.7 SELF-ASSEMBLY OF DROPLETS

In the crossover stage, a thin polymer layer prevents the droplets from coalescing, leading to the self-assembly of droplets into a hexagonal array. There are three possible explanations for this self-assembly. The first idea is that Marangoni convection; an effect caused by the gradient of surface tension at the interface between two phases (liquid and vapor) causes self-assembly of droplets. In the case of polymer films and water, a layer of polymer is formed around the water droplet due to instant precipitation. This layer prevents coalescence of water droplets.[37–40] The long-range order of these droplets is then driven by the long-range attractive (and repulsive) forces, leading to a surface filled with water droplets stabilized by solvated polymer layer. This makes droplets to levitate on the solid surface until the later stages of breath figure formation process.

In a phenomenological theory proposed by Pitios and Francois the droplets do not levitate on the surface but touch the surface with the interfacial polymer layer keeping the droplets apart.[18] A third idea contrary to both of the above mechanisms was put forth by Nishikawa et al.[42] who stated that the initial water droplets neither levitate nor are ordered on the surface, but thermocapillary and Marangoni forces cause the water droplets to submerge into the organic solution. The thermocapillary and Marangoni forces that are exerted on the submerged droplets are then responsible for the observed hexagonal order.[42]

1.2.8 COALESCENCE OF DROPLETS

The last stage is the coalescence of droplets. The force of attraction between the droplets depends on the diameter of the droplets and is given as follows:[93,94]

$$F = \frac{4\pi R^6 \omega^2}{3l\sigma}\left(\frac{1}{d} + 0.25(1-p^2)^{1.5} - 0.75(1-p^2)^{0.5}\right)^2, \tag{1.24}$$

where R is the radius of the droplet (m), ω is the specific weight of the liquid ($kgm^{-2}s^{-1}$), σ is the surface tension of the liquid (N/m), d is the liquid–air relative density (kg/m^3), and p is the ratio of the apparent droplet radius above the surface to the droplet radius below the surface.

1.2.9 EVAPORATION OF DROPLETS

The evaporation rate of pinned droplets is proportional to the initial drop volume and occurs in two different modes.[95,96] They are constant contact angle mode and constant contact radius mode.

1.2.9.1 CONSTANT CONTACT RADIUS MODE

When the contact line is pinned, the contact radius remains constant during the evaporation of a droplet whereas the contact angle is initially reduced. This mode of evaporation is called constant contact radius mode. This mode corresponds to pinned triple line. The rate of change of volume of the droplet in the constant contact radius mode is given[96,97]

$$-\frac{dV}{dt} = constant. \tag{1.25}$$

1.2.9.2 CONSTANT CONTACT ANGLE MODE

When the pinning of the line is reduced, the contact radius starts to vary keeping the contact angle constant. This mode of evaporation is called constant contact angle mode.[95,97] This corresponds to moving triple line. For this mode, the volume of the droplet is given as

$$V \propto (t_p - t)^{3/2}, \tag{1.26}$$

where t_p is the pinning time (s), that is, the time between the pinned mode and moving triple line mode. There will be a change in Gibb's free energy with different evaporation modes. The droplet in a constant contact radius mode initially shows a decrease in Gibb's free energy than the moving droplet. But when the transition from constant contact radius to constant contact angle takes place, this Gibb's free energy in constant contact radius (CCR) mode is smaller than constant contact angle (CCA) mode,[98] that is, for constant contact radius mode

$$\Delta G_{\text{CCR}} > \Delta G_{\text{CCA}} ; G_{n+1} = G_n - \Delta G_{\text{CCR}} , \tag{1.27}$$

and for constant contact angle mode

$$\Delta G_{\text{CCR}} < \Delta G_{\text{CCA}} ; G_{n+1} = G_n - \Delta G_{\text{CCA}} . \tag{1.28}$$

1.3 EXPERIMENTAL PROCEDURE

The preparation of self-assembled droplet pattern begins with spin coating the polymer solution on the substrate. This can be done either in the chamber maintained with the humid atmosphere or outside the chamber. The way the humid air is passed favors two different kinds of process, namely, static and dynamic process. In the static process, the substrate is placed in a vessel containing saturated water vapor, whereas in the dynamic process, the moist air is blown over the substrate. In both the process, non-aqueous vapor can also be used.[23]

1.4 FACTORS AFFECTING THE SELF-ASSEMBLED-DROPLET PATTERNS

1.4.1 POLYMER

The weight, structure, and concentration of the polymers affect the formation of the films. The properties of the film formed will depend on the functional groups attached to the polymer.[99] For example, fluorinated polymers will give super-hydrophobic films. If thermos-responsive amphiphilic copolymers are used, they will yield a polymer film with thermo-responsive properties.[100] When liquid crystal polymers are used as building blocks, electro-responsive films are obtained.[101]

With the increase in polymer concentration, the honeycomb pore size decreases.[102] This can be explained by Henry's law:

$$P = P_0 (1 - X_B), \quad (1.29)$$

where P_0 is the vapor pressure of the pure solvent (kPa), P is the vapor pressure of the solvent in solution (kPa), and X_B is the molar fraction of the solute (mol/L). If the molar fraction of the solute is high, then the vapor pressure of the solvent in solution decreases. This leads to a slower solvent evaporation and in turn higher surface temperature leading to a small temperature difference between the environment and the surface. The smaller temperature difference decreases the growth of droplets during nucleation process. As a result, smaller pores are formed.

Perry's law states that higher molecular weight leads to higher vapor pressure, which in turn produces larger pores. Higher molecular weight of polymers reduces the depth of the pores (i.e., the depth to which the condensing droplets penetrate into the surface of the polymer). The pores are formed as a consequence of evaporation of the thus penetrated droplets.[102–104]

1.4.2 SOLVENT

The main requirement for the honeycomb film formation is the higher evaporation rate of the solvent. Solvents with low water solubility, high interfacial tension with water, and high thermodynamic abilities are required for the formation of highly ordered polymer films. Solvents, such as chloroform and dichloromethane enable highly ordered pore formation, whereas acetone, tetrahydrofuran, and toluene do not favor the formation of ordered film.[34] The properties of some widely used solvents are given in Table 1.1.

TABLE 1.1 Properties of Some Widely Used Solvents.

Solvent	ρ (g/cm^3)	η (m Pa.s)	S (g/100 mL)	ST (mN/m)	γ (mN/m)	BP (°C)	VP (kPa)	H (kJ/mol)
Acetone	0.7899	0.306	Miscible	23.46	–	56	30.8	30.99
Carbon disulfide	1.2632	0.352	0.22	31.58	46	48.1	48.2	27.51
Chloroform	1.4833	0.537	0.8	26.5	33.5	61	26.2	31.28
Ethyl acetate	0.9003	0.423	7.9	23.39	6.8	77	12.6	35.60
Dichloromethane	1.3266	0.413	2	27.2	28.3	40	58.2	28.82
Tetrahydrofuran	0.8892	0.456	Miscible	26.4	–	65	21.6	31.99
Toulene	0.8669	0.560	0.05	27.93	36.1	111	3.79	38.01

1.4.3 SUBSTRATE

Honeycomb films were seen to be formed on inorganic and organic substrates, such as glass, silicon, polyvinylchloride, polyethylene, fluorinated glass, etc. Formation of a good polymer film on a substrate depends on the extent of wettability of the substrate by the polymer. Higher the wettability of the substrate by polymer solution, better will be the structures of honeycomb pattern formed.[105]

Pore size is directly proportional to the surface energy of the solid for a given solvent. For example, the substrate with higher surface energy promotes the formation of long-range order of breath figure films with polystyrene-chloroform solution, whereas polystyrene-carbon disulfide solution on the same substrate does not yield any ordered films. The solvent involved in the process is important even though the substrate has all the properties required for ordered formation of pores.[34]

1.4.4 RELATIVE HUMIDITY

It is seen that higher relative humidity (RH) leads to bigger pore formation. The lower humidity leads to reduced vapor pressure of the solvent. This causes the solvent to evaporate very fast resulting in a smaller array of pores.[106] If the RH is low (RH <45%), then there will be a lack of water molecules to form ordered array. For RH >90%, there are a large number of water droplets which result in a higher vapor pressure and slower solvent evaporation. So, there will be a larger number of droplets which try to coalesce with each other forming bigger droplets. In both the cases, there is no possibility of forming ordered array of pores.[34,106]

Thus to get an ordered array of pores, relative humidity of around 70% has to be maintained in the chamber. In a dynamic process of self-assembled pattern formation, gas flow can be adjusted to vary the humidity and thus increase the solvent evaporation to create a temperature gradient between the solution surface and the bulk solution.[106]

1.4.5 AIRFLOW

Airflow velocity and direction also plays an important role in the formation of breath figure arrays. If the airflow velocity is very high, then pore sizes get decreased. When the airflow is perpendicular to the substrate, circular pores are formed. When the airflow is at some direction to the substrate, the pore shape also changes. For example, airflow at 15° to the substrate yields elliptical pores.[107,108]

1.4.6 VAPOR CONDITION

Apart from water, breath figure formation can be formed in non-aqueous environment like ethanol, methanol, etc. With proper choice of solvent, polymer, and substrate, it is possible to get an ordered array of pores using water vapor. In the case of ethanol and methanol, the pores are slightly distorted from its circular shape and are shallow. This is due to the low surface tension of the liquids. Alcohols immiscible in polymer solution produce larger holes in polymer solution.[109]

1.5 BREATH FIGURE FILMS ON NON-PLANAR SUBSTRATE

In addition to all the above mentioned properties, morphology of the substrate also plays a vital role in the formation of an ordered array of pores. The substrates that are usually used to study the self-assembly of droplets over constrained surfaces are commercial TEM grids, spectroscopic gratings, and square/hexagonal gratings. These gratings come with inbuilt 3D patterns, with typical grating separation (constraint dimension) being of the order of a few tens of micrometer. Formation of breath figures on such non-planar substrate enables to form hierarchical structures. Hierarchical structures exhibit unique properties, such as super hydrophobicity and can be used as a master for soft lithography process.

Connal et al. were the pioneers of forming breath figures on non-planar substrates using various polymers.[16,29] TEM grids were used as substrate with a variety of polymers like poly dimethyl siloxane (PDMS), poly(ethyl acrylate), poly(methyl acrylate), and poly(methyl methacrylate) as solvent. TEM grids have uniform rectangular meshes/holes of specific dimensions of the order of 50 μm and were chosen for the same. When the TEM grid is placed on glass slide, an ordered non-planar substrate is obtained.[110] When a polymer solution is coated over this non-planar substrate followed by passage of humid air, it results in the formation of breath figures both inside the rectangular area of the grids and over the grid area. As a result of this, a hierarchical pattern is obtained.

Following them, Gong et al. used TEM grid to prepare a two-level micro-porous film from polystyrene-b-poly acrylic acid solution.[111] One level of ordered surface features originates from the contour of the hard templates (TEM grids), while the other level originates from the condensation of water droplets, that is, breath figure arrays. The patterns formed are shown in Figure 1.10.

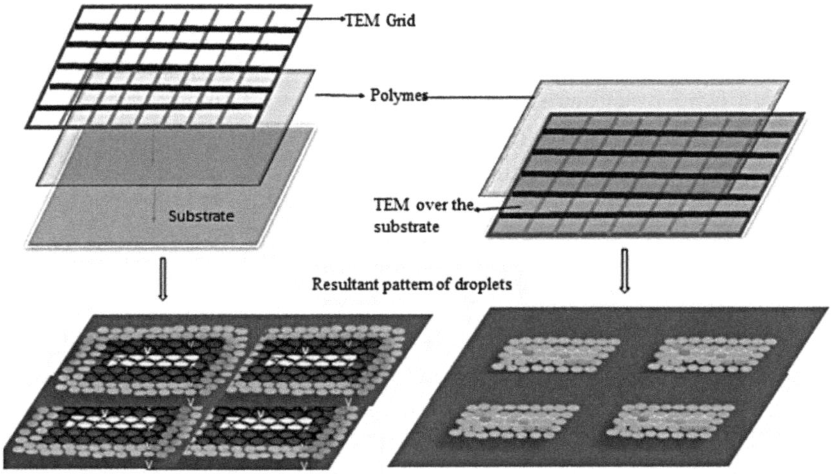

FIGURE 1.10 Patterns of droplets when the grid is placed over the polymer and below the substrate are shown.

Galleoti et al. used the honeycomb pattern obtained by breath figure method as a master for soft lithography and a positive imprint was obtained using PDMS. TEM grid was placed over the positive imprint and PDMS was poured onto it to obtain two-level structure.[112]

Qiao et al. prepared hierarchically ordered films without using photolithography. They used core cross-linked star polymer with different glass transition temperature (*T*g) to fabricate honeycomb films. TEM grids were placed over the glass slides before coating the polymer film. They obtained crack free polymer films on non-planar substrates using high *T*g polymers.[113–115]

They also found that Young's modulus of the substrate is the most influential factor in determining the occurrence of cracks on honeycomb films while using non-planar substrates. They also tried placing a grid template on the surface of the evaporating solution, leading to the formation of BFAs only in the mesh space of the grid.[14] In all the above works, they showed that the shape of the underlying surface (grating/grids) plays a very important role in the orientation of the pores in the honeycomb film through physical constraints.

For instance, the hexagonal grid/grating induces perfect hexagonal honeycomb structure. This is because the hexagonal grating shares all the principal axis with hexagonal array of pores.[116]

Parallel bar grating with a dimension 50 μm is also used as substrates for creating hierarchical structures. In the case of parallel grating, the pores are linearly arranged along the edge of the grating resulting in highly ordered two-dimensional structure.[116] Attempt was also made on using a square grating to improve the degree of ordering of pores. But in this case, there is an inconsistency between the grating and pore array axis.[14,116] Figure 1.11 shows the droplet pattern on a different type of gratings.

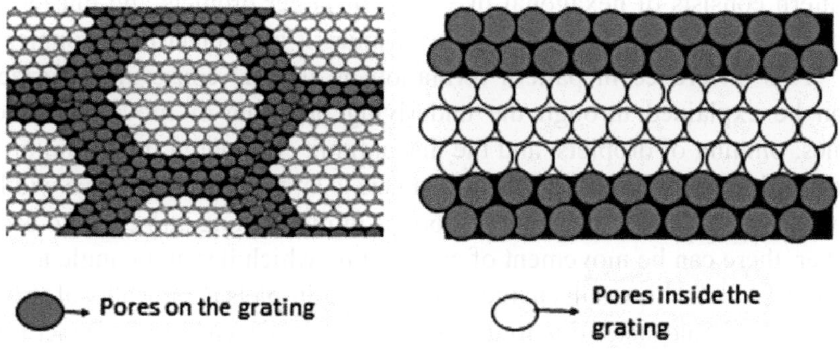

FIGURE 1.11 The assembly of droplets on various gratings.

Gu et al. used micro-pipette with curvature gradient as a substrate for breath figure formation using poly(L-lactic acid) (PLA) as the film-forming polymer and di-oleoylphosphatidyl ethanolamine (DOPE) as the surfactant. They showed that the honeycomb structures including pore size and regularity change remarkably with gradually increasing surface curvature. At high curvatures, large number of pores are closer together and this leads to coalescence of droplets causing the formation of hemispherical pore strings.[117]

In stainless steel gauze, using dip coating technique with polycarbonate as polymer hybrid honeycomb films were obtained.[118] 1D constraints have been introduced by the use of optical gratings of separation 50 μm. A similar pattern was seen with the difference that in the case of TEM grid, the outer structure is that of a square while in the case of grating the outer structure is that of straight lines.[116]

In all the above-discussed substrates, the size of the grating measures around 50 μm. The presence of 2D constraints when polymer is deposited over a TEM grid, square grating, or parallel grating yielded a double pattern of the breath figures deposited. Honeycomb structures were formed at a separation of 50 Å. These correspond to the underlying grid lines of the same dimension. The intermediate space is filled with a second honeycomb structure of a smaller dimension.[14]

An attempt was made to verify the formation of same hierarchical pattern in 1D constrained surface. The space between the grating measures around 2.5 μm which is much lesser when compared to the dimension of TEM grids. The surfaces with 1D constraints showed an interesting pattern consists of hexagonal like rings of larger droplets and cluster of smaller droplets inside the ring in both aqueous and ethanol environment.

This difference in pattern formation on the 1D constrained surface can be explained through the underlying principle of moving contact lines, pinning of droplets, and the discontinuity of energy minimum on a grooved surface. Provided the droplet size is larger than the groove width, for 1D constrained surface with grooves infinitely long and close to each other, there can be movement of contact line which lies at an angle to the grooves.[30,79,80] This results in the formation of droplets having a local structure (metastable state) due to the continuous displacement of contact line without any pinning.

The ring formation is attributed to the moving three-phase contact line and change in its direction due to pinning of droplets at certain points and successive evaporation of droplets. On the 1D constrained surface, there are spaces for occupying new droplets. This favors re-nucleation at the spaces inside the ring. This confirms that the difference in pattern is attributed completely to the geometry of the underlying surface. The lines in the 1D constrained surface are comparable to a capillary with a number of grooves normal to the axis and as indicated in the theory, the argument of contact line moving is valid for this order of length scale.[30,31] Such a pattern has not been observed with 1D grooves separated by 50 μm.

1.6 APPLICATIONS

Polymer surfaces patterned by self-assembled liquid droplets find a wide variety of applications. Some of them are discussed below.

1.6.1 STIMULI RESPONSE MATERIALS

The honeycomb patterns with stimuli-responsive properties find applications in drugdelivery, gene delivery, sensing, wettability switching, etc. These surfaces respond to external stimuli like temperature, solvent, pH, stress, and light.[14] Using block co-polymer polystyrene-block-poly(N-isopropylacrylamide) (PSb-PNIPAM) for the preparation of thin dense films, Stenzel et al. fabricated a pin-cushion-like films which show thermo-responsive properties, whereas ordinary honeycomb pattern using the same polymer does not show any thermo-responsive behavior. This is because the polymer segregates at the water-solution interface forming a hydrophobic end at the top and hydrophilic pores.[119,120] A similar phenomena was observed with the blend of polystyrene and copolymer containing PNIPAM as polymer to form honeycomb films by Yabu et al.[121] Stenzel et al. also fabricated thermo-responsive honeycomb films by grafting PNIPAM from the PS-co-PHEMA film surface. The surface created by Stenzel et al. is slightly different from the surfaces created using PS b-PNIPAM. Therefore, it can be concluded that the surfaces grafted with PS-b-PNIPAM show a switchable hydrophilic/hydrophobic behavior.[119]

pH-sensitive devices are also fabricated using honeycomb films made out of PS-b-P4VP surface. These honeycomb films are composed of pores at the micrometer scale and phase separated structures at the nanometer scale. It is also confirmed that the pin-cushion like structures enhance the pH sensitivity.[14,122]

1.6.2 SUBSTRATES AND TEMPLATES

Breath figure arrays are used as templates. Breath figure arrays can be used to form honeycomb structure on materials which cannot be directly utilized in breath figure process. Compared to other templates like colloidal crystals and block co-polymer patterns, breath figure process is versatile of its pore size adjustability.[23]

Galeotti et al. showed that the polystyrene honeycomb patterns generated by breath figure method can be reproduced using PDMS. The positive mold obtained using the above process consists of regularly arranged micro-metric protuberances. This was inked with a biomolecule solution and subsequently printed on the desired substrate. In this way, they showed that spots of a few microns can be easily produced without the need of sophisticated apparatus.[112]

Similarly, Lei Li et al. patterned surfaces with breath figure arrays using amphiphilic diblock copolymer polystyrene-b-poly(acrylic acid) (PS-b-PAA). They sputter-coated the top surface of the film with gold. Large etching rate selectivity between gold mask and underlying substrate plays an important role in the effective transfer of the patterns. The surface features can be adjusted by adjusting the pore sizes in the micro porous films. These transferred surfaces are then used as templates for soft lithography process.[14,83,118,123]

The breath figure arrays are used as masks to transfer the honeycomb pattern to the substrates. Hirai et al. peeled off the top layer of the breath figure film, and used it as a mask for lithography process. After peeling, the nano spike arrays were then transferred to the silicon substrates. The silicon surfaces are chosen in such a way that they have various crystal facets. The resultant surfaces showed both anti-reactive and super-hydrophobic properties. Using sputter coating of breath figure array film, a metal mask with a honeycomb pattern can film can be obtained.[102,124–126]

Li et al. have demonstrated that by using 3-amino propyltrimethoxysilane (APETS), zinc acetylacetonate ($Zn(acct)_2$), and ferrocene as precursors, silica, ZnO, and ferric oxide (Fe_2O_3) honeycomb patterns can be prepared by pyrolyzing corresponding cross-linked polystyrene block PAA/precursor composite.[83]

The honeycomb films formed by surfactant-modified Au or Ag nanoparticles could be used as the substrate of surface-enhanced Raman spectroscopy (SERS).[83] Wan et al. showed an alternative approach to prepare highly sensitive SERS substrates with breath figure array films decorated with Ag nano particles. Their results confirmed that the enhancement factor of the substrates will be high depending on the number of analytes to silver nano particles.[127]

Hao et al. prepared the SERS substrates using the blends of polymer and gold nano particles modified by mercaptan. The relationship between the performance of SERS substrates and the morphology of the film was studied. They showed that the more regular honeycomb films were able to produce very strong SERS signals. This is due to the fact that more regular breath figure array films have large surface areas which allow them to adsorb more analytes from the surrounding solution.[128]

1.6.3 BIOMEDICAL APPLICATIONS

Breath figure arrays were used in cell culture and other biomedical applications. The breath figure films have the properties like surface chemical heterogeneity and adjustable surface topography. These properties of breath figure films both as prepared and chemically modified made them a potential candidate for cell culture and for the investigation of cell behaviors.[129–131] Bio-compatible and biodegradable polymers like poly(e-caprolactone), poly(l-lactide), polyalkylcyanoacrylate, glycopolymer, DNA, and their copolymers are also used as building blocks for fabricating breath figure arrays. On all the above-said polymers, the behavior of various types of cells was investigated.[132–137]

These breath figure arrays are found to be enhancing the cell attachment and spreading compared to unpatterned polymer films. These films are also used to control the morphology and differentiation of cultured cells, which was a crucial thing in tissue engineering. The cells like hepatocytes,[23] breast epithelial cells,[138] osteoblasts,[23] stem cells,[23] endothelial cells,[139,140] etc. The pattern geometry was greatly influenced the morphology of cells because the pore walls guided the spreading of the cells.[23] The properties of structured porous films like hydrophilicity enhance the adhesion of cells on the pores. Breath figure arrays of protein were also produced after selectively grafting protein recognition moieties on the surface of the pores.[23,141] Beattie et al. prepared a breath figure array film using soluble PPy composite. They found that these were suitable as scaffolds for tissue engineering.[142] Du et al. prepared breath films and cultured human embryonic skin fibroblasts (ESFs) on them. These cultured cells adhered, spread, and grew better on breath figure array films rather than organogels. This enhancement is due to the honeycomb structure and surface roughness of the films.[143]

Titanium n-butoxise as starting material, they prepared breath figures of TiO_2 consisting of anatase nanocrystals. The pore size was varied and verified for the cell adhesion property. The enhancement in the cell adhesion and spreading were found when the pore size ranges about 4:6 m.[144]

It was also found by Tang et al. that breath figure films prepared using polymers were not suitable for growth of cancer cells, because these films are hydrophobic in nature, and there is air in the pores of the films.[145] Effective inhibition to bacterial adhesion, growth, and biofilm formation were found in polystyrene breath figure films with a pore size of 5–11 μm by Manabe et al.[146]

The improved anti-bacterial effect was observed by in the breath figure films containing Mn(III) meso-tetra(4-sulfonatophenyl) porphine chloride (MnTPPS). The same was also observed in breath figure films made of cellulose with a quaternary ammonium (QA). Glucose-sensing films based on phenylboronic acid are used to find the glucose sensitivity.[14,23] The ordered structures of breath figure films were used to fabricate micro-optical device and the rough surface of breath figure arrays was employed to reduce the light reaction and increase the light harvest.[23]

1.6.4 OPTICAL APPLICATIONS

The spherical pores of breath figure arrays are good templates for the fabrication of micro lens arrays. Micro lens arrays are used in light modulation, which is affected by the shape and size of the micro-lens in the micro-lens arrays.[23] Yabu and Shimomura prepared breath figure arrays under standard conditions using PDMS as prepolymer. These spherical pores after curing at 300° became an array of spherical lenses. The upper bubble layer was peeled out and formed hemispherical lenses. The optical performance of both the spherical and hemispherical lens arrays was compared and it was found that hemispherical lens arrays showed higher optical performance than the spherical one.[147–150]

The micro lens arrays are usually prepared using PDMS. To prepare them with other materials, a low-surface energy non-stick surface on BFA film was created by Chari et al. via a surface reconstruction process after heating the BFA based on a PDMS-containing co polymer in dry air. They observed that during the above process of surface reconstruction, the low energy PDMS segments migrated to the surface of the breath figure array film yielding a non-stick surface. This non-stick surface was used as a master/template for preparing micro lens array. Using polyurethane micro-lens arrays were prepared by which showed a relatively high refractive index of 1.54.[23,151]

Pintani et al. have demonstrated the use of breath figure pattern as a means to produce PDMS masters possessing arrays of micron-size protrusions that can subsequently be used to pattern the active layer of organic thin-film device structures. They fabricated micro LED using the stamp that was created by breath figure method to pattern an insulating PI layer on top of a PEDOT: PSS covered ITO-coated glass substrate. This defines an array of micro-receptacles. The higher efficiency and higher current and power efficiency were obtained with the resultant devices. The features of the master can be tuned by varying the conditions of the breath figure approach.[23,152]

1.6.5 OTHER APPLICATIONS

Microfiltration and sterile filtration for protein concentration are the most important application of porous honeycomb patterned polymer films. Breath figure arrays are also good candidates for micro-sieves. By adjusting the solution concentration, pore sizes can be varied from 4.5 to 1.0 μm using brominated poly(phenylene oxide) as polymer which can serve as micro sieves.[23] The films prepared hydrophilic polymers with highly volatile, non-polar solvents, such as carbon disulfide and chloroform are highly hydrophilic. That is why non-polar or amphiphilic polymers are used in breath figure process to reduce the hydrophilic blocks. Therefore, in contrast to hydrophobic or super-hydrophobic surfaces, hydrophilic honeycomb films are scarcely reported. The surfaces created using breath figure processing can show interesting wetting properties.[11]

1.7 CONCLUSIONS

It is seen that self-assembly of liquid droplets on polymeric surfaces has enormous potential for technological applications. While the ideal pattern of such assemblies is a honeycomb structure, deviations from it, including a hierarchical drop size distribution can be achieved by constraining the underlying surface. Use of polymeric substrates for this purpose retains the inherent simplicity of the self-assembly process. A large class of applications for self-assembled droplet-based patterning of polymers can be envisaged.

KEYWORDS

- **honeycomb**
- **self-assembly**
- **nucleation**
- **thermocapillary effect**
- **air flow**
- **constraints**
- **wetting**
- **depinning**

REFERENCES

1. Menz, W.; Mohr, J.; Paul, O.; *Microsystem Technology*, 2nd Ed.; Wiley-VCH: Weinheim, Germany, 2001.
2. Rosi, N. L.; Mirkin, C. A. Nanostructures in Bio-diagnostics. *Chem. Rev.* **2005,** *105* (4), 1547–1562.
3. Munoz-Bonilla, A.; Emmanuel, I.; Eric, P.; Rodriguez-Hernandez, J. Self-organized Hierarchical Structures in Polymer Surfaces: Self-assembled Nanostructures within Breath Figures. *Langmuir* **2009,** *25* (11), 6493–6499.
4. Vincent Rotello, M. Nanoparticles, In *Building Blocks for Nanotechnology*; Springer: Berlin, Germany, 2004; p 32.
5. Zhiqun, L. *Evaporative Self-assembly of Ordered Complex Structures;* World Scientific Publishing Company: Singapore, 2012.
6. Aranzazu Del, C.; Eduard, A. *Chem. Rev.* **2008,** *108,* 911.
7. Gates, B. D.; Xu, Q. B.; Stewart, M.; Ryan, D.; Willson, C. G.; Whitesides, G. M. New Approaches to Nanofabrication: Molding, Printing, and Other Techniques. *Chem. Rev.* **2005,** *105,* 1171–1196.
8. Geissler, M.; Xia, Y. N. Patterning: Principles and Some New Developments. *Adv. Mater.* **2004,** *16,* 1249–1269.
9. Whitesides, G. M.; Grzybowski, B. Self-assembly at All Scales. *Science.* **2002,** *295,* 2418–2421.
10. Yanqiong, Z.; Yuki, K.; Makoto, K.; Koji, M. *J. Mech. Sci. Technol.* **2011,** *25* (1), 33.
11. Bunz, U. H. F. *Adv. Mater.* **2006,** 18, 973–989.
12. Stenzel, M. H. *Aust. J. Chem.* **2002,** 55, 239–243.
13. Ma, H.; Hao, H. *Chem. Soc. Rev.* **2011,** 40, 5457–5471.
14. Pierre, E.; Laurent, R.; Laurent, B.; Maud, S. Recent Advances in Honeycomb-structured Porous Polymer Films Prepared via Breath Figures. *Eur. Polym. J.* **2012,** *48,* 1001–1025.

15. Hernandez-Guerrero, M.; Stenzel, M. H. *Polym. Chem.* **2012,** *3,* 563–577.
16. Steyer, A.; Guenoun, P.; Beysens, D.; Knobler, C. *Phys. Rev. B.* **1990,** *42,* 1086–1089.
17. Kuo, C. T.; Lin, Y. S.; Liu, T. K.; Liu, H. C.; Hung, W. C.; Jiang, I. M.; Tsai, M. S.; Hsu, C. C.; Wu, C. Y. *Opt. Exp.* **2010,** *18,* 18464–18470.
18. Pitois, O.; Francois, B. *Colloid Polym. Sci.* **1999,** *277,* 574–578.
19. Zhao, B.; Zhang, J.; Wang, X.; Li, C. *J. Mater. Chem.* **2006,** *16,* 509–513.
20. Ma, H.; Kong, L.; Guo, X.; Hao, J. *RSC Adv.* **2011,** *1,* 1187–1189.
21. Liping, H.; Rongrong, H.; Sijie, C.; Jie, L.; Lei, J.; BenZhong, T. Patterned Honeycomb Structural Films with Fluorescent and Hydrophobic Properties. *J. Nanomater.* **2013,** *853154,* 1–8.
22. Lei, L.; Zhong, Y.; Jianliang, G.; Jian, L.; Jin, H.; Zhi, M. Fabrication of Robust Micro-patterned Polymeric Films via Static Breath-figure Process and Vulcanization. *J. Colloid Interface Sci.* **2011,** *354,* 758–764.
23. Hua, B.; Can, D.; Aijuan, Z.; Lei, L. Breath Figure Arrays: Unconventional Fabrications, Functionalizations, and Applications. *Angew. Chem. Int. Ed.* **2013,** *52,* 12240–12255.
24. Lord, R. *Nature* **1911,** *86,* 416.
25. Xiaopeng, X.; Mingfeng, L.; Weiwei, Z.; Xinyu, L. Kinetic Control of Preparing Honeycomb Patterned Porous Film by the Method of Breath Figure. *React. Funct. Polym.* **2011,** *71,* 964–971.
26. Xiaofeng, L.; Yang, W.; Liang, Z.; Shuaixia, T.; Xiaolan, Y.; Ning, Z.; Guoqiang, C.; Jian, X. Fabrication of Honeycomb-patterned Polyalkylcyanoacrylate Films from Monomer Solution by Breath Figures Method. *J. Colloid Interface Sci.* **2010,** *350,* 253–259.
27. Elisa, F.; Paola, F.; Francesco, P. *Langmuir* **2011,** *27* (5), 1874.
28. Zhiguang, L.; Xiaoyan, M.; Duyang, Z.; Beirong, S.; Xiu, Q.; Qing, H.; Xinghua, G. *RSC Adv.* **2014,** *4,* 49655.
29. Luke Connal, A.; Robert, V.; Paul Gurr, A.; Craig Hawker, J.; Greg Qiao, G. *Langmuir* **2008,** *24,* 556.
30. de Gennes, P. G. Wetting and Spreading. *Rev. Mod. Phys.* **1985,** *57* (3), 827.
31. de Gennes, P. G. *Soft Interfaces, The 1994 Dirac Memorial Lecture;* Cambridge University Press: Cambridge, England, 1994.
32. Widawski, G.; Rawieso, M.; Francois, B. Self-organized Honeycomb Morphology of Star-polymer Polystyrene Films. *Nature* **1994,** *369* (6479), 387–389.
33. Bolognesi, A.; Mercogliano, C.; Yunus, S.; Civardi, M.; Comoretto, D.; Turturro, A. Self-organization of Polystyrenes into Ordered Microstructured Films and Their Replication by Soft Lithography. *Langmuir* **2005,** *21,* 3480–3485.
34. Yingying, D.; Mingliang, J.; Guofu, Z.; Lingling, S. Breath Figure Method for Construction of Honeycomb Films. *Membranes* **2015,** *5,* 399–424.
35. Ling-Shu, W.; Liang-Wei, Z.; Yang, O.; Zhi-Kang, X. Multiple Interfaces in Self-assembled Breath Figures. *Chem. Commun.* **2014,** *50,* 4024–4039.
36. Briscoe, B. J.; Galvin, K. P. *J. Phys. D. Appl. Phys.* **1990,** *23,* 422.
37. Lulu, S.; Vivek, S.; Jung Park, O.; Srinivasarao, M. *Soft Matter.* **2011,** *7,* 1890.
38. Srinivasarao, M.; David, C.; Alan, P.; Sanjay, P. Three-dimensionally Ordered Array of Air Bubbles in a Polymer Film. *Science.* **2001,** *292,* 79–82.

39. Barrow, M. S.; Jones, R. L.; Park, J. O.; Srinivasarao, M.; Williams, P. R.; Wright, C. *J. Spectroscopy*. **2004,** *18,* 577–585.
40. Rayleigh, L. *Nature* **1912,** *90,* 436.
41. Schatz, M. F.; Neitzel, G. P. *Annu. Rev. Fluid Mech.* **2001,** *33,* 93.
42. Nishikawa, T.; Ookura, R.; Nishida, J.; Arai, K.; Hayashi, J.; Kurono, N.; Sawadaishi, T.; Hara, M.; Shimomura, M. *Langmuir* **2002,** *18,* 5734–5740.
43. Zdenek, K.; Kyotaka, S.; Pavel, D.; Alexei, S. Homogeneous Nucleation of Droplets from Supersaturated Vapor in a Closed System. *J. Chem. Phys.* **2004,** *120,* 6660.
44. Kalikmanov, V. I. *Nucleation Theory, Lecture Notes in Physics;* Springer: New York, NY, 2012; Vol. 860.
45. Nepomnyashchy, A.; Golovin, A.; Tikhomirova, A.; Volpert, V. Nucleation and Growth of Droplets at a Liquid–Gas Interface. *Phys. Rev. E.* **2006,** *74,* 021605.
46. Fletcher, N. Size Effect in Heterogeneous Nucleation. *J. Chem. Phys.* **1958,** *29,* 572–576.
47. Lazaridis, M. *J. Colloid Interface Sci.* **1993,** *155,* 386.
48. Narhe, R. D.; Beysens, D. A. Nucleation and Growth on a Super-hydrophobic Grooved Surface. *Phys. Rev. Lett.* **2004,** *93* (7), 076103.
49. Adamson, A. W. *Physical Chemistry of Surfaces,* 5th ed.; John Wiley and Sons Inc.: Hoboken, NJ, 1990.
50. de Gennes, P. G.; Brochard Wyart, F.; Quere, D. *Capillarity and Wetting Phenomenon, Drops, Bubbles, Pearls and Waves;* Springer: New York, NY, 2004.
51. Tak-Sing, W.; Taolei, S.; Lin, F.; Joanna, A. Interfacial Materials with Special Wettability. *MRS Bull.* **2013,** *38,* 366–371.
52. Mittal, K. L. *Contact Angle, Wettability and Adhesion;* CRC Press: Boca Raton, FL, 2009; Vol. 6.
53. Bhushan, B. *Principles and Applications of Tribology;* Wiley: New York, NY, 1999.
54. Bhushan, B. *Introduction to Tribology;* Wiley: New York, NY, 2002.
55. Young, T. An Essay on the Cohesion of Fluids. *Phil. Trans. R. Soc. Lond.* **1805,** *95,* 65–87.
56. Johnson, Jr. R. E.; Dettre, R. H. Wettability and Contact Angle. In *Surface and Colloid Science;* Matijevic, E., Ed.; Wiley-Interscience: New York, NY, 1969; Vol. 2, pp 85–153.
57. Huh, C.; Mason, S. G. Effects of Surface Roughness on Wetting (Theoretical). *J. Colloid Interface Sci.* **1977,** *60,* 11–38.
58. Good, R. J. Contact Angles and the Surface Free Energy of Solids. In *Surface and Colloid Science;* Good, R. J., Stromberg, R. R., Eds.; Plenum Press: New York, NY, 1979; Vol. 11, pp 1–29.
59. Wenzel, R. N. Resistance of Solid Surfaces to Wetting by Water. *Ind. Eng. Chem.* **1936,** *28,* 988–994.
60. Cassie, A. B. D.; Baxter, S. Wettability of Porous Surfaces. *Trans. Faraday Soc.* **1944,** *40,* 546–551.
61. Cassie, A. B. D. Contact Angles, Disc. *Faraday Soc.* **1948,** *3,* 11–16.
62. Marmur, A. Equilibrium and Spreading of Liquids on Solid Surfaces. *Adv. Colloid Interface Sci.* **1983,** *19,* 75–102.
63. Nosonovsky, M.; Bhushan, B. Multiscale Friction Mechanisms and Hierarchical Surfaces in Nano- and Bio-tribology. *Mater. Sci. Eng. R.* **2007,** *58,* 162–193.

64. Nosonovsky, M.; Bhushan, B. Hierarchical Roughness Makes Superhydrophobic Surfaces Stable. *Microelectron. Eng.* **2007,** *84,* 382–386.
65. Nosonovsky, M.; Bhushan, B. Hierarchical Roughness Optimization for Biomimetic Superhydrophobic Surfaces. *Ultramicroscopy* **2007,** *107,* 969–979.
66. Yevgeniy Kalinin, V.; Viatcheslav, B.; Robert Thorne, E. Contact Line Pinning by Microfabricated Patterns: Effects of Microscale Topography. *Langmuir* **2009,** *25* (9), 5391–5397.
67. Gibbs, J. W. *The Collected Works of Willard Gibbs, J;* Yale University Press: New Haven, CT, 1961; Vol. 1.
68. Oliver, J. F.; Huh, C.; Mason, S. G. Resistance to Spreading of Liquids by Sharp Edges. *J. Colloid Interface Sci.* **1977,** *59* (3), 568–581.
69. Dyson, D. C. Contact Line Stability at Edges—Comments on Gibbs Inequalities. *Phys. Fluids* **1988,** *31* (2), 229–232.
70. Shuttleworth, R.; Bailey, G. L. J. The Spreading of a Liquid over a Rough Solid. *Disc. Faraday Soc.* **1948,** *3,* 16–22.
71. Bikerman, J. J. Surface Roughness and Contact Angle. *J. Phys. Colloid Chem.* **1950,** *54* (5), 653–658.
72. Furmidge, C. G. *J. Colloid Sci.* **1962,** *17,* 309.
73. David, L. Wood, Slopes of Films and Drops on Substrates with Disjoining Pressure, Dissertation, University of Washington, 2008.
74. Nikolayev, V. S.; Gavrilyuk, S. L.; Gouin, H. Modelling of the Moving Deformed Triple Contact Line: Influence of the Fluid Inertia. *J. Colloid Interface Sci.* **2006,** *302,* 605–612.
75. Ondarcuhu, T.; Piednoir, A. Pinning of a Contact Line on Nanometric Steps during the Dewetting of a Terraced Substrate. *Nano Lett.* **2005,** *5* (9), 1744–1750.
76. Abbott, N. L.; Folkers, J. P.; Whitesides, G. M. Manipulation of the Wettability of Surfaces on the 0.1-Micrometer to 1-Micrometer Scale Through Micromachining and Molecular Self-assembly. *Science.* **1992,** *257* (5075), 1380–1382.
77. Dejonghe, V.; Chatain, D. Experimental-study of Wetting Hysteresis on Surfaces with Controlled Geometrical and/or Chemical Defects. *Acta Mater.* **1995,** *43* (4), 1505–1515.
78. Dettre, R.; Johnson, R. *Advances in Contact Angle, Wettability and Adhesion* (Chapter 8); Chemistry Series, No. 43; Fowkes, F. M., Ed.; American Chemical Society: Washington, DC, 1964; Vol. 43, p 136.
79. Cox, R. G. *J. Fluid Mech.* **1983,** *131,* 1.
80. Mason, S. G. In *Wetting, Spreading and Adhesion;* Padday, J. F., Ed.; Academic: New York, NY, 1978; p 321.
81. Garoff, S.; Schwartz, L. Contact Angle Hysteresis on Heterogeneous Surfaces. *Langmuir* **1985,** *1,* 219–230.
82. Marmur, A. Contact-angle Hysteresis on Heterogeneous Smooth Surfaces: Theoretical Comparison of the Captive Bubble and Drop Methods. *Colloids Surf. A.* **1998,** *136,* 209–215.
83. Zhang, X.; Mi, Y. Dynamics of a Stick-Jump Contact Line of Water Drops on a Strip Surface. *Langmuir* **2009,** *25,* 3212–3218.
84. Dobbs, H.; Bonn, D. Predicting Wetting Behavior from Initial Spreading Co-efficients. *Langmuir* **2001,** *17,* 4674–4676.

85. Aijuan, Z.; Hua, B.; Lei, L. Breath Figures: A Nature Inspired Preparation Method for Ordered Porous Films. *Chem. Rev.* **2015,** *115,* 9801–9868.
86. Beysens, D.; Steyer, A.; Guenoun, P.; Fritter, D.; Knobler, C. M. How does Dew Form. *Phase Transit.* **1991,** *31,* 219–246.
87. Kolb, M. Comment on "Scaling of the Droplet-size Distribution in Vapor Deposited Thin-films." *Phys. Rev. Lett.* **1989,** *62,* 1699.
88. Brilliantov, N. V.; Andrienko, Y. A.; Krapivsky, P. L.; Kurths, J. Polydisperse Adsorption: Pattern Formation Kinetics, Fractal Properties, and Transition to Order. *Phys. Rev. E.* **1998,** *58,* 3530–3536.
89. Family, F.; Meakin, P. Scaling of the Droplet-size Distribution in Vapor-deposited Thin Films. *Phys. Rev. Lett.* **1988,** *61,* 428–431.
90. Jean Louis, V.; Daniel, B.; Charles Knobler, M. Scaling Description for the Growth of Condensation Patterns on Surfaces. *Phys. Rev. A.* **1988,** *37,* 4965–4970.
91. Stricker, L.; Vollmer, J. Impact of Microphysics on the Growth of One-dimensional Breath Figures. *Phys. Rev. E.* **2015,** *92,* 042406. arXiv:1502.00566v2.
92. Haupt, M.; Miller, S.; Sauer, R.; Thonke, K.; Mourran, A.; Moelle, M. Breath Figures: Self-organizing Masks for the Fabrication of Photonic Crystals and Dichroic Filters. *J. Appl. Phys.* **2004,** *96* (6), 3065–3069.
93. Steyer, A.; Guenoun, P.; Beysens, D. Hexatic and Fat Fractal Structures for Water Droplets Condensing on Oil. *Phys. Rev. E. Stat. Phys. Plasmas Fluids Relat. Interdiscip. Top.* **1993,** *48,* 428–431.
94. Chan, D. Y. C.; Henry, J. D.; Jr. White, L. R. The Interaction of Colloidal Particles Collected at Fluid Interfaces. *J. Colloid Interface Sci.* **1981,** *79,* 410–418.
95. Picknett, R.G.; Bexon, R. The Evaporation of Sessile or Pendant Drops in Still Air. *J. Colloid Interface Sci.* **1977,** *61,* 336–350.
96. Ying-Song, Y.; Ziqian, W.; Ya-Pu, Z. Experimental and Theoretical Investigations of Evaporation of Sessile Water Droplet on Hydrophobic Surfaces. *J. Colloid Interface Sci.* **2012,** *365* (1), 254–259.
97. Rowan, S. M.; Newton, M. I.; McHale, G. *J. Phys. Chem.* **1995,** *99,* 13268.
98. Dong Yu, I.; Ho Jae, K.; Seung Woo, D.; Seon Ahn, H.; Hyun Sun, P.; Moriyama, K.; Moo hwan, K. Dynamics of Contact Line Depinning during Droplet Evaporation Based on Thermodynamics. *Langmuir* **2015,** *31,* 1950–1957.
99. Dong, R.; Yan, J.; Ma, H.; Fang, Y.; Hao, J. Dimensional Architecture of Ferrocenyl-based Oligomer Honeycomb-patterned Films: From Monolayer to Multilayer. *Langmuir* 2011, *27,* 9052–9056.
100. Yabu, H.; Hirai, Y.; Kojima, M.; Shimomura, M. Structured Films Containing Thermoresponsive Polymers and Their Surface Wettability. *Chem. Mater.* **2009,** *21,* 1787–1789.
101. Yabu, H.; Akagi, K.; Shimomura, M. Micropatterning of Liquid Crystalline Polyacetylene Derivative by Using Self-organization Processes. *Synth. Met.* **2009,** *159,* 762–764.
102. Xu, Y.; Zhu, B.; Xu, Y. A Study on Formation of Regular Honeycomb Pattern in Polysulfone Film. *Polymer* **2005,** *46,* 713–717.
103. Matsuyama, H.; Ohga, K.; Maki, T.; Teramoto, M. The Effect of Polymer Molecular Weight on the Structure of a Honeycomb Patterned Thin Film Prepared by Solvent Evaporation. *J. Chem. Eng. Jpn.* **2004,** *37,* 588–591.

104. Bormashenko, E.; Pogreb, R.; Stanevsky, O.; Bormashenko, Y.; Gendelman, O. Formation of Honeycomb Patterns in Evaporated Polymer Solutions: Influence of the Molecular Weight. *Mater. Lett.* **2005,** *59,* 3553–3557.
105. Ferrari, E.; Fabbri, P.; Pilati, F. Solvent and Substrate Contributions to the Formation of Breath Figure Patterns in Polystyrene Films. *Langmuir* **2011,** *27,* 1874–1881.
106. Peng, J.; Han, Y.; Yang, Y.; Li, B. The Influencing Factors on the Macroporous Formation in Polymer Films by Water Droplet Templating. *Polymer* **2004,** *45,* 447–452.
107. Li, J.; Peng, J.; Huang, W.; Wu, Y.; Fu, J.; Cong, Y.; Xue, L.; Han, Y. Ordered Honeycomb-structured Gold Nanoparticle Films with Changeable Pore Morphology: From Circle to Ellipse. *Langmuir* **2005,** *21,* 2017–2021.
108. Wu, X.; Wang, S. Integration of Photo-crosslinking and Breath Figures to Fabricate Biodegradable Polymer Substrates with Tunable Pores that Regulate Cellular Behavior. *Polymer* **2014,** *55,* 1756–1762.
109. Ding, J. Y.; Zhang, A. J.; Bai, H.; Li, L.; Li, J.; Ma, Z. *Soft Matter.* **2013,** *9,* 506–514.
110. Connal, L. A.; Qiao, G. G. *Adv. Mater.* **2006,** *18* (22), 3024.
111. Jianliang, G.; Lichao, S.; Yawen, Z.; Chunyin, M.; Lei, L.; Suyuan, X.; Vladimir, S. *Nanoscale.* **2012,** *4,* 278.
112. Galeotti, F.; Chiusa, I.; Morello, L.; Giani, S.; Breviario, D.; Hatz, S.; Damin, F.; Chiari, M.; Bolognesi, A. *Eur. Polym. J.* **2009,** *45* (11), 3027.
113. Connal, L. A.; Qiao, G. G. Preparation of Porous Poly (Dimethylsiloxane)-Based Honeycomb Materials with Hierarchal Surface Features and Their Use as Soft-lithography Templates. *Adv. Mater.* **2006,** *18* (22), 3024–3028.
114. Connal, L. A.; Gurr, P. A.; Qiao, G. G.; Solomon, D. H. From Well-defined Star Microgels to Highly Ordered Honeycomb Films. *J. Mater. Chem.* **2005,** *15* (12), 1286–1292.
115. Connal, L. A.; Vestberg, R.; Gurr, P. A.; Hawker, C. J.; Qiao, G. G. Patterning on Nonplanar Substrates: Flexible Honeycomb Films from a Range of Self-assembling Star Copolymers. *Langmuir* **2008,** *24* (2), 556–562.
116. Park, J. S.; Lee, S. H.; Han, T. H.; Kim, S. O. Hierarchically Ordered Polymer Films by Templated Organization of Aqueous Droplets. *Adv. Funct. Mater.* **2007,** *17* (14), 2315–2320.
117. Xiaoli, J.; Tianzhu, Z.; Lina, X.; Changling, W.; Xuefeng, Z.; Ning, G. Surfactant-induced Formation of Honeycomb Pattern on Micropipette with Curvature Gradient. *Langmuir* **2011,** *27* (9), 5410–5419.
118. Ling-Shu, W.; Liang-Wei, Z; Yang, O.; Zhi-Kang, X. Multiple Interfaces in Self-assembled Breath Figures. *Chem. Commun.* **2014,** *50,* 4024–4039.
119. Hernandez-Guerrero, M.; Min, E.; Barner-Kowollik, C.; Muller, A. H. E.; Stenzel, M. H. Grafting Thermoresponsive Polymers onto Honeycomb Structured Porous Films Using the RAFT Process. *J. Mater. Chem.* **2008,** *18* (39), 4718–4730.
120. Min, E. H.; Ting, S. R. S.; Billon, L.; Stenzel, M. H. Thermoresponsive Glycopolymer Chains Grafted onto Honeycomb Structured Porous Films via RAFT Polymerization as a Thermo-dependent Switcher for Lectin Concanavalin a Conjugation. *J. Polym. Sci. A. Polym. Chem.* **2010,** *48,* 3440–3455.
121. Yabu, H.; Hirai, Y.; Kojima, M.; Shimomura, M. Simple Fabrication of Honeycomb- and Pincushion-structured Films Containing Thermoresponsive Polymers and Their Surface Wettability. *Chem. Mater.* **2009,** *21* (9), 1787–1789.

122. Escale, P.; Rubatat, L.; Derail, C.; Save, M.; Billon, L. PH Sensitive Hierarchically Self-organized Bioinspired Films. *Macromol. Rapid Commun.* **2011,** *32* (14), 1072–1076.
123. Hai-yan, C.; Ting, Z.; Wei-chao, X.; Ming, Z. *Preparation of Photoluminescent Microporous Hybrid Films by Breath Figure Method,* Proceedings of the 3rd International Conference on Industrial Application Engineering, 2015.
124. Hirai, Y.; Yabu, H.; Matsuo, Y.; Ijiro, K.; Shimomura, M. *J. Mater. Chem.* **2010,** *20,* 10804–10808.
125. Li, L.; Zhong, Y.; Li, J.; Gong, J.; Ben, Y.; Xu, J.; Chen, X.; Ma, Z. *J. Colloid Interface Sci.* **2010,** *342,* 192–197.
126. Hirai, Y.; Yabu, H.; Matsuo, Y.; Ijiro, K.; Shimomura, M. *Macromol. Symp.* **2010,** *295,* 77–80.
127. Ou, Y.; Wang, L.; Zhu, L.; Wan, L.; Xu, Z. In-situ Immobilization of Silver Nanoparticles on Self-assembled Honeycomb-patterned Films Enables Surface-enhanced Raman Scattering (SERS) Substrates. *J. Phys. Chem. C.* **2014,** *118,* 11478–11484.
128. Kong, L.; Dong, R.; Ma, H.; Hao, J. Au NP Honeycomb-patterned Films with Controllable Pore Size and Their Surface-enhanced Raman Scattering. *Langmuir* **2013,** *29,* 4235–4241.
129. Chen, S.; Lu, X.; Huang, Z.; Lu, Q. In Situ Growth of a Polyphosphazene Nanoparticle Coating on a Honeycomb Surface: Facile Formation of Hierarchical Structures for Bio Application. *Chem. Commun.* **2015,** *51,* 5698–5701.
130. Chen, S.; Lu, X.; Hu, Y.; Lu, Q. Biomimetic Honeycomb Patterned Surface as the Tunable Cell Adhesion Scaffold. *Biomater. Sci.* **2015,** *3,* 85–93.
131. Mongkhontreerat, S.; Walter, M. V.; Cai, Y.; Brismar, H.; Hult, A.; Malkoch, M. Functional Porous Membranes from Amorphous Linear Dendritic Polyester Hybrids. *Polym. Chem.* **2015,** *6,* 2390–2395.
132. Tanaka, M.; Takebayashi, M.; Shimomura, M. *Macromol. Symp.* **2009,** *279,* 175–182.
133. Ponnusamy, T.; Lawson, L. B.; Freytag, L. CF.; Blake, D. A.; Ayyala, R. A.; John, V. T. *Biomatter.* **2012,** *2,* 77–86.
134. Fukuhira, Y.; Kitazono, E.; Hayashi, T.; Kaneko, H.; Tanaka, M.; Shimomura, M.; Sumi, Y. *Biomaterials* **2006,** *27,* 1797–1802.
135. Sunami, H.; Ito, E.; Tanaka, M.; Yamamoto, S.; Shimomura, M. *Colloids Surf. A.* **2006,** 548–551, 284–285.
136. Sun, H.; Li, W.; Wu, L. *Langmuir* **2009,** *25,* 10466–10472.
137. Li, X.; Wang, Y.; Zhang, L.; Tan, S.; Yu, X.; Zhao, N.; Chen, G.; Xu, J. *J. Colloid Interface Sci.* **2010,** *350,* 253–259.
138. Ponnusamy, T.; Chakravarty, G.; Mondal, D.; John, V. T. Novel Breath Figure Based Synthetic PLGA Matrices for in Vitro Modelling of Mammary Morphogenesis and Assessing Chemotherapeutic Response. *Adv. Healthcare Mater.* **2014,** *3,* 703–713.
139. Nishikawa, T.; Nishida, J.; Ookura, R.; Nishimura, S.; Wada, S.; Karino, T.; Shimomura, M. Honeycomb-patterned Thin Films of Amphiphilic Polymers as Cell Culture Substrates. *Mater. Sci. Eng. C.* **1999,** *8–9,* 495–500.
140. Sunami, H.; Ito, E.; Tanaka, M.; Yamamoto, S.; Shimomura, M. Effect of Honeycomb Film on Protein Adsorption, Cell Adhesion and Proliferation. *Colloids Surf. A.* **2006,** *284–285,* 548–551.

141. Huh, M.; Jung, M. H.; Park, Y. S.; Kang, T. B.; Nah, C.; Russell, R. A.; Holden, P. J.; Yun, S. I. Fabrication of Honeycomb Structured Porous Films from Poly (3-Hydroxybutyrate) and poly(3-Hydroxybutyrate-co-3-Hydroxyvalerate) via the Breath Figures Method. *Polym. Eng. Sci.* **2012,** *52,* 920–926.
142. Beattie, D.; Wong, K. H.; Williams, C.; Poole-Warren, L. A.; Davis, T. P.; Barner-Kowollik, C.; Stenzel, M. H. Honeycombstructured Porous Films from Polypyrrole-containing Block Copolymers Prepared Via RAFT Polymerization as a Scaffold for Cell Growth. *Biomacromolecules.* **2006,** *7,* 1072–1082.
143. Du, M. C.; Zhu, P. L.; Yan, X. H.; Su, Y.; Song, W. X.; Li, J. B. Honeycomb Self-assembled Peptide Scaffolds by the Breath Figure Method. *Chem. Eur. J.* **2011,** *17,* 4238–4245.
144. Li, H.; Jia, Y.; Du, M.; Fei, J.; Zhao, J.; Cui, Y.; Li, J. Self-organization of Honeycomb-like Porous TiO2 Films by Means of the Breath-figure Method for Surface Modification of Titanium Implants. *Chem. Eur. J.* **2013,** *19,* 5306–5313.
145. Heng, L. P.; Hu, R. R.; Chen, S. J.; Li, M. C.; Jiang, L.; Tang, B. Z. Ordered Honeycomb Structural Interfaces for Anticancer Cells Growth. *Langmuir* **2013,** *29,* 14947–14953.
146. Manabe, K.; Nishizawa, S.; Shiratori, S. Porous Surface Structure Fabricated by Breath Figures that Suppresses Pseudomonas Aeruginosa Biofilm Formation. *ACS Appl. Mater. Interface* **2013,** *5,* 11900–11905.
147. Yabu, H.; Shimomura, M. *Langmuir* **2005,** *21,* 1709.
148. Wu, M. C. *Proc. IEEE.* **1997,** *85,* 1833.
149. Jones, C. D.; Serpe, M. J.; Schroeder, L.; Lyon, L. A. *J. Am. Chem. Soc.* **2003,** *125,* 5292.
150. Kim, J.; Serpe, M. J.; Lyon, L. A. *Angew. Chem. Int. Ed.* **2005,** *44,* 1333.
151. Chari, K.; Lander, C. W.; Sudol, R. J. *Appl. Phys. Lett.* **2008,** *92,* 111916.
152. Pintani, M.; Huang, J.; Ramon, M. C.; Bradley, D. D. C. *J. Phys. Condens. Matter.* **2007,** *19,* 016203.

CHAPTER 2

AN OVERVIEW OF TECHNIQUES AND PARAMETERS AFFECTING SYNTHESIZATION OF SILVER NANOWIRES

AYŞE ÇELIK BEDELOĞLU[1*], GÖKÇENUR SAĞLAM[1], and SUKHWINDER K. BHULLAR[2,3]

[1]*Department of Fiber and Polymer Engineering, Bursa Technical University, Bursa 16310, Turkey*

[2]*Department of Mechanical Engineering, Bursa Technical University, Bursa 16310, Turkey*

[3]*Department of Mechanical Engineering, University of Victoria, Finnerty Road, Victoria, BC, Canada*

[*]*Corresponding author. E-mail: ayse.bedeloglu@btu.edu.tr*

CONTENTS

ABSTRACT

Over the last decades, the applications of biomaterial and nanotechnology have been growing interest in healthcare and also helping in understanding and controlling these materials at nanoscale. Nanostructures such as rods, wires, belts, ribbons, and tubes have been intensively investigated for years due to their unique features including small sizes and high surface-to-volume ratios. Among these, nanowires in particular have been the focus of many recent studies in a variety of different disciplines such as biotechnology, nanobiotechnology, and bioengineering. Since bulk silver has the highest electrical, thermal conductivity, and low-cost among all noble metals, therefore, by combining optical transparency and flexibility, silver nanowire (AgNW) networks are useful in the development of electronic and optoelectronic devices. Overall, the focus of this chapter is to study synthesis techniques and affecting parameters in AgNWs synthesization. In addition, biomedical applications of AgNWs are highlighted.

2.1 GENERAL OVERVIEW

Nanotechnology has become an important area of modern research related to design, synthesis, and manipulation of particle structure ranging approximately from one to hundred nano or micrometers. Also, the dimensional features such as one-, two-, and three-dimensional nanostructures play a very important role in determining the critical properties of materials.[1] In addition, the size of the structures provides an important role to control over many of the physical and chemical properties of nanoscale materials. Different morphologies of nanostructures in the forms of spheres, discs, rods, wires, belts,[2] rice,[3] ribbons,[4] stars, prisms,[5,6] and cubes[7–9] actively investigated for years as shown in Figure 2.1.

Besides all of the nanostructures, in particular, metal nanowires[10–13] have been the recent focus of many researchers due to their unique electronic, optical,[14] physicochemical, thermal, and catalytic properties.[15,16] They have promising applications in areas such as flexible transparent electrodes, highly stretchable conductors,[17–19] plasmonic fibers, photonic crystals,[15] optical polarizers,[20] catalysts, biomolecular sensing, antimicrobials, batteries, water purification, gas sensors, and potential technological applications[15,18] because of their dual plasmon resonance bands,[19,20] good storage[21] superior in electrical (6.3 × 107 S m^{-1}) and thermal (429 $Wm^{-1}K^{-1}$) conductivities.[18–22] Furthermore, silver nanowires (AgNWs) are characterized by using different techniques including the ultraviolet–visible (UV–Vis) spectroscopy, scanning electron microscopy (SEM), transmission electron microscopy (TEM), Fourier-transform infrared (FTIR) spectroscopy, X-ray diffraction (XRD), and nuclear magnetic resonance (NMR).[23]

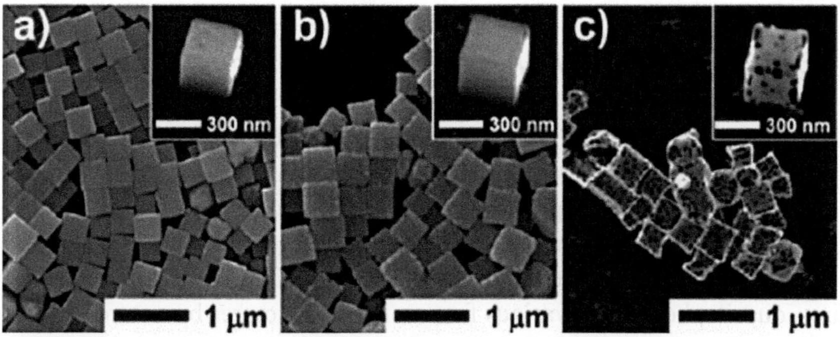

FIGURE 2.1 SEM images: (a) PVP-Ag nanocubes, (b) gold (Au)-electrodeposited PVP-Ag nanocubes, and (c) Au nanoboxes obtained by chemical etching of (b). Reproduced from Ref. [8] with permission of The Royal Society of Chemistry.

2.2 SYNTHESIS TECHNIQUES FOR SILVER NANOWIRES

Chemical synthesis,[24,25] microemulsion, liquid-phase,[26] template,[27,30] hydrothermal synthesis,[28] polyol process,[29] photoinduced reduction, electrospinning,[31] solution-phase,[32] etc., have been investigated for years to synthesis nanowires with uniform and controllable dimensions.

A variety of techniques to synthesis AgNWs with uniform and controllable dimensions such as chemical synthesis,[24,25] microemulsion, liquid-phase,[26] template,[27,30] hydrothermal synthesis,[28] polyol process,[29] photoinduced reduction, electrospinning,[31] and solution-phase[32] have been studied. A number of researchers as listed in Table 2.1 analyzed these different synthesis techniques for AgNWs techniques and they are described in Sections 2.1–2.7.

TABLE 2.1 Different Synthesis Techniques for Silver Nanowires.

Synthesis techniques	References
Polyol process	[29–36]
Chemical reduction synthesis	[24–25, 37–41]
Microemulsion	[26, 27, 42–48]
Photoinduced reduction	[49, 50]
UV-initiated photoreduction	[51–54]
Electrochemical synthetic synthesis	[55–59]
Template-directed synthesis	[60–71]

2.2.1 POLYOL SYNTHESIS

Polyol synthesis was developed by Lin et al.[34] for synthesis of colloidal particles made of metals and alloys. The typical reaction of this process involves the reduction agent, precursor metal, solvent, and stabilizer at a promoted temperature. In the polyol synthesis, the nucleation and growth of silver atoms are directed in the solution to produce nanocubes, nanowires, quasi-spheres, and various forms of nanostructures by controlling the ratio of the capping agent and silver precursor.[35] It was reported by Wiley et al.[32] that the polyol synthesis of AgNWs involves heating ethylene glycol (EG) (polyol) with $AgNO_3$ (salt precursor) and PVP (polymeric capping agent) to form metal colloids. Also, polyol synthesis to synthesize AgNW without metal salts was reported by Jian et al.[36] The authors in this study analyzed the relationship between the molar ratio and different PVP molecular weights to determine importance of PVP in the synthesis of AgNWs with a high aspect-ratio. They concluded that the addition of PVP with molecular weight of 360 K enabled the synthesis of optimal AgNWs with high aspect ratios and PVP/$AgNO_3$ molar ratio 2.5.[36]

2.2.2 CHEMICAL REDUCTION SYNTHESIS

In the synthesis of nanoparticles or nanowires via chemical reduction synthesis, organic and inorganic reducing agents such as sodium citrate, ascorbate, sodium borohydride ($NaBH_4$), elemental hydrogen, polyol process, tollens reagent, nitrogen, N-dimethylformamide (DMF), and poly(ethylene glycol)-block copolymers are used for reduction of silver ions to metallic silver[37–39] Polymeric compounds such as poly(vinyl alcohol)

(PVA), poly(vinylpyrrolidone), poly(ethylene glycol), poly(methacrylic acid), and poly(methyl methacrylate) have been used as protective agents to stabilize dispersive nanoparticles.[40] The synthesis of silver nanorods and nanowires were prepared by chemical reduction of a metal precursor ($AgNO_3$) using the strong reducing agent $NaBH_4$ in the presence of surfactant cetyltrimethyl ammonium bromide (CTAB) to stabilize the nanoparticles was studied by Nikhil et al.[41]

2.2.3 MICROEMULSION TECHNIQUES

Microemulsion synthesis has admitted appreciable attention in recent years. This synthesis technique allows size-controllable and uniform preparation of nanoparticles. It was reported by authors that reverse micelles are good candidates for fabrication of nanostructures.[42] Also, it was recommended by Pileni[43] that sodium bis(2-ethylhexyl) sulfosuccinate (AOT) is the most common surfactant used to form reverse micelles which are composed of two hydrophobic chains and a sulfonate hydrophilic head.[44] They found the ternary systems types of alkane/AOT/water show enormous advantages. Alkane is used due to its low toxicity but high viscosity as oil solvent in microemulsion. In another study, authors proposed different synthesis: In one aspect, silver ions were reduced by $NaBH_4$ or hydrazine (N_2H_4) instead of AOT surfactant in their reaction system.[45] In the other aspect, silver nanoparticles were formed in AOT microemulsion[43] when they used AOT microemulsion for fabricating silver bromide nanoparticles since the solubility of potassium bromide (KBr) is higher than that of the other salt in the AOT/alkene/water microemulsions.[46] This technique requires high toxic organic solvents and large amounts of surfactant and organic solvent must be separated and removed from the final product.[47] Advantages of this technique can be found as catalysts to catalyze most organic reactions.[48]

2.2.4 PHOTOINDUCED REDUCTION

It is reported by Shchukin[49] that in photoinduced reduction technique, the particle growth process was controlled using dual-beam illumination of nanoparticles by using poly(styrene sulfonate)/poly(allylamine hydrochloride) polyelectrolyte capsules as microreactors. Also, it is studied by authors that the photoinduced synthesis can be used to alter silver nanospheres to triangular silver nanocrystals (nanoprisms) and the citrate and poly(styrene sulfonate) may be used as stabilizing agents.[50]

2.2.5 UV-INITIATED PHOTOREDUCTION

Several studies are reported in literature describing that silver nanoparticles such as nanosphere, nanowire, and dendrite have also been synthesized by an ultraviolet irradiation photoreduction technique at room temperature using the PVA (as protecting and stabilizing agent), poly(acrylic acid),[51] citrate[52], and collagen.[52,53] The concentration, and rate of both PVA and citrate play significant role in the size of the nanorods and dendrites as a weak reducing agent and a capping agent which stabilize the particles.[54]

2.2.6 ELECTROCHEMICAL SYNTHETIC SYNTHESIS

The first study on electrochemical synthetic synthesis was reported by authors Reetz et al.[55] with the development of size-selective metal particles in an organic phase. Later, it was reported by Sanchez et al.,[56] Reetz & Helbig,[57a] and Reetz et al.[57b] that these particles could be prepared electrochemically by using tetraalkylammonium served as the supporting electrolyte and stabilizer for the metal nanoclusters. Also, it was reported by Sanchez et al.,[58] and Reetz and Helbig[59] that tetraalkylammonium salts have one head group to bond with an inorganic ion and carbon chain is not long enough to form an effective hydrophobic domain around the metal clusters to stabilize them in the aqueous phase. The researchers investigated that easier and more convenient electrochemical synthesis to prepare silver nanoparticles or nanowires in aqueous phase under the protection of PVP. The role of PVP is found to accelerate the silver particle formation and may be directly applied in the electroplating system. This electrochemical synthesis will make possible large-scale preparation of the size-controlled silver nanoparticles, as well as, other metals, such as gold and platinum.

2.2.7 TEMPLATE-DIRECTED SYNTHESIS

A number of research studies are presented by authors in literature on the template-directed synthesis technique. The hard and soft templates have been used for synthesis of AgNWs by researchers, especially, hard templates that include channels contained in membrane of alumina[60–62] or track-etched polycarbonate,[63] pores within zeolite[64] or mesoporous silica,[65] carbon nanotubes,[66,67] and solid materials. Furthermore, studies on a range of soft templates were investigated. It was found soft templates including polymer film of PVA,[17] deoxyribonucleic acid (DNA) chains,[63,68] meso-structures self-assembled from diblock copolymers,[19] rod-shaped micelles of cetyltrimethylammonium bromide,[69] liquid crystalline phases of oleate,[70] or sodium bis(2-ethylhexyl) sulfosuccinate/p-xylene/water[71] and arrays of calix[4] hydroquinone nanotubes[67] were effective in synthesis of nanowires with uniform and controllable dimensions.

2.3 PARAMETERS EFFECTING SYNTHESIS OF SILVER NANOWIRES

The morphology, yield, and dimension of AgNWs can be changed by altering preparation procedure of materials and conditions.[72] Several studies have been reported by authors on parameters as given in Table 2.2 which affect the synthesis conditions of AgNWs, properties of the product and yield. A brief description of these is given in Sections 2.3.1–2.3.5.

TABLE 2.2 Parameters Effecting Synthesis of Silver Nanowires.

Effective parameters	References
Reducing and capping agents	[73–81]
Molar ratios	[82–87]
Molecular weight	[24, 26, 27, 88–90]
Base materials	[20, 76, 91–97]
Temperature	[39, 98]

2.3.1 EFFECT OF REDUCING AND CAPPING AGENTS

In many studies, polymeric compounds such as polyol especially; poly (vinylpyrrolidone),[73] poly(ethylene glycol), glycerol,[73] CTAB,[74] oleyl-amine,[75] KBr,[76] sodium citrate, polymethyl methacrylate, etc.,[77] have been reported as the effective reductants and capping agents for the synthesis of AgNWs. Glycerol with three hydroxyl groups and secondary hydroxyl group shows powerful reduction ability compared to EG. Since, glycerol has a higher viscosity (1412 mPa s at 20°C) compared to EG (25.66 mPa s at 16°C, boiling point ~195–198°C) and also, viscosity decreases with a reaction temperature increase, the reduction rate in EG is faster than that in glycerol. Besides, the reduction rate and nucleation of the Ag^0 are also relevant to the concentration and the volume ratio of EG and glycerol.[78] The effect of glycerol and EG on the formation of AgNW structures was studied by Jia et al.[79] mixing glycerol as solvents at different ratios. Also, it was proved by Jia et al.[79] that the formation of uniform AgNWs is the result of the synergetic effect between EG and glycerol. The roles of citrate and PVP as the capping agents in the formation of Ag nanowires have been also studied by Zhang et al.,[5] Caswell et al.,[80] and Korte.[81] The authors described that at room temperature (25°C), citrate serves as the capping agent, since it cannot reduce silver salt, and it strongly combines with Ag^+ to form Ag^-citrate complex (pH = 7.1). It was also mentioned by authors that because of the rapid reaction, more nuclei will be formed quickly quasi-spherical particles and resulting in decreased average particle size. The authors also discussed further that the citrate concentration exceeds limit detection which prompt to the formation of quasi-spherical particles rather than AgNWs.[80] In the case of PVP as the capping agent, it was reported by authors that the concentration of PVP is important in the formation spherical Ag particles.[5,81]

2.3.2 EFFECT OF PVP AND PVP/$AGNO_3$ RATIO

The effect of PVP and PVP/$AgNO_3$ ratio is reported in several studies[82–87] as protecting agent in polyol synthesis technique due to its high efficiency. It was studied by Lin and Hsueh[82] that PVP is a homopolymer with a long and soft polyvinyl chain and contains an amide group. Also, it was discussed that the N and O atoms of this polar group probably have a strong affinity for the silver ions and metallic silver.[82] In another research study, it was found that PVP is a reducing agent toward silver ion and silver nucleation hence, has a great impact on the morphology and structure of AgNWs.[83] It was also further discussed by authors that PVP helps to stabilize the growing particles with various faces and generally used as a stabilizer to prevent agglomeration is the main role of the surfactant.[84] Zongtao et al.[85] discussed that the backbone of the vinyl polymer forms a hydrophobic domain, which surrounds metal particles, whereas the hydrophilic pendant groups of the polymer interact with water or polar solvent. In other words, the steric effect of the polymer in the surface of metal particles prevents the particles from agglomeration. The researchers investigated in a further study[86] that the average diameter of AgNWs depends on the average number of repeating unit in one PVP macromolecule and molar ratios between the capping agent (PVP) and the precursor salt ($AgNO_3$). It was shown by Sahin et al.[87] that with the molar ratio 6 of PVP/$AgNO_3$ the mostly results in the production of silver nanoparticles while the PVP/$AgNO_3$ molar ratio 9 composed AgNWs.[87] Furthermore, it was discussed with the molar ratio 3, the resulted product contains nanorods as illustrated in Figure 2.2.

2.3.3 EFFECT OF MOLECULAR WEIGHT OF PVP

Yunxia et al.[88] and Jones & Tamplin[89] suggested that the PVP with higher molecular weight should more easily induce the PVP-Ag coordination compound to arrange in a one-dimensional nanowires.[88] It was found that both yields and aspect ratios of the obtained amount of Ag nanowires increased when the molecular weight (Mw) of PVP increase. The different average Mws of PVP (15,000, 38,000, 58,000, 200,000, and 800,000) were analyzed by Tornblom and Henriksson.[90] They concluded that with 15,000 molecular weight of PVP nanoparticles (~60%) and nanowires (~40%) were formed while with Mw 800.000 obtained nanowires were ~99%.

2.3.4 EFFECT OF BASE MATERIALS

According to research studies cation and the anion play an important role in the formation and growth of AgNWs.[5,91] It was also investigated by authors that the addition of a trace amount of salt, such as sodium chloride (NaCl),[92] Iron(III) nitrate ($Fe(NO)_3$), Platinum(II) chloride($PtCl_2$),[93,94] Silver bromide (AgBr),[95] KBr,[76] and copper(I) chloride (CuCl) has been affected influence the morphology of the final metal products. There are many researchers investigating the effect of silver nitrate and silver chloride. It was reported by Liu et al.[96,] and Korte et al.[97] upon using silver nitrate instead of silver chloride, the reduction of Ag^+ ions to metallic silver was faster, thus silver nanoparticles instead of AgNWs were obtained.

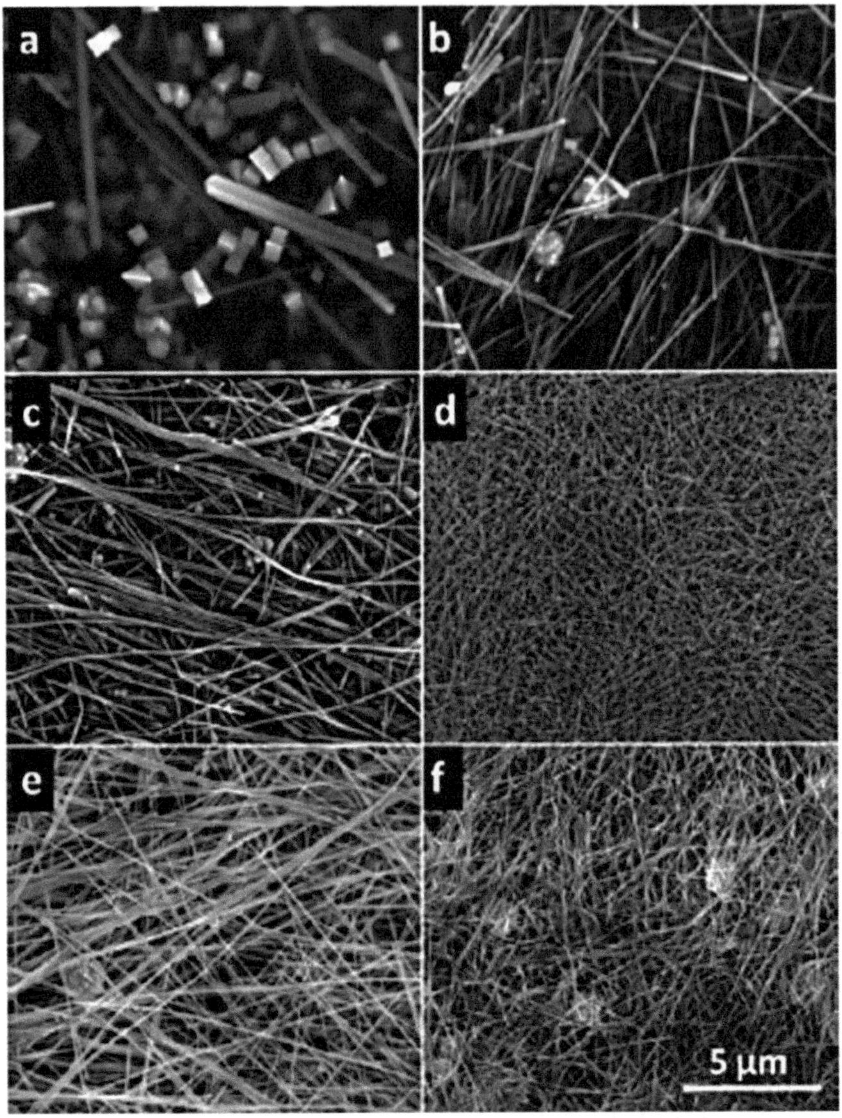

FIGURE 2.2 SEM images of Ag nanowires synthesized at different PVP:$AgNO_3$ molar ratios of (a) 3:1, (b) 4.5:1, (c) 6:1, (d) 7.5:1, (e) 9:1, and (f) 11:1. Reprinted with permission from Ref. [87]. Copyright (2011) American Chemical Society.

2.3.5 EFFECT OF TEMPERATURE

It was investigated by Coskun et al.[98] that temperature affects the reduction ability of polyol alcohol and temperature can be adapted to control the nucleation and growth processes of metals and synthesize various forms of nano-particles.[39] Also, researcher investigated that temperature is an important parameter for Ag nanowire synthesis. The change in nanowire length and diameter with respect to the growth temperature at 110, 130, 150, 170, 190, and 200°C shows that, below a critical temperature, high aspect ratio Ag nanowire synthesis is not possible. High temperatures, upon the 150°C, are also crucial for the conversion of EG to glycolaldehyde, which reduces Ag^+ ions to Ag.[98]

To summarize AgNWs synthesis, techniques, and effective parameters in synthesization, polyol, and chemical reduction synthesis are promising, since these techniques provide simple, effective, low-cost, and high yield production compared to others. Moreover, Bi and Ye[33] reported that by controlling the parameters such as reaction time, addition of control agent, temperature, and molar ratio between capping agent and metallic precursor, a reasonable controlled growth of AgNWs may be achieved as illustrated in Figure 2.3.

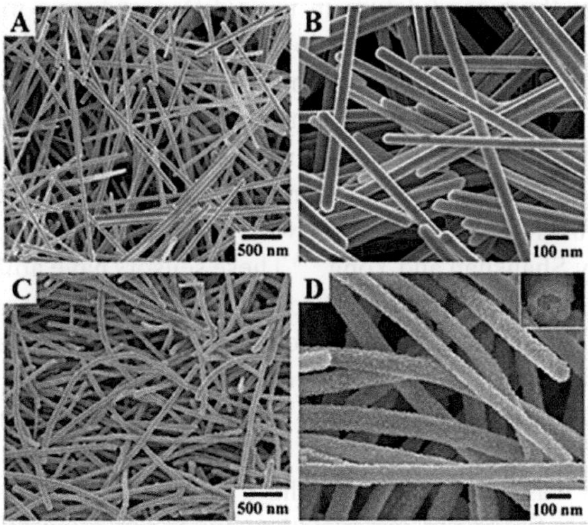

FIGURE 2.3 (A, B) SEM images of Ag nanowires and (C, D) SEM images of as-prepared Ag/AgCl core–shell nanowires. Reprinted from Ref. [33] with permission of The Royal Society of Chemistry.

2.4 APPLICATIONS

Silver nanomaterials in a variety of forms such as nanoparticles, nanotubes, nanorods, nanospheres, nanodisks, nanoplates, and nanowires offer their potentials for biomedical and many other applications. As our main focus in this review was on AgNWs, therefore, some of their biomedical applications are discussed in the following section.

2.4.1 BIOMEDICAL APPLICATIONS

Nanomaterials such as nanotubes, nanoparticles, and nanowires are currently emerging materials for the application of electrodes for a variety of sensors, batteries and supercapacitors, and photovoltaics.[99–104] Also, their unique properties such as high surface to volume ratio, excellent electrical transport behavior, and high surface area promote their candidature in biomedical applications. For example, biomedical devices industries have benefited from the nanowires, nanopores, and nanoneedles such as biosensors for the detection of biological targets-DNA;[105–109] neural activity;[110,111] cancer therapeutics;[112,113] antimicrobial susceptibility,[114,115] and contrast agents in imaging.[116] Silver nanomaterials promote their candidature in their different forms in biomedical industry for a number of applications such as antibacterial, antimicrobial, antiviral, biosensors, antitumor, optical imaging, and imaging intensifier.[117] To detect multiple stimuli simultaneously like stretch, pressure, temperature, or touch utilizing highly sensitive and wearable sensors is recent interest of researcher. Using screen-printed AgNW electrode multifunctional sensors which are highly stretchable were developed by Shanshan and Yong.[118] These sensors have ability to detect strain up to 50% and pressure up to 1.2 MPa. Also, it was reported by Po-Chun et al.[119] that AgNW electrode sensors can be used for several wearable applications for human motions, prosthetic and healthcare as shown in Figure 2.4. Furthermore, they have finger touch sensitivity, fast response time equal to 40 ms and good pressure mapping function. Also, it was mentioned by authors that AgNWs are promising for medical textile fibers and smart clothes due to their flexibility, conductivity, and heat capacity.

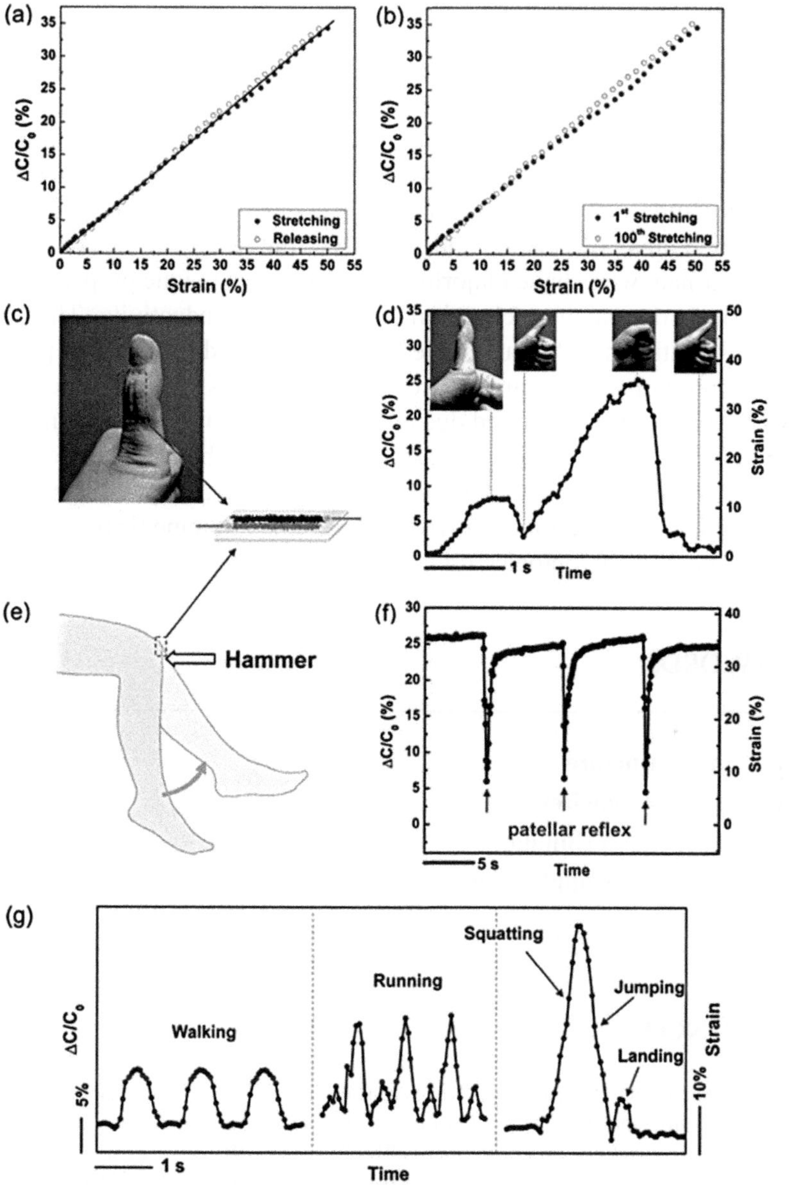

FIGURE 2.4 Silver nanowire electrode sensors for healthcare and their strain sensing capability Reprinted with permission from Ref. [119]. Copyright (2015) American Chemical Society.

2.5 CONCLUSIONS

In this chapter, mainly, a review of synthesization techniques of the AgNWs and the different parameters affecting AgNWs morphology were presented. An overview of AgNWs was also presented by giving recent applications of AgNWs especially in biomedical field. One-dimensional nanostructures (wires, belt, rod, etc.) possess many important applications in diverse areas of science and industry. Studies about AgNWs among the other nanowires have majority, since silver has unique properties such as optical transparency, electrical, and thermal conductivity. Therefore, various synthesis techniques have been investigated for years to produce nanowires with uniform and controllable dimensions. Among these, polyol synthesis is widely preferred due to its scalability, highest efficiency, and easy processing conditions. Since AgNWs are flexible, suitable for large-scale manufacturing and have low-cost processes, they can be a very promising candidate for industrial applications in biomedical, electronic, electromechanical, optoelectronic, and photonic devices.

KEYWORDS

- **nanostructure**
- **silver nanowires**
- **synthesis techniques**
- **biomedical applications**

REFERENCES

1. Miller, M. S.; O'Kane, J. C.; Niec, A.; Carmichael, R. S.; Carmichael, T. B. Silver Nanowire/Optical Adhesive Coatings as Transparent Electrodes for Flexible Electronics. *ACS Appl. Mater Interfaces* **2013,** *5,* 10165–10172.
2. Liu, B.; Zhao, X. A Facile Synthesis of Ordered Ultralong Silver Nanobelts. *Mater. Res. Bull.* **2009,** *44,* 682–687.
3. Yongyun, M.; Chuan, W.; Hongwei, Y. Rapid and Uniform Synthesis of Silver Nanowires via Rice-shaped Silver Nucleant. *Mater. Lett.* **2015,** *142,* 102–105.
4. Marchal-Roch, C.; Mayer, C. M.; Michel, A.; Dumas, E.; Liu, F. X.; Séchersse, F. Facile Synthesis of Silver Nano/Micro-ribbons or Saws Assisted by Polyoxomolybdate as

Mediator Agent and Vanadium(iv) as Reducing Agent. *Chem. Commun.* **2007,** *36,* 3750–3752.

5. Zhang, J.; Liu, H.; Zhan, P.; Wang, Z.; Ming, N. Controlling the Growth and Assembly of Silver Nanoprisms. *Adv. Funct. Mater.* **2007,** *17,* 1558–1566.
6. Skrabalak, S. E.; Au, L.; Li, X.; Xia, Y. Facile Synthesis of Ag Nanocubes and Au Nanocages. *Nat. Prot.* **2007,** *2,* 2182–2190.
7. Sun, Y.; Xia, Y. Shape-controlled Synthesis of Gold and Silver Nanoparticles. *Science* **2002,** *298,* 2176–2179.
8. Okazaki, K.; Yasui, J.; Torimoto, T. Electrochemical Deposition of Gold Frame Structure on Silver Nanocubes. *Chem. Commun.* **2009,** *0,* 2917–2919.
9. Zebin, R.; Xiaoyi, L.; Jingxia, G.; Ruibo, W.; Yanni, W.; Mingdi, Z. Solution-based Metal Enhanced Fluorescence with Gold and Gold/silver Core–shell Nanorods. *Opt. Commun.* **2015,** *357,* 156–160.
10. Comini, E.; Faglia, G.; Sberveglieri, G.; Pan, Z.; Wang, Z. L. Stable and Highly Sensitive Gas Sensors Based on Semiconducting Oxide Nanobelts. *Appl. Phys. A.* **2002,** *81,* 1869.
11. Fields, L. L.; Zheng, J. P.; Cheng, Y.; Xiong, P. Room-temperature Low-power Hydrogen Sensor Based on a Single Tin Dioxide Nanobelt. *Appl. Phys. Lett.* **2006,** *88,* 263–302.
12. Kwon, W. J.; Kim, J. H.; Jin, S.; Lee, S. U.; Lim, Y. S. Metal Nanobelt and Method of Manufacturing the Same, and Conductive Ink Composition and Conductive Film Comprising the Same. US20120128996 A1, 2012.
13. Wana, Y.; Guob, Z.; Jianga, X.; Fanga, K.; Lub, X.; Zhanga, Y.; Gua, N. Quasi-spherical Silver Nanoparticles: Aqueous Synthesis and Size Control by the Seed-mediated Lee–Meisel Method. *J. Colloid Interface Sci.* **2013,** *394,* 263–268.
14. Kamat, P. V. Photophysical, Photochemical and Photocatalytic Aspects of Metal Nanoparticles. *J. Phys. Chem.* **2002,** *106,* 7729.
15. Angmo, D.; Krebs, F. C.; Lieber. Flexible ITO-free Polymer Solar Cells. *J. Appl. Polym. Sci.* **2013,** *129,* 1–14.
16. Yang, C.; Wong, C. P.; Yuen, M. M. F. Printed Electrically Conductive Composites: Conductive Filler Designs and Surface Engineering. *J. Mater. Chem. C.* **2013,** *1,* 4052–4069.
17. Gao, J. J.; Qu, R. J.; Tang, B.; Wang, C. H.; Ma, Q. L.; Sun, C. M. Preparation and Property of Polyurethane/Nanosilver Complex Fibers. *J. Nanopart. Res.* **2011,** *13,* 5289–5299.
18. Yu, Y. H.; Ma, C. C. M.; Teng, C. C.; Huang, Y. L.; Lee, S. H.; Wang, I.; Wei, M. H. Electrical, Morphological, and Electromagnetic Interference Shielding Properties of Silver Nanowires and Nanoparticles Conductive Composites. *Mater. Chem. Phys.* **2012,** *136,* 334–340.
19. De, S.; Higgins, T. M.; Lyons, P. E.; Doherty, E. M.; Nirmalraj, P. N.; Blau, W. J.; Boland, J. J.; Coleman, J. N. Silver Nanowire Networks as Flexible, Transparent, Conducting Films: Extremely High DC to Optical Conductivity Ratios. *ACS Nano.* **2009,** *28,* 1767–1774.
20. Tang, X.; Tsuji, M. Syntheses of Silver Nanowires in Liquid Phase. In *Nanowires Science and Technology;* Nicoleta Lupu., Ed.; ISBN 978-953-7619-89-3, InTech: London, 2010.

21. Shao, J.; Zhang, X. J.; Tian, H. K.; Geng, Y. H.; Wang, F. S. Donor–acceptor–donor Conjugated Oligomers Based on Isoindigo and Anthra[1,2-b]Thieno[2,3-d]Thiophene for Organic Thin-film Transistors: The Effect of the Alkyl Side Chain Length on Semiconducting Properties. *J. Mater. Chem. C.* **2015,** *3,* 7567–7574.
22. Ge, J.; Yao, H. B.; Wang, X.; Ye, Y. D.; Wang, J. L. Stretchable Conductors Based on Silver Nanowires: Improved Performance Through a Binary Network Design. *Angew. Chem. Int. Ed.* **2013,** *125,* 1698.
23. Borodko, Y.; Habas, S. E.; Koebel, M.; Yang, P.; Frei, H.; Somorjai, G. A. Probing the Interaction of Poly(Vinylpyrrolidone) with Platinum Nanocrystals by UV-Raman and FTIR. *Phys. Chem. B.* **2006,** *110,* 23052–23059.
24. Jun, H.; Feng, Z.; Jing, W.; John Xiao, Q. Synthesis of Single-crystalline Fe Nanowires Using Catalyst-assisted Chemical Vapor Deposition. *Mater. Lett.* **2015,** *160,* 529–532.
25. Kim, S. H.; Choi, B. S.; Kang, K.; Choi, Y. S.; Yang, S. I. Low Temperature Synthesis and Growth Mechanism of Ag Nanowires. *J. Alloy. Compd.* **2007,** *433,* 261–264.
26. Xu, J.; Hu, J.; Peng, C.; Liu, H.; Hu, Y. A Simple Approach to the Synthesis of Silver Nanowires by Hydrothermal Process in the Presence of Gemini Surfactant. *J. Colloid Interface Sci.* **2006,** *2,* 689–693.
27. Keren, K.; Krueger, M.; Gilad, R.; Ben-Yoseph, G.; Sivan, U.; Braun, E. Sequence-specific Molecular Lithography on Single DNA Molecules. *Science* **2002,** *297,* 72–75.
28. Wang, Z.; Zhao, Z.; Qiu, J. A General Strategy for Synthesis of Silver Dendrites by Galvanic Displacement Under Hydrothermal Conditions. *J. Phys. Chem. Solids.* **2008,** *69,* 1296–1300.
29. Sun, Y.; Xia, Y. Large-Scale Synthesis of Uniform Silver Nanowires through a Soft, Self-seeding, Polyol Process. *Adv. Mater.* **2002,** *14,* 833.
30. Yang, R.; Sui, C.; Gong, J.; Qu, L. Silver Nanowires Prepared by Modified AAO Template Method. *Mater. Lett.* **2007,** *61,* 900.
31. Li, D.; Ouyang, G.; McCann, J. T.; Xia, Y. N. Collecting Electrospun Nanofibers with Patterned Electrodes. *Nano Lett.* **2005,** *5,* 913.
32. Wiley, B.; Sun, Y.; Mayers, B.; Xia, Y. Shape-controlled Synthesis of Metal Nanostructures: The Case of Silver. *Chem. Eur. J.* **2005,** *11,* 454–463.
33. Bi, Y.; Ye, J. In Situ Synthesis of Ag/AgCl Core–shell Nanowires and Their Photocatalytic Properties. *Chem. Commun.* **2009,** *0,* 6551–6553.
34. Lin, J. Y.; Hsueh, Y. L.; Huang, J. J.; Wu, J. R. Effect of Silver Nitrate Concentration of Silver Nanowires Synthesized Using a Polyol Method and Their Application as Transparent Conductive Films. *Thin Solid Films.* **2015,** *584,* 243–247.
35. Lin, J. Y.; Hsueh, Y. L.; Huang, J. J. The Concentration Effect of Capping Agent for Synthesis of Silver Nanowire by Using the Polyol Method. *J. Solid State Chem.* **2014,** *214,* 2–6.
36. Jian, L. Y.; Yu, L. H.; Jung, H. J. The Concentration Effect of Capping Agent for Synthesis of Silver Nanowire by Using the Polyol Method. *J. Solution Chem.* **2014,** *214,* 2–6.
37. Merga, G.; Wilson, R.; Lynn, G.; Milosavljevic, B.; Meisel, D. Redox Catalysis on "Naked" Silver Nanoparticles. *J. Phys. Chem. C.* **2007,** *111,* 12220–12206.
38. Oliveira, M.; Ugarte, D.; Zanchet, D.; Zarbin, A. Influence of Synthetic Parameters on the Size, Structure, and Stability of Dodecanethiol-stabilized Silver Nanoparticles. *J. Colloid Interface Sci.* **2005,** *292,* 429–435.

39. Yongyun, M.; Chuan, W.; Hongwei, Y. Rapid and Uniform Synthesis of Silver Nanowires Via Rice-shaped Silver Nucleant. *Mater. Lett. J.* **2015,** *142,* 102–105.
40. Oliveira, M.; Ugarte, D.; Zanchet, D.; Zarbin, A. Influence of Synthetic Parameters on the Size, Structure, and Stability of Dodecanethiol-stabilized Silver Nanoparticles. *J. Colloid Interface Sci.* **2005,** *292,* 429–435.
41. Nikhil, R. J.; Latha, G.; Catherine, J. M. One-dimensional Colloidal Gold and Silver Nanostructures. *Chem. Commun.* **2001,** *105,* 617–618.
42. May, S.; Shaul, A. B. Molecular Theory of the Sphere-to-rod Transition and the Second CMC in Aqueous Micellar Solutions. *J. Phys. Chem. B.* **2001,** *105,* 630.
43. Pileni, M. P., Ed.; *Structure and Reactivity in Reverse Micelles;* Elsevier: Amsterdam, the Netherlands, 1989.
44. Wongwailikhit, K.; Horwongsakul, S. The Preparation of Iron (III) Oxide Nanoparticles Using W/O Microemulsion. *Mater. Lett.* **2011,** *65,* 2820–2822.
45. Andersson, M.; Pedersen, J. S.; Palmqvist, A. E. C. Silver Nanoparticle Formation in Microemulsions Acting Both as Template and Reducing Agent. *Langmuir* **2005,** *21,* 11387.
46. Chero, C. H.; Gan, L. M.; Shah, D. O. *J. Dispers. Sci. Technol.* **1990,** *11,* 593.
47. Zhang, W.; Qiao, X.; Chen, J. Synthesis of Nanosilver Colloidal Particles in Water/oil Microemulsion. *Colloids Surf. A Physicochem. Eng. Asp.* **2007,** *299,* 22–28.
48. Cozzoli, P.; Comparelli, R.; Fanizza, E.; Curri, M.; Agostiano, A.; Laub, D. Photocatalytic Synthesis of Silver Nanoparticles Stabilized by TiO_2 Nanorods: A Semiconductor/metal Nanocomposite in Homogeneous Nonpolar Solution. *J. Am. Chem. Soc.* **2004,** *126,* 3868–3879.
49. Shchukin, D. G.; Radtchenko, I. L.; Sukhorukov, G. Photoinduced Reduction of Silver Inside Microscale Polyelectrolyte Capsules. *Chem. Phys. Chem.* **2003,** *4,* 1101–1103.
50. Chen, H. J.; Jia, J. B.; Dong, S. J. Direct Electrochemistry and Electrocatalysis of Horseradish Peroxidase Immobilized in Sol-gel-derived Ceramic-carbon Nanotube Nanocomposite Film. *Nanotechnology.* **2007,** *22* (8), 1811–1815.
51. Tian, X. L.; Chen, K.; Cao, G. Y. Seedless, Surfactantless Photoreduction Synthesis of Silver Nanoplates. *Mater. Lett.* **2006,** *60,* 828–839.
52. Wei, G.; Wang, L.; Sun, L. L.; Song, Y. H.; Sun, Y. J.; Guo, C. L.; Yang, T. *J. Phys. Chem. C.* **2007,** *111,* 1976.
53. Huang, H.; Yang, Y. Preparation of Silver Nanoparticles in Inorganic Clay Suspensions. *Compos. Sci. Technol.* **2008,** *68,* 2948–2953.
54. Zhou, Y.; Yu, S. H.; Wang, C. Y.; Li, X.G.; Zhu, Y. R.; Chen, Z. Y. A Novel Ultraviolet Irradiation Photoreduction Technique for the Preparation of Single-crystal Ag Nanorods and Ag Dendrites. *Adv. Mater.* **1999,** *11,* 850–852.
55. Reetz, M. T.; Helbig, W. J. Size-selective Synthesis of Nanostructured Transition Metal Clusters. *Am. Chem. Soc.* **1994,** *116,* 7401.
56. Reetz, M. T.; Winter, M.; Breinbauer, R.; Thurn-Albrecht, T.; Vogel, W. Size-selective Electrochemical Preparation of Surfactant-stabilized Pd-, Ni- and Pt/Pd Colloids. *Chem. Eur. J.* **2001,** *7,* 1084.
57. (a) Reetz, M. T.; Helbig, W. J. Size-selective Synthesis of Nanostructured Transition Metal Clusters. *Am. Chem. Soc.* **1994,** *116,* 7401.

(b) Reetz, M. T.; Helbig, W.; Quaiser, S. A.; Stimming, U.; Breuer, N.; Vogel, R. Visualization of Surfactants on Nanostructured Palladium Clusters by a Combination of STM and High-resolution TEM. *Science* **1995,** *267,* 367.

58. Sanchez, L. R.; Blanco, M. C.; Lopez-Quintela, M. A. Electrochemical Synthesis, Characterisation and Phytogenic Properties of Silver Nanoparticles. *J. Phys. Chem. B.* **2000,** *104,* 9683.
59. Reetz, M. T.; Helbig, W. Size-selective Synthesis of Nanostructured Transition Metal Clusters. *J. Am. Chem. Soc.* **1994,** *116,* 7401.
60. Zhuo, S.; Zhang, J.; Shi, Y.; Huang Robinson, Y. Self-template-directed Synthesis of Porous Perovskite Nanowires at Room Temperature for High-performance Visible-light Photodetectors. *Angew. Chem. Int. Ed. Engl.* **2015,** *54,* 5693–5696.
61. Wang, J. L.; Liu, J. W.; Lu, B. Z.; Lu, Y. R. Ge, J. Recycling Nanowire Templates for Multiplex Templating Synthesis: A Green and Sustainable Strategy. *Eur. J.* **2015,** *22,* 35–40.
62. Li, Y.; Han, Z.; Jiang, L.; Su, Z.; Liu, F.; Lai, Y. Liu, Y. Template-Directed Synthesis of Ordered Iron Pyrite (FeS2) Nanowires and Nanotubes Arrays. *J. Sol-Gel Sci.* **2014,** *72,* 100–105.
63. (a) Martin, C. R.; Mitchell, D. T. Template-synthesized Nanomaterials in Electrochemistry. *Electroanal. Chem.* **1999,** *21,* 1.

    (b) Cepak, V. M.; Martin, C. R. Preparation of Polymeric Micro- and Nanostructures Using a Template-based Deposition Method. *Chem. Mater.* **1999,** *11,* 1363.
64. (a) Romanov, S. Electronic Structure of the Minimum-diameter Tl, Pb and Bi Quantum Wire Superlattices. *J. Phys. Condens. Matter.* **1993,** *5,* 1081.

    (b) Anderson, P. A.; Armstrong, A. R.; Porch, A.; Edwards, P. P.; Woodall, L. J. Structure and Electronic Properties of Potassium-loaded Zeolite. *J. Phys. Chem. B.* **1997,** *101,* 9892.
65. (a) Ryoo, R.; Kim, J. M.; Ko, C. H.; Shin, C. H. An HREM Study of Channel Structures in Mesoporous Silica SBA-15 and Platinum Wires Produced in the Channels. *J. Phys. Chem.* **1996,** *100,* 718.

    (b) Ko, C. H.; Ryoo, R. Imaging the Channels in Mesoporous Molecular Sieves with Platinum. *Chem. Commun.* **1996,** *21,* 2467.
66. Iijima, S. Helical Microtubules of Graphitic Carbon. *Nature* **1991,** *354,* 56–58.
67. Kaifer, M. G.; Reddy, P. A.; Gutsche, C. D.; Echegoyen, L. Electroactive Calixarenes. 1. Redox and Cation Binding Properties of Calixquinones. *J. Am. Chem. Soc.* **1994,** *116,* 3580.
68. Eichen, Y.; Braun, E.; Sivan, U.; Ben-Yoseph, G. DNA-templated Assembly and Electrode Attachment of a Conducting Silver Wire. *Nature* **1998,** *39,* 775–778.
69. Yu, Y. Y.; Chang, S. S.; Lee, C. L.; Wang, C. R. Gold Nanorods: Electrochemical Synthesis and Optical Properties. *Phys. Chem. B.* **1997,** *101,* 6661.
70. Jiang, X.; Xie, Y.; Lu, J.; Zhu, L.; He, W.; Qian, Y. J. Oleate Vesicle Template Route to Silver Nanowires. *Mater. Chem.* **2001,** *11,* 1775.
71. Huang, L.; Wang, H.; Wang, Z.; Mitra, A.; Bozhilov, K. N.; Yan, Y. Nanowire Arrays Electrodeposited from Liquid Crystalline Phases. *Adv. Mater.* **2002,** *14,* 61.
72. Gou, L.; Chipara, M.; Zaleski, J. M. Convenient, Rapid Synthesis of Ag Nanowires. *Chem. Mater.* **2007,** *19,* 1755–1760.

73. Fritzche, W.; Porwol, H.; Wiegand, A.; Bornmann, S.; Köhler, J. M. In-situ Formation of Ag-containing Nanoparticles in Thin Polymer Films. *Nanostruct. Mater.* **1998,** *10,* 89.
74. Nagasawa, H.; Maruyama, M.; Komatsu, T.; Isoda, S.; Kobayashi, T. *Phys. Status Solidi. A.* **2002,** *67,* 191.
75. Wang, C.; Peng, S.; Chan, R.; Sun, S. Synthesis of AuAg Alloy Nanoparticles from Core/Shell-Structured Ag/Au. *Small.* **2009,** *5,* 567–570.
76. Liangbing, H.; Sun, K. H.; Jung-Yong, L.; Peter, P.; Yi, C. Scalable Coating and Properties of Transparent, Flexible, Silver Nanowire Electrodes. *ACS Nano.* **2010,** *5,* 1936–0851.
77. Gaddy, G. A.; McLain, J. L.; Steigerwalt, E. S.; Broughton, R.; Slaten, B. L. Photogeneration of Silver Particles in PVA Fibers and Films. *Mills Cluster Sci.* **2001,** *12,* 457.
78. Yang, C.; Gu, H. W.; Lin, W.; Yuen, M. M.; Wong, C. P.; Xiong, M. Y.; Gao, B. Silver Nanowires: From Scalable Synthesis to Recyclable Foldable Electronics. *Adv. Mater.* **2011,** *23,* 3052–3056.
79. Jia, C.; Yang, P.; Zhang, A. Glycerol and Ethylene Glycol Co-mediated Synthesis of Uniform Multiple Crystalline Silver Nanowires. *Mater. Chem. Phys.* **2014,** *143,* 794–800.
80. Caswell, K.; Bender, C.; Murphy, C. Seedless, Surfactantless Wet Chemical Synthesis of Silver Nanowires. *Nano Lett.* **2003,** *3,* 667.
81. Korte, K.; Skrabalak, S.; Xia, Y. Rapid Synthesis of Silver Nanowires through a $CuCl^-$ or $CuCl^{2-}$ Mediated Polyol Process. *J. Mater. Chem.* **2008,** *18,* 437.
82. Lin, J. Y.; Hsueh, Y. L. J. J. Effect of Silver Nitrate Concentration of Silver Nanowires Synthesized Using a Polyol Method and Their Application as Transparent Conductive Films. *Eur. Phys. J. B.* **2015,** *584,* 243–247.
83. Ducamp-Sanguesa, C.; Herrera-Urbina, R.; Figlarz, M. Synthesis and Characterization of Fine and Monodisperse Silver Particles. *J. Solid State Chem.* **1992,** *100,* 272.
84. Zeng, X.; Zhou, B.; Gao, Y.; Wang, C.; Li, S.; Yeung, C. Y.; Wen, W. *Nanotechnology.* **2014,** *16,* 1–16.
85. Zongtao, Z.; Bin. Z.; Liming, H. Poly (Vinyl pyrrolidone): A Dual Functional Reductant and Stabilizer for the Facile Synthesis of Noble Metal Nanoplates in Aqueous Solutions. *J. Solid State Chem.* **1996,** *121,* 105–110.
86. Silvert, P. Y.; Urbina, R. U.; Tekaia, K. E. Preparation of Colloidal Silver Dispersions by the Polyolprocess. *J. Mater. Chem.* **1997,** *7,* 293.
87. Sahin, C.; Burcu, A.; Husnu, U. E. Polyol Synthesis of Silver Nanowires: An Extensive Parametric Study. *Cryst. Growth Des.* **2011,** *11,* 4963–4969.
88. Yunxia, R.; Weiwei, H.; Me, W. K.; Shulin, J.; Changhui, Y. A One-step Route to Ag Nanowires with a Diameter below 40 nm and an Aspect Ratio above 1000†. *Chem. Commun.* **2014,** *50,* 14877–14880.
89. Jones, W. S.; Tamplin, W. S. *Glycols*; Curme, Jr. G. O., Johnston, F., Eds.; Reinhold: New York, NY, 1952; p 38.
90. Tornblom, M.; Henriksson, U. Effect of Solubilization of Aliphatic Hydrocarbons on Size and Shape of Rodlike C16TABr Micelles Studied by 2H NMR Relaxation. *J. Phys. Chem. B.* **1997,** *101,* 6028–6040.
91. Korte, K. E.; Skrabalak, S. E.; Xia, Y. Rapid Synthesis of Silver Nanowires through a CuCl-or CuCl2-mediated Polyol Process. *J. Mater.Chem.* **2008,** *18,* 437–441.

92. Tsuji, M.; Matsumoto, K.; Jiang, P.; Matsuo, R.; Tang, X. L.; Kamarudin, K. S. Roles of Pt Seeds and Chloride Anions in the Preparation of Silver Nanorods and Nanowires by Microwave-polyol Method. *Colloids Surf. A.* **2008,** *316,* 266–277.
93. Sun, Y.; Yin,Y.; Mayers, B. T.; Herricks, T.; Xia, Y. Uniform Silver Nanowires Synthesis by Reducing $AgNO_3$ with Ethylene Glycol in the Presence of Seeds and Poly (Vinyl pyrrolidone). *Chem. Mater.* **2002,** *14,* 4736–4745.
94. Tsuji, M.; Nishizawa, Y.; Hashimoto, M.; Tsuji, T. Uniform Silver Nanowires Synthesis by Reducing $AgNO_3$ with Ethylene Glycol in the Presence of Seeds and Poly (Vinyl pyrrolidone). *Chem. Lett.* **2004,** *33,* 370–371.
95. Liu, S.; Yue, J.; Gedanken, A. Synthesis and Characterization of Monoclinic ZrO_2 Nanorods by a Novel and Simple Precursor Thermal Decomposition Approach. *Adv. Mater.* **2001,** *13,* 656–659.
96. Liu, S.; You, J.; Gedanken, A. Synthesis and Characterization of Monoclinic ZrO_2 Nanorods by a Novel and Simple Precursor Thermal Decomposition Approach. *Adv. Mater.* **2001,** *13,* 656–659.
97. Korte, K. E.; Skrabalak, S. E.; Xia, Y. Rapid Synthesis of Silver Nanowires through a CuCl-or CuCl2-mediated Polyol Process. *J. Mater. Chem.* **2008,** *18,* 437–44.
98. Coskun, S.; Aksoy, B.; Unalan, H. E. Polyol Synthesis of Silver Nanowires: An Extensive Parametric Study. *Cryst. Growth Des.* **2011,** 11, 4963–4969.
99. Nair, G. K.; Jayaseelan, D.; Biji, P. Direct-writing of Circuit Interconnects on Cellulose Paper Using Ultra-long, Silver Nanowires Based Conducting Ink. *RSC Adv.* **2009,** *5,* 76092–76100.
100. Siegel, C.; Phillips, S. T.; Dickey, M. D.; Lu, N.; Suo, Z.; Whitesides, G. M. *Adv. Funct. Mater.* **2010,** *20,* 28–35.
101. MacKenzie, R.; Fraschina, C.; Dielacher, B.; Sannomiya, T.; Dahlin, A. B.; Voros, J. *Nanoscale.* **2013,** *5,* 4966–4975.
102. Vlad, A.; Reddy, A. L. M.; Ajayan, A.; Singh, N.; Gohy, J. F.; Melinte, S.; Ajayan, P. M. *Proc. Natl. Acad. Sci. USA.* **2012,** *109,* 15168–15173.
103. Jung, H. Y.; Karimi, M. B.; Hahm, M. G.; Ajayan, P. M.; Jung, Y. J. *Sci. Rep.* **2012,** *2,* 1–5.
104. Mostafalu, P.; Sonkusale, S. Paper-based Super-capacitor Using Micro and Nano Particle Deposition for Paper-based Diagnostics; IEEE Sensors: Baltimore, MD, 3–6 Nov. 2013.
105. Esfandyarpour, R.; Esfandyarpour, H.; Harris, J. S.; Davis, R. W. *Nanotechnology.* **2013,** *24,* 465301.
106. Esfandyarpour, R.; Esfandyarpour, H.; Javanmard, M.; Harris, J. S.; Davis, R. W. *Sens. Actuators B.* **2013,** *177,* 848–855.
107. Tyagi, P.; Postetter, D.; Saragnese, D. L.; Randall, C. L.; Mirski, M. A.; Gracias, D. H. *Anal. Chem.* **2013,** *81,* 9979–9984.
108. Wang, Y.; Zhu, Y.; Chen, J.; Zeng, Y. *Nanoscale.* **2012,** *4,* 6025–6031.
109. Xie, P.; Xiong, Q.; Fang, F.; Qing, Q.; Lieber, C. M. *Nat. Nanotechnol.* **2012,** *7,* 119–112.
110. Heim, M.; Yvert, B.; Kuhn, A. *J. Physiol.* **2012,** *106,* 137–145.
111. Seker, E.; Berdichevsky, Y.; Begley, M. R. Reed, M. L.; Staley, K. J.; Yarmush, M. L. *Nanotechnology.* **2010,** *21,* 125504.

112. Lim, Z. Z. J.; Li, J. E. J.; Ng, C. T.; Yung, L. Y. L.; Bay, B. H. *Acta Pharmacol. Sin.* **2011,** *32,* 983–990.
113. Shen, H.; You, J.; Zhang, G.; Ziemys, A.; Li, Q.; Bai, L.; Deng, X.; Erm, D. R.; Liu, X.; Li, C.; Ferrari, M. *Adv. Healthcare Mater.* **2012,** *1,* 84–89.
114. Ren, G.; Hu, D.; Cheng, E. W. C.; Vargas-Reus, M. A.; Reip, P.; Allaker, R. P. *Int. J. Antimicrob. Agents.* **2009,** *33,* 587–590.
115. Samuel, U.; Guggenbichler, J. P. *Int. J. Antimicrob. Agents.* **2004,** *23* (suppl. 1), 75–78.
116. Chen, J.; Saeki, F.; Wiley, B. J.; Cang, H.; Cobb, M. J.; Li, Z. Y.; Au, L.; Zhang, H.; Kimmey, M. B.; Li, X.; Xia, Y. *Nano Lett.* **2005,** *5,* 473–477.
117. Huang, Z.; Jiang, X.; Guo, D.; Gu, N. Controllable Synthesis and Biomedical Applications of Silver Nanomaterials. *J. Nanosci. Nanotechnol.* **2011,** *11,* 9395–9408.
118. Shanshan, Y.; Yong, Z. Wearable Multifunctional Sensors Using Printed Stretchable Conductors Made of Silver Nanowires. *Nanoscale.* **2014,** *6,* 2345–2352. DOI: 10.1039/C3NR05496A
119. Po-Chun, H.; Xiaoge, L.; Chong, L.; Xing; X.; Hye Ryoung, L.; Welch Alex, J.; Zhao, T.; Cui, Y. Personal Thermal Management by Metallic Nanowire-Coated Textile. *Nano Lett.* **2015,** *15* (1), 365–371.

CHAPTER 3

FABRICATION OF METAL NANOPARTICLES ON POLY(GLYCIDYL METHACRYLATE): SYNTHESIS, CHARACTERIZATION, AND CATALYTIC APPLICATION

S. MOHAMMED SAFIULLAH, K. ABDUL WASI, and K. ANVER BASHA*

P.G. and Research Department of Chemistry, C. Abdul Hakeem College, Melvisharam 632509, Vellore District, Tamil Nadu, India

**Corresponding author. E-mail: kanverbasha@gmail.com*

CONTENTS

ABSTRACT

Metal nanoparticles were immobilized on the poly(glycidyl methacrylate) (PGMA) to yield PGMA core/metal nanoparticle shell nanocomposite (PGMA/Cu nanohybrid 1% and PGMA/Ni nanohybrid 1%) by deposition method. The deposition method consists of two simple steps: (1) the synthesis of PGMA beads by suspension polymerization followed by (2) direct deposition of metal nanoparticles on activated PGMA beads. The PGMA beads were used as a soft template to host nanoparticles without surface modification. Two different metal nanoparticles were employed to know the viability and variability of the designed method. The catalytic activity of resultant nanocomposites was tested for the reduction of 4-nitrophenol (4NP) to 4-aminophenol (4AP) with an excess amount of sodium borohydride ($NaBH_4$). The X-ray diffraction (XRD) study results showed that the metal nanoparticles were embedded on the surface of the PGMA matrix. The scanning electron microscopy (SEM) images revealed that the fabrication of metal nanoparticles on the PGMA matrix possesses different shapes and changes the morphology and nature of PGMA beads significantly. The catalytic studies have demonstrated that the PGMA/Cu nanohybrid 1% exhibits better catalytic activity than PGMA/Ni nanohybrid 1%.

3.1 INTRODUCTION

The hybridization of inorganic nanoparticles with the functional polymer has attracted significant interest in the research communities. Metal nanoparticles have gained considerable attention due to its useful properties such as catalytic and antibacterial activity and good thermal and electrical conductivity. However, the applications of metal nanoparticles usually suffer from irreversible aggregation and difficulty to recover the metal nanoparticles due to the high surface energy and large surface area.[1–3] Immobilization of metal nanoparticles on a polymer support has been proven to be capable of addressing this issue and broadening the applications.[4–8]

However, the reported polymer substrates are mainly based on chemically inert materials including polystyrene, polypropylene, or poly(methyl methacrylate), which limits the development of new novel functional materials.[8] The poly(glycidyl methacrylate) (PGMA) belongs to the class of functional polymers. The facile reaction of the epoxy groups with a large variety of reagents provides a novel route for preparing various multifunctional polymers through chemical modifications of this polymer. Recently, Zhang et al. immobilized Ag nanoparticles on the sulfhydryl functionalized PGMA microspheres.[9] Nam et al. used amine functionalized PGMA microspheres as the template to fabricate gold crystals with high catalytic activity in converting 4-nitrophenol (4NP) to 4-aminophenol (4AP).[10] Li et al. deposited gold nanoparticles (12 ± 3 nm) on the surface of poly(allylamine hydrochloride)-modified PGMA spheres.[11]

However, to the best of our knowledge, the immobilization of copper and nickel nanoparticles on the PGMA support that used as a catalyst has not been reported so far. In this chapter, an effective and facile method was developed to fabricate metal nanoparticles on PGMA support without surface modification. This method facilitates the incorporation of metal nanoparticles on the surface of PGMA matrix. The elemental, structural, morphological, and thermal studies have been carried out to characterize PGMA/Cu nanohybrid and PGMA/Ni nanohybrid. The PGMA/Cu nanohybrid and PGMA/Ni nanohybrid were used as a catalyst in the reduction of 4NP to 4AP, and were investigated.

3.2 EXPERIMENTAL

3.2.1 MATERIALS

Glycidyl methacrylate (GMA) purchased from Sigma-Aldrich was freed from inhibitor by distillation at 60–70°C under reduced pressure. Benzoyl peroxide (BPO) was recrystallized from chloroform–methanol (1:1) and used as an initiator. Used disodium hydrogen phosphate (DSHP), sodium borohydride ($NaBH_4$), copper sulfate pentahydrate ($CuSO_4$), hydrazine (N_2H_4), nickel sulfate ($NiSO_4$), and sodium hydroxide (NaOH) were purchased from Sigma-Aldrich. $NaBH_4$ and N_2H_4 were used as a reducing agent, whereas NaOH was added to adjust the pH and to accelerate the reduction reaction in water. Polyvinyl alcohol [PVA (14000) (Sigma-Aldrich)] was used as a surfactant and polyethylene glycol [PEG (8000) (Sigma-Aldrich)] as a size controller and capping agent. Ascorbic acid (Qualichems) employed as an anti-oxidant of colloidal copper in an aqueous medium.

3.2.2 SYNTHESIS OF PGMA/CU NANOHYBRID

The PGMA beads were prepared by the procedure reported earlier.[12] The experimental procedure is presented schematically for the preparation of PGMA/Cu nanohybrid in Figure 3.1a. Required quantity of monodisperse PGMA beads was activated by soaking in chloroform for 24 h and immersed in sol containing CuNPs overnight, to synthesize the PGMA/Cu nanohybrid. The red color PGMA/Cu nanohybrid solid beads were filtered, washed, and dried at 100°C under vacuum. The sol was prepared as follows: required quantity of ascorbic acid was dissolved in 15 mL of deionized water; calculated amount of $CuSO_4$ and PEG was added under stirring. NaOH solution was added dropwise into the reaction media to adjust the pH to 12; a yellow color solution was obtained. After 30 min, $NaBH_4$ was added slowly into the reaction media under stirring to avoid agglomeration. Finally, an aqueous dispersion of red color CuNPs was obtained.

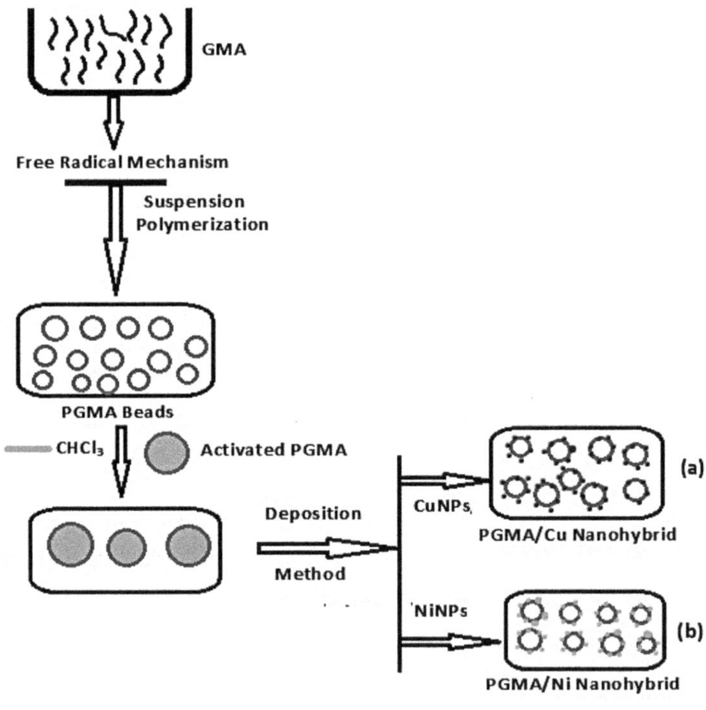

FIGURE 3.1 Synthetic procedure of (a) PGMA/Cu nanohybrid and (b) PGMA/Ni nanohybrid.

3.2.3 SYNTHESIS OF PGMA/NI NANOHYBRID

The schematic representation of PGMA/Ni nanohybrid synthesis is presented in Figure 3.1b. For this, 100 mg of PGMA beads was activated by soaking in chloroform for 24 h and it was dispersed in sol containing nickel nanoparticles (NiNPs). The violet color PGMA/Ni nanohybrid solid beads were filtered, washed, and dried at 100°C under vacuum. The sol was prepared by dissolving $NiSO_4$ (0.1 g) and PEG (1%) in water under stirring and the solution becomes pale green. To the pale green solution, calculated amounts of NaOH and N_2H_4 were added slowly under stirring to avoid agglomeration; the solution turns into dark green. After a period of time, colloidal dispersion of pale blue color NiNPs in water was obtained.

3.2.4 CATALYTIC HYDROGENATION OF 4-NITROPHENOL

Catalytic hydrogenation reactions of 4NP were conducted at room temperature (25°C) in the presence of PGMA/Cu and Ni nanohybrid catalysts. Typically, aqueous solutions of 4NP (1 mmol) and $NaBH_4$ (0.1 M) were freshly prepared. An amount of 50 mg of catalyst was added in 0.5 mL of 4NP solution with 5 mL of deionized water, and 0.5 mL of $NaBH_4$ solution was injected into the mixture under continuous stirring. To evaluate the reaction progress, 3 mL aliquots were taken out of the reaction mixture at specified time intervals and monitored by UV–Vis spectroscopy. The catalysts were easily recovered by filtration and reused without any treatment.

3.2.5 CHARACTERIZATION

3.2.5.1 FOURIER-TRANSFORM INFRARED SPECTROSCOPY

The PGMA/Cu and Ni nanohybrid were structurally characterized by Fourier-transform infrared (FTIR) spectroscopy. The FTIR spectra of the PGMA/Cu and Ni nanohybrid were recorded with KBr-pressed pellet over the range of 4000–400 cm^{-1} using Bruker Tensor-27 FTIR spectrophotometer with OPUS software.

3.2.5.2 X-RAY DIFFRACTION ANALYSIS

The structures of PGMA/Cu and Ni nanohybrid were characterized by using X-ray diffraction (XRD). The XRD (D8-Advance, Bruker AXS) with a Cu $K\alpha$ radiation source (0.15418 nm). The Bragg angular region (2θ) was fixed between 0 and 90°C at a scan rate of 0.3 s^{-1}. The sizes of the metal nanoparticles were calculated from the XRD patterns by using Scherrer formula.

$$D = \frac{0.89\lambda}{\beta \cos \theta}$$

3.2.5.3 SCANNING ELECTRON MICROSCOPY (SEM)

The morphology of the PGMA/Cu and Ni nanohybrid were investigated using scanning electron microscopy (SEM, CARL ZEISS EVO 18).

3.2.5.4 ENERGY DISPERSIVE X-RAY SPECTROSCOPY (EDAX)

The EDAX elemental composition analyzer (Bruker) was used to study the elemental composition of PGMA/Cu and Ni nanohybrid.

3.2.5.5 THERMOGRAVIMETRIC ANALYSIS (TGA)

Thermal stability of the PGMA/Cu and Ni nanohybrid were determined by thermogravimetric analysis (TGA, Model: TGA Q500 V20. 13 Build 39 thermal analyzer) at a heating rate of 20°C min^{-1} under dry nitrogen atmosphere.

3.2.5.6 DIFFERENTIAL SCANNING CALORIMETRY

Differential scanning calorimetry (DSC) of the PGMA/Cu and Ni nanohybrid were recorded by using TGA, Model: TGA Q500 V20. 13 Build 39 thermal analyzer at a heating rate of 20°C min^{-1} under static air atmosphere to measure the glass transition temperature (T_g). The measurements are carried out from 30 to 300°C.

3.2.5.7 UV–VIS SPECTROSCOPY

The UV–Vis spectra were recorded over time in order to monitor the change in color using UV–Vis spectroscopy, SHIMADZU TCC-240A.

3.3 RESULTS AND DISCUSSION

3.3.1 FTIR CHARACTERIZATION

The FTIR spectra of the PGMA/Cu and Ni nanohybrid were recorded to study the interaction of CuNPs with PGMA matrix (Fig. 3.2). The weak band at 524 cm^{-1} is assigned to the metal–oxygen stretching vibration.[13] The asymmetrical stretching of epoxy ring was observed at 992 cm^{-1}. The C—H symmetry and asymmetry stretching due to the methyl and methylene groups were observed at around 3000 cm^{-1} and 2935 cm^{-1}, respectively. The carbonyl stretching of ester group was observed at 1731 cm^{-1}. The appearance of broad peak at 3440 cm^{-1} indicates the presence of water molecules in PGMA/metal nanohybrid and reveals the hydrophilic nature of the nanocomposites. No shift/disappearance of carbonyl and epoxy band confirmed that the interaction between metal nanoparticles and PGMA matrix was insignificant, due to the poor interaction of nanoparticle with the polymer matrix.[14] Since the reduction of metal ions was carried out in the aqueous medium, the hydrophilic metal nanoparticles were produced. Thereby the fabrication of hydrophilic nanoparticles induced the hydrophilicity in the PGMA matrix.

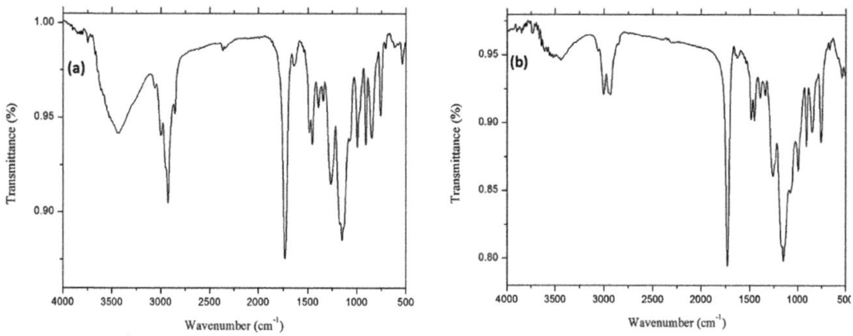

FIGURE 3.2 FTIR spectrum of (a) PGMA/Cu nanohybrid and (b) PGMA/Ni nanohybrid.

3.3.2 XRD ANALYSIS

The XRD patterns of the PGMA/metal nanohybrids are presented in Figure 3.3. The presence of the broad peak at 20°C in Figure 3.3 confirms the amorphous nature of the PGMA/metal nanohybrid. The addition of the nanoparticles did not induce any crystallinity in the polymer. It was also confirmed that the diffraction peaks with high intensities appeared at angles corresponding to (111), (200), and (220) planes of FCC structure of copper with the space group of FM3M. This is in agreement with the standard data from JCPDS card no. 85-1326. However, the structure provides sufficient stability and no structural changes were observed in the XRD spectra taken after 1 month. This suggested that the PGMA matrix provides sufficient stability to the CuNPs. Since the characteristics peaks of CuNPs were observed in Figure 3.3a, it was confirmed that the CuNPs were doped on the surface of the PGMA matrix. But Tian et al. reported that the characteristics diffraction peaks of the CuNPs in the XRD spectra were absent because it was coated with polymer.[15] The size of the CuNPs was calculated using Scherrer formula and it was found to be 11 nm.

Figure 3.3b presents the XRD pattern of PGMA/Ni nanohybrid. All the peaks except broad peak around 20°C in the XRD patterns can be indexed to an FCC Ni in Figure 3.3b. It was also confirmed that diffraction peaks with high intensities appear at angles corresponding to (111), (200), and (222) planes of the FCC structure of nickel corresponds to $2\theta = 44.5$, 51.8, and 76.4, respectively. The appearance of the characteristic diffraction peaks of the NiNPs in the XRD pattern (Fig. 3.4b) confirmed that the nanoparticles were present on the surface of PGMA. The nickel nanoparticle's size was found to be 82 nm.

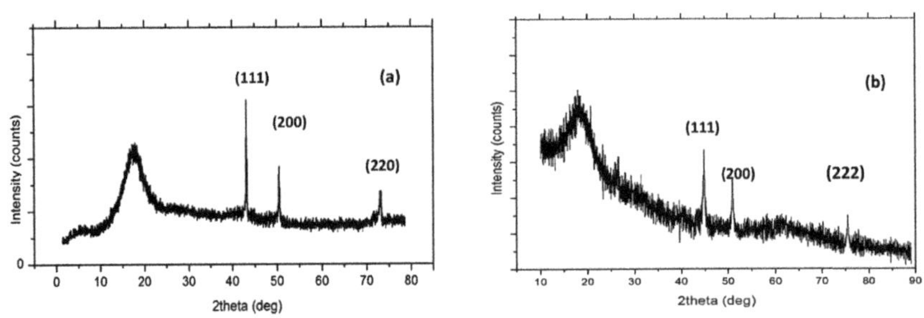

FIGURE 3.3 XRD pattern of (a) PGMA/Cu nanohybrid and (b) PGMA/Ni nanohybrid.

3.3.3 SEM AND EDAX ANALYSIS

Figures 3.4a and 3.5a demonstrate spherical shape of the PGMA/Cu and Ni nanohybrids. It was observed that the fabrication of nanoparticles on the surface of porous PGMA beads leads to the formation of a rough surface (Figs. 3.4b and 3.5b).[14] The dramatically different morphology suggests a 3D heterogeneous nucleation and growth rather than a layer-by-layer epitaxial growth of CuNP on the PGMA core.[16] In the case of 1% (w/w) $CuSO_4$, the fabricated copper nanoparticles on the PGMA surface exist in two shapes, namely nanocubes and nanospheres (Fig. 3.4c,d). The conformal epitaxial growth, heterogeneous nucleation, and island growth of CuNPs lead to the formation of copper nanosphere on the PGMA.[16] As depicted in Figure 3.5c, Ni nanoplates deposited on the PGMA surface lead to the formation of the rough surface. The formation of Ni nanoplates is attributed to the selective deposition of Ni species on the PGMA surface during the synthetic process.[17]

As depicted in Figures 3.4c and 3.5c, the self-assemblies of nanoparticles were not uniform because it is not functionalized to locate in a controlled way.[18] It should be noted that the nanoparticles formed in this study were in aqueous suspension form; therefore, the nanoparticles are relatively hydrophilic. The doping of nanoparticles stays on the PGMA surface, leads to the formation of a rough surface, and induces the hydrophilicity on it.[19] This is in agreement with FTIR data. Some cracks are shown in Figures 3.4d and 3.5d; this may be due to the following reasons: the non-equilibrium contraction arises due to either the presence of nanoparticles in the resultant nanohybrid or the absence of atoms to form internal surfaces.[12] Thus it can be concluded that the doping of the nanoparticles on PGMA matrix by the deposition method changes the morphology and nature of PGMA significantly. The EDAX spectra (Figs. 3.4e and 3.5e) confirm the presence of CuNPs and NiNPs, respectively, in the PGMA matrix.

3.3.4 THERMAL ANALYSIS

The thermal stability of PGMA/metal nanohybrids was studied by using TGA; the TG curves are shown in Figure 3.6. The thermal degradation temperatures $T_{-10\%}$ and $T_{-50\%}$ for different nanoparticle loadings of PGMA/metal nanohybrids are listed in Table 3.1. It can be seen that the prepared PGMA/metal nanohybrids exhibit lower $T_{-10\%}$ value than PGMA. This is

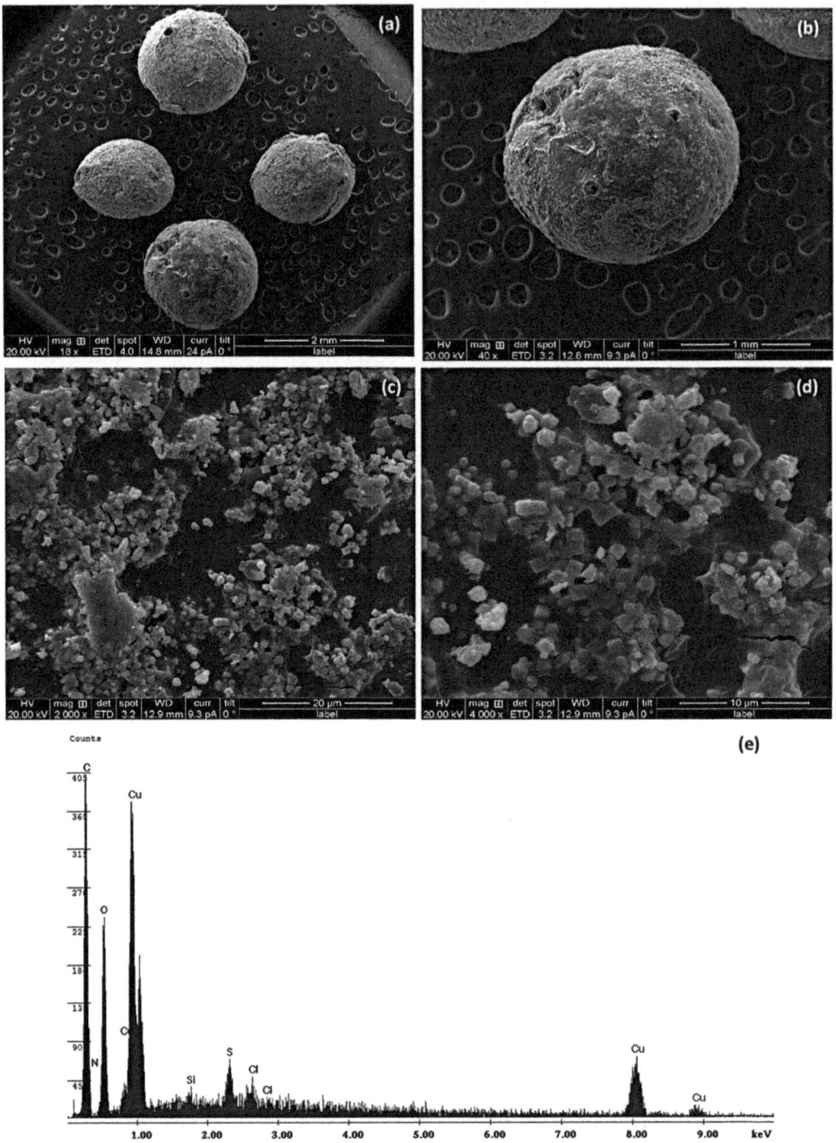

FIGURE 3.4 Scanning electron microscopic images of (a) 2 mm, (b) 1 mm, (c) 20 μm, (d) 10 μm, and (e) EDAX spectra of PGMA/Cu nanohybrid.

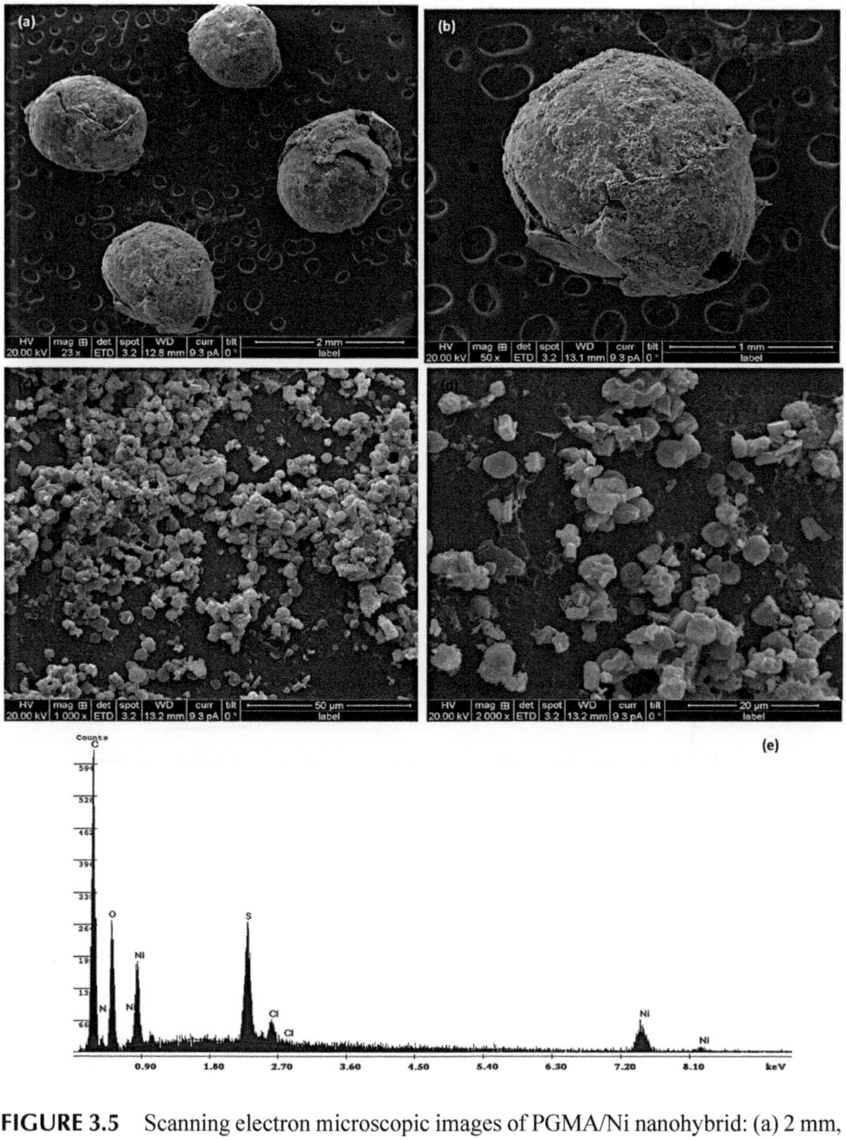

FIGURE 3.5 Scanning electron microscopic images of PGMA/Ni nanohybrid: (a) 2 mm, (b) 1 mm, (c) 20 μm, (d) 10 μm, and (e) EDAX spectra of PGMA/Ni nanohybrid.

because the doping of nanoparticles on the activated PGMA surface leads to weakening of the rigid surface of it. The thermal studies show that the thermal stability of PGMA is higher than that of PGMA/metal nanohybrids; it may be due to the fast thermal degradation through the cracks formed on the surface of PGMA/metal nanohybrid.

The thermograms exhibit a continuous weight loss with the same rate up to 450°C. Both the TG curves (Fig. 3.6a,b) indicate that the overall decomposition process consists of two reactions: namely, depolymerization to monomer and ester decomposition. It was proposed by Zulfiqar et al. that the monomer formation at the lower temperature is due to initiation

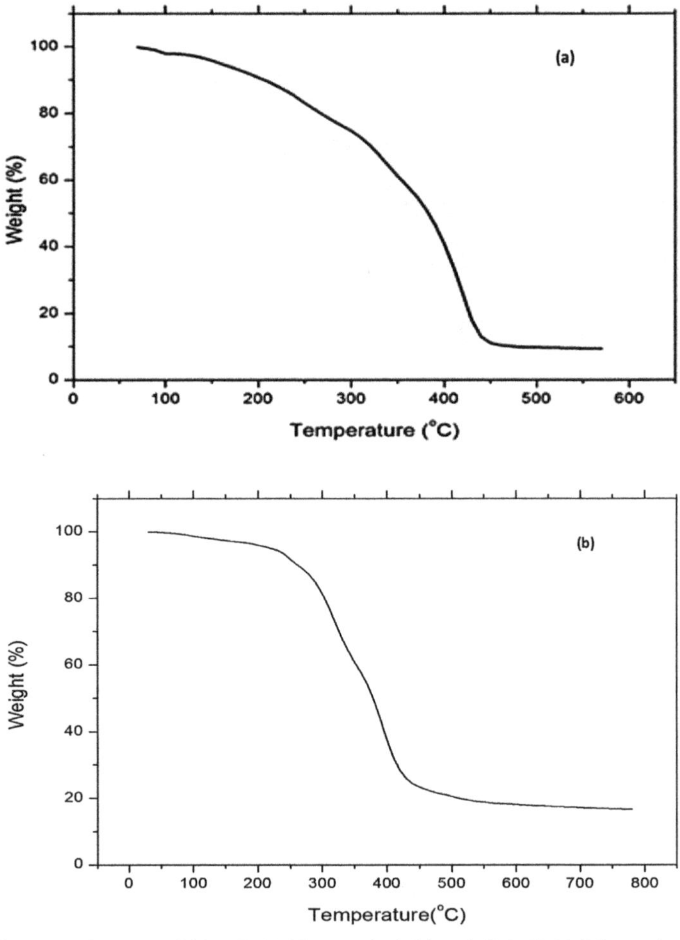

FIGURE 3.6 TG curve of (a) PGMA/Cu nanohybrid and (b) PGMA/Ni nanohybrid.

at unsaturated chain ends, whereas degradation in the region of 300–400°C is associated with the initiation by random scission.[20] From Figure 3.6a,b the thermal degradation of PGMA/metal nanohybrid is continuous at the lower temperature, hence the thermal degradation due to monomer formation was not observed. This observation confirmed that the immobilization of nanoparticles on PGMA surface blocks the unsaturated chain ends. This study confirmed that the doping of nanoparticles on the activated PGMA bead leads to a decrease in the thermal stability.

The DSC curves are shown in Figure 3.7a,b. The corresponding T_g values of the PGMA/Cu and Ni nanohybrid are tabulated in Table 3.1. The considerable increases in T_g of PGMA/Ni nanohybrid (1%) than PGMA/Cu nanohybrid (1%) were due to the insignificant deterioration of rigid surface structure upon loading of the NiNPs on the PGMA matrix.

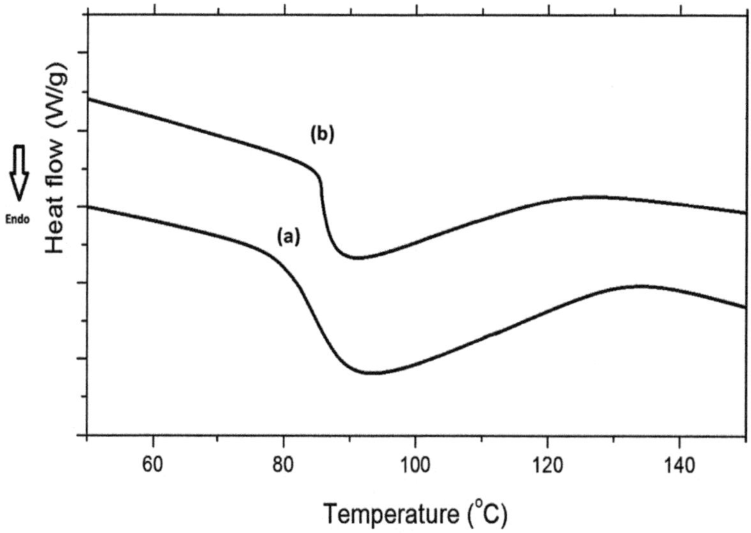

FIGURE 3.7 DSC curves of (a) PGMA/Cu nanohybrid and (b) PGMA/Cu nanohybrid.

TABLE 3.1 Properties of PGMA/Cu and Ni Nanohybrids.

Particulars	$T_{-10\%}$ (°C)	$T_{-50\%}$ (°C)	Tg (°C)	% of CuNPs Loading (%) (EDAX)	Size of the NPs (nm)
PGMA/Cu nanohybrid	212	384	99	33.94	11
PGMA/Ni nanohybrid	260	378	110	23.30	14

3.3.5 CATALYTIC HYDROGENATION OF 4-NITROPHENOLS

The catalytic reduction of 4NP by $NaBH_4$ was chosen as a model reaction to evaluate the catalytic activities of the PGMA/metal nanohybrid catalyst. The UV–Vis absorption spectra were recorded over time to monitor the change in precursor color. As shown in Figure 3.8a, the original adsorption peak of 4NP was centered at 317 nm and shifted to 400 nm after the addition of a freshly prepared $NaBH_4$ aqueous solution (Fig. 3.8b; 0 s), and the color of the solution changed from light yellow to bright yellow immediately. This red shift was just due to the formation of 4-nitrophenolate ions in alkaline conditions caused by the addition of $NaBH_4$.[21] Figure 3.8b depicts the UV–Vis absorption spectra of the reduction of 4NP over PGMA/Cu nanohybrid

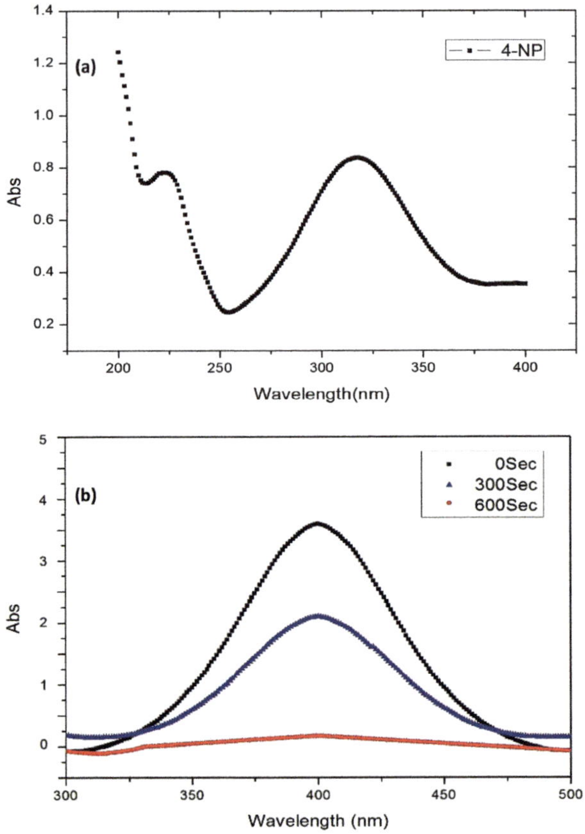

FIGURE 3.8 UV–Vis absorption spectra of the reduction of 4NP over PGMA/Cu nanohybrid (1%) catalyst in aqueous media at room temperature.

catalyst. It was noticed that after the addition of PGMA/Cu nanohybrid catalyst the peak at 400 nm sharply decreased. As a result, the initially light yellow solution consequently underwent fading to become colorless.[22,23]

Likewise, PGMA/Ni nanohybrid catalyst was also used, and the completion of the reaction by changing its color was noted and tabulated in Table 3.2. It is evident from the table that the PGMA/Cu nanohybrid 1% and PGMA/Ni nanohybrid 1% catalyst exhibit quite similar catalytic activity.

TABLE 3.2 Catalytic Activities of PGMA/Cu and Ni Nanohybrids.

Catalyst	NPs:4NP wt:wt	Time to reach 100% conversion (s)
PGMA/Cu nanohybrid	1:10	600
PGMA/Ni nanohybrid	1:10	680

3.4 CONCLUSIONS

An effective deposition of metal nanoparticles on the PGMA matrix was achieved by simple methodology. In this method, nanoparticles have been stabilized on the surface of PGMA by non-covalent interaction and confirmed by FTIR spectroscopy. The XRD analysis results also confirmed the immobilization of nanoparticles on the surface of PGMA. The deposition of nanoparticles had a significant effect on the morphology of polymer. Thermal stability of PGMA/metal nanohybrid was determined by TGA and DSC. The significant decrease in the thermal stability of PGMA/metal nanohybrid than its PGMA confirms that the deposited nanoparticles weaken the rigid structure of PGMA. The obtained PGMA/metal nanohybrid proved to be a catalyst as it successfully catalyzed the 4NP reduction.

KEYWORDS

- **metal nanoparticle**
- **poly(glycidyl methacrylate)**
- **polymer-supported catalyst**
- **nanocomposite**
- **thermal properties**

REFERENCES

1. Wu, Y.; Wang, D.; Li, Y. *Chem. Soc. Rev.* **2014,** *43,* 2112.
2. Zhang, J.; Zhang, M.; Tang, K.; Verpoort, F.; Sun, T. *Small* **2014,** *10,* 32.
3. Kumar, K. S.; Kumar, V. B.; Paik, P. *J. Nanopart.* **2013,** *1,* 1.
4. Chen, G. F.; Lu, J. R.; Lam, C.; Yu, Y. *Analyst* **2014,** *139,* 5793.
5. Taheri, S.; Baier, G.; Majewski, P.; Barton, M.; Förch, R.; Landfester, K.; Vasilev, K. *Nanotechnology* **2014,** *25,* 305102.
6. Wang, R. Z.; Wang, Z.; Lin, S.; Deng, C.; Li, F.; Chen, Z.; He, H. *RSC Adv.* **2015,** *5,* 40141–40147. DOI: 10.1039/C5RA03288A
7. Wang, S.; Zhang, J.; Yuan, P.; Sun, Q.; Jia, Y.; Yan, W.; Chen, Z.; Xu, Q. *J. Mater. Sci.* **2015,** *50,* 1323.
8. Kuroda, K.; Ishida, T.; Haruta, M. *J. Mol. Catal. A Chem.* **2009,** *298,* 7–11.
9. Zhang, W.; Sun, Y.; Zhang, L. *Ind. Eng. Chem. Res.* **2015,** *54,* 6480–6488.
10. Oh, J. S.; Dang, L. N.; Yoon, S. W.; Nam, J. D. *Macromol. Rapid Commun.* **2013,** *34,* 504.
11. Li, M.; Chen, G. *Nanoscale* **2013,** *5,* 11919.
12. Mohammed Safiullah, S.; Abdul Wasi, K.; Anver Basha, K. *Polymer* **2015,** *66,* 29–37.
13. Olad, A.; Mohammad, B.; Hamidreza, S. *Prog. Org. Coat.* **2011,** *72,* 599–604.
14. Mohammed Safiullah, S.; Abdul Wasi, K.; Anver Basha, K. *Mater. Lett.* **2014,** *133,* 60–63.
15. Tian, K.; Liu, C.; Yang, H.; Ren, X. *Colloids. Surf. A. Physicochem. Eng. Asp.* **2012,** *397,* 12–15.
16. Fan, F. R.; Liu, D. Y.; Wu, Y. F.; Duan, S.; Xie, Z. X.; Jiang, Z. Y. *J. Am. Chem. Soc.* **2008,** *130,* 6949–6951.
17. Seo, D.; Park, J. C.; Song, S. *J. Am. Chem. Soc.* **2006,** *128,* 14863.
18. Rycenga, M.; McLellan, J. M.; Xia, Y. *Adv. Mater.* **2008,** *20,* 2416–2420.
19. Yeum, J. H.; Sun, Q.; Deng, Y. *Macromol. Mater. Eng.* **2005,** *290,* 78–84.
20. Zulfiqar, S.; Zulfiqar, M.; Nawaz, M.; McNeill, I. C.; Gorman, J. G. *Polym. Degrad. Stab.* **1990,** *30,* 195–203.
21. Maksod, I. H. A. E.; Saleh, T. S. *Green Chem.* **2010,** *3,* 127–134.
22. Ohashi, M.; Beard, K. D.; Ma, S.; Blom, D. A.; St-Pierre, J.; Zee, J. W. V.; Monnier, J. R. *Electrochim. Acta.* **2010,** *55,* 7376–7384.
23. Feng, L.; Si, F.; Yao, S.; Cai, W.; Xing, W.; Liu, C. *Catal. Commun.* **2011,** *12,* 772–775.

CHAPTER 4

STRUCTURAL (NEUTRON DIFFRACTION) AND PHYSICAL PROPERTY STUDIES IN Sb_2Se_3-(CuI) CHALCOHALIDE GLASSES AND NANO-CRYSTALS

RASHMI M. JOGAD[1], P. S. R. KRISHNA[2], G. P. KOTHIYAL[3], and MAHANTAPPA S. JOGAD[4,1*]

[1]*Department of Physics, Karnataka State Women's University, Vijayapura, India*

[2]*Solid State Physics Division, Bhabha Atomic Research Center, Mumbai 400085, India*

[3]*Glass and Advanced Ceramics Division, Bhabha Atomic Research Center, Mumbai 400085 (Superannuated), India*

[4]*Department of Physics, School of Physical Sciences, Central University, Karnataka, Kadaganchi, Aland Road, Kalaburagi 585367, India*

[*]*Corresponding author. E-mail: jogad1952@rediffmail.com*

CONTENTS

ABSTRACT

Halide and chalcogenide glasses have received a great deal of interest as potential candidates for infrared transmitting materials. Chalcohalide glasses such as Sb_2Se_3-(CuI) are good candidates for application in CO_2 laser fibers and infrared windows. We report here the preparation of chalcohalide $x$$Sb_2Se_3$-(1 − x)CuI glass (where x is varied from 0.7 to 0.3) in vacuum by melt quench technique. These glasses have been characterized using X-ray diffraction (XRD), neutron diffraction (ND), Fourier-transform infrared spectroscopy (FTIR), Field emission scanning electron microscope (FESEM), etc. Glassy nature was confirmed by XRD while the presence of nanoparticles/nano-strips in the glassy matrix was identified by FESEM. FTIR measurement on these glasses exhibited a high transmission of about 75% in the range 10–30 μm. Neutron diffraction was studied on the sample 0.4Sb_2Se_3-0.6CuI glass to understand the short-range order (SRO) and the bonding nature. Magnetically controlled growing rod (MCGR) technique was employed for analysis of neutron diffraction patterns. We concluded the presence of intermediate range order (IRO) in 0.4Sb_2Se_3-0.6CuI glass by the use of neutron diffraction in combination with X-ray measurements. In addition, glasses were found to be consisting of mainly Sb—Se as well as Cu—Se bonds. Some investigations on the electrical switching memory for chalcohalide glasses of 0.4Sb_2Se_3-0.6CuI and 0.6Sb_2Se_3-0.4CuI composition were also carried out and the results are discussed.

4.1 INTRODUCTION

Glass is one of the most ancient materials known and used by humankind. However, where and when glasses first appeared is not exactly known. Glass, existing for millions of years, has fascinated and attracted enormous interest in both scientific and technological[11,82] areas. The common usage of the term glass follows from the definition of Morey.[45] According to this definition, glass is an inorganic material and is analogs to a liquid with very high viscosity typically of the order of 10^{15} poise; owing to such high viscosity, the material is rigid and often glass is referred to as supercooled liquid and is said to be thermodynamically unstable. Glasses are generally obtained by quenching the melt at a sufficiently rapid rate to avoid nucleation and crystallization. This definition is, however, not complete as in subsequent developments polymeric glasses have been obtained. Glass exhibits SRO and is also defined as amorphous solid exhibiting a characteristic glass transition temperature. Though they lack in the long-range order (LRO) found in crystals, these materials preserve a SRO with bond lengths and bond angles fluctuating narrowly around the values found in corresponding crystals.[83] As the amorphous materials possess crystal-like SRO and liquid-like LRO, their properties are generally influenced by both these features. Despite the importance of the glassy materials in modern life, from a physical point of view, our understanding of this state of matter is poor and the transition from the liquid to the glassy state is one of the greatest unresolved problems of condensed matter physics.[3,15] The dynamical aspects of disordered materials, including liquids, and glasses were recently reviewed by Price et al.[54] However, during recent years significant theoretical developments and experimental advances have led to a renewed interest in the physics of glasses.[9,11,17,39,69]

The absence of LRO in glassy chalcogenides provides the convenience of changing the elemental ratios and hence the properties over a wide range. The interesting properties exhibited by chalcogenide glasses make them suitable materials for phase-change memories (PCM) and other applications such as infrared optical devices, photoreceptors, sensors, and waveguides.[48] One of the most remarkable properties of chalcogenides is their electrical switching behavior.[47] Reversible (threshold type) or irreversible (memory type) switching from a high resistance OFF state to a low resistance ON state in glassy chalcogenides occurs at a critical voltage called the threshold/switching voltage (V_T). Investigations on the switching behavior and its composition dependence throw light on the local structural effects of amorphous chalcogenide semiconductors and X also help us in identifying suitable samples for PCM applications.[4,5]

Transmission in chalcogenide glasses is found to show high loss due to their large refractive indices.[31] However, it is reported by Baldwin et al.,[6] Poulain[53] Xiujian Z et al.[79] and Cornet[13] that this loss can be reduced by adding metal halides in chalcogenide glasses. This also helps in increasing the infrared transmission range. Therefore, chalcohalides are a new class of materials having high potential in infrared transmission materials, optical fiber materials for CO_2 laser radiation, and switching memory applications.

In this work, we report the preparation and characterization of $x\text{Sb}_2\text{Se}_3$-$(1-x)$CuI ($0.3 \leq x \leq 0.7$) chalcohalide glasses. The glasses were prepared by melt quench technique in a sealed (in vacuum) ampoule. They were studied for thermophysical, optical properties and submicron structural features. These glasses have been characterized using X-ray diffraction, neutron diffraction, FTIR, and FESEM, etc. Glassy nature was confirmed by XRD while the presence of nanoparticles/nano-strips in the glassy matrix was identified by FESEM. FTIR measurements on these glasses exhibited a high transmission of about 75% in the wavelength range of 10–30 μm. Neutron diffraction studies on the sample $0.4Sb_2Se_3$-0.6CuI glass have been carried out to understand the SRO and the bonding nature. We concluded the presence of IRO in $0.4Sb_2Se_3$-0.6CuI glass by the use of neutron diffraction in combination with X-ray measurements. In addition, glasses were found to be consisting of mainly Sb—Se as well as Cu—Se bonds. Some investigations on the electrical switching memory for chalcohalide glasses of $0.4Sb_2Se_3$-0.6CuI and $0.6Sb_2Se_3$-0.4CuI composition have also been carried out. Switching memory behavior was found to have dependence on composition as well as on microstructure of these materials. It seems that during the process of switching Se atoms easily move because of smaller atomic radii and suitable bond angles.

4.2 THEORETICAL BACKGROUND

4.2.1 STRUCTURAL THEORIES OF GLASS FORMATION

As mentioned earlier that glasses are normally prepared by melt quenching. Different chemical systems require specific cooling rates to be met in order to form glasses. This fact has led to several attempts to produce a complete atomic theory of glass formation based on the nature of the chemical bonds and the shape of the structural units involved. The glass formation ability (GFA) has been a subject of great interest and various factors are believed to decide the GFA of a particular material. Several models based on structural, thermodynamic, and kinetic factors have been developed to understand the origin of glass formation. Glass forming ability in chalcogenide glasses is related to the number of lone pair electrons.[81] The existence of lone pair electrons is related to the presence of specific network defects.

In the case of chalcogenide materials, the compound formation and the resulting atomic structure, viscosity of the melt, speed of quenching, and the frustration in a multi-component melt are some of the important issues, which influence the GFA[18] that need to be addressed. The GFA is also understood to increase with covalency of the additive element and decrease with its size.[13] There have also been efforts to understand the compositional dependency of GFA on the basis of rigidity percolation using the constraint theory.[50]

Glass formation in non-oxide system could not be explained by a simple model. A number of theories have been suggested by different authors,[65] based on different criteria like bonding strength between the constituent elements, and their electronegativity values. The presence of deep eutectics existing in the phase diagram of certain systems is favorable for its glass formation. Sun[70] has given the criterion that glass formation can be related to the single bond energy E_b, which is defined as (E_d/CN), where E_d is the dissociation energy into gaseous atoms. According to Rawson,[59] the glass formation is closely related to the (E_b/T_m) and it is greater than 0.05 for glass forming system. This condition, known as the Sun–Rawson condition, has been questioned recently by Minaev et al.[43] who has shown that the condition is violated in As—Te, Ga—Te, and Al—Te glasses. In these materials, the liquidus line moves to higher temperatures owing to covalent ion bonding, leading to a violation of the Sun–Rawson condition. Lower coordination number of the constituent elements favors glassy phase formation. Soga and Ota,[66] have proposed a criterion based on kinetic approach according to them, higher viscosity and lower cooling rate favors glass formation criteria and conclude that the electronegativity criterion applicable to oxide system can be equally extended to fluoride and halide system. Further, another parameter of interest in the discussion of glass formation is a specific bond strength (SBS). It is a measure of the thermal energy available to break the bond. If SBS > 1 then, the glass formation is favored. The important structural models are discussed in the following subsections with an emphasis on GFA.

4.2.2 GLASS STRUCTURE MODELS

A good structural model should explain (1) property-compositional relationship; (2) atomic arrangements in LRO as well as SRO; and (3) general applicability for as many systems as possible. The concepts of a continuous random network of Zachariasen,[80] and the polymeric crystallite concept of Porai–Koshits were partially successful in explaining the glass structure. A number of models have been suggested by Zachariasen[80] and Kingery et al.[27,28] to describe the structure of glasses which are: (1) crystallite model,

(2) random network model, (3) chemically ordered continuous random network model, (4) polymeric polymorphous-crystalloid structure model, and (5) topological model based on the bond constraint theory. We shall briefly mention about these models here, however, readers may refer to details in several articles in the literature.[27,55,56]

4.2.2.1 CRYSTALLITE MODEL

X-ray diffraction patterns from glasses generally exhibit broad peaks centered in the range in which strong peaks are also seen in the diffractions pattern of the corresponding crystals. Such observation leads to the suggestion that glasses are composed of assemblages of very small crystals, termed crystallite, with the observed breadth of the glass diffraction pattern resulting from particle size broadening. This model was applied to both single-component and multi-component glasses, but the model is not acceptable presently in its original form.

4.2.2.2 RANDOM NETWORK MODEL (RNM)

In the random network model, glasses are viewed as three-dimensional networks or arrays, lacking symmetry, and periodicity. Based on the idea given by Warren[76] early controversies between proponents of the crystallite and random network models of glass structure were in general resolved in favor of the random network model.

4.2.2.3 CHEMICALLY ORDERED CONTINUOUS RANDOM NETWORK MODEL (COCRN)

Based on continuous random network by Zachariasen,[80] chemically ordered continuous random network by Lucovsky and Hayes,[38] stereo chemically defined structure by Gaskell,[21] and structural role of constituents (i.e., based on columbic interaction by network formers, network modifiers, and intermediates), a new model called COCRN was given. It is widely used as a realistic description of the structure of both covalent glasses and amorphous solids. In short, we may say that the continuous random network model has been given valuable information in the understanding of covalent solids such as amorphous silicon (a-Si), silica (a-SiO_2), and a-As_2Se_3, and the random close packing (RCP) model, also sometimes referred to as the dense random packed (DRP) model has been quite useful in studying metallic glasses such as nickel–phosphorus, gold–silicon, and copper–zirconium alloys. The random coil model is primarily applicable to polymer chain organic glass such as polystyrene.

4.2.2.4 POLYMERIC POLYMORPHOUS-CRYSTALLOID STRUCTURE MODEL

Based on the early concept of microcrystallite by Lebedev, polymeric crystallite concept by Porai–Koshits, polymeric polymorphous crystalloid structure by Minaev[40] and nano-paracrystallite by Popescu[52] formation of glass is understood as follows. It is the process of generation, mutual transformation, and copolymerization of structural fragments of various polymorphs of crystal substance without an LRO (crystalloids). Crystalloid is a fragment of crystal structure consisting of a group of atoms connected by chemical bonds. In every non-crystalline substance there are two or more SRO's, two or more IROs, and there is no LRO. Glass structure is not absolutely continuous, and there are separate broken chemical bonds and other structural defects.

4.2.2.5 TOPOLOGICAL MODEL

Topological models are based on constraint theory and on structural dimensionality considerations.[50] In these models, some of the properties can be discussed in terms of the average coordination number Z, which is indiscriminate of the species or valence bond. Two topological threshold values, defined at $Z = 2.4$ and 2.67, are present in several glassy systems such as Ge–Se and Sb (As)–Se, which are typical examples. The idea of mechanical constraint counting by Philips,[50] floppy mode and mean field rigidity threshold at $Z = 2.40$ by Thorpe,[73] structural transition from 2D to 3D at $Z = 2.67$ by Tanaka,[72] intermediate phase in addition to floppy and rigid phases by Boolchand et al.[8] have been developed. Network topological thresholds also play an important role in determining the structural aspects of chalcogenide glasses. Generally, the covalent network of chalcogenide glasses is known to be influenced by two topological effects namely, the rigidity percolation threshold (also known as stiffness or mechanical threshold) and the chemical threshold.

4.2.3 NEUTRON DIFFRACTION FORMALISM

Neutrons have a spin (1/2), zero charge and a magnetic moment $\mu_{n=} -1.913$ μN. Neutrons with thermal energies between 1 and 100 meV have de Broglie wavelengths of 2.86–0.905 Å, comparable to atomic separation in condensed matter. Having no charge but a magnetic moment makes them a useful tool to detect atomic magnetic moments with no concomitant coulomb effects.[23,78] Neutron being a chargeless particle, it can penetrate deep into matter and, therefore allows one to study bulk materials. Unlike X-ray scattering, neutron nuclear scattering cross section is governed by short-range nuclear interaction and it does not vary monotonically. This allows one to locate lighter atoms in presence of heavier atoms. Nuclear scattering even allows one to detect various isotopes of an atom.

The neutron matter interaction has two major parts, which form the basis of most of the thermal neutron scattering investigations in condensed matter. The first is the interaction of neutrons with the nuclei of the target via short-range ($\sim 10^{-12}$ cm) nuclear forces. This is known as nuclear scattering.[36] In addition, there are other very weak interactions which do not contribute significantly to the scattered intensities. The diffusive and vibrational motions of the atoms in a liquid or a glass can be measured by both inelastic neutron scattering (INS) and inelastic X-ray scattering (IXS).

For a monatomic system, a single probabilistic function called the pair distribution function is used to describe the spatial correlations between pairs of atoms. In a polyatomic liquid or glass, this concept is extended such that the probability of finding one type of atom at a given distance from another is related to the partial pair distribution function for those chemical species in that system. In all materials, the existence of chemical bonding will give rise to short range and intermediate-range structural order but not too long-range periodic order as in the case of crystals.[62] The determination of this order in terms of partial pair distribution functions is therefore fundamental to an understanding of many physical properties of liquids and glasses.

The Fourier transform of a (partial) pair distribution function leads to a reciprocal space function known as the (partial) structure factor. Neutron and X-ray diffraction techniques measure the differential scattering cross-section that is essentially proportional to the structure factor of a monatomic system, or to a weighted sum of partial structure factors for a polyatomic system. From a series of diffraction experiments on a given polyatomic system for which the scattering power of the atoms is varied (by isotopic substitution in neutron diffraction), it is often possible to determine, with accuracy, its partial structure factors and thereby its partial distribution functions. The need for high accuracy and absolute normalization of the measured intensities makes diffraction measurements on liquids and glasses particularly challenging.

As neutrons and X-rays have fundamentally different interactions with matter, diffraction experiments made using these methods can provide complementary information.

The formalism used here holds for neutron diffraction and starts with a system that comprises point-like scattering centers. The results thus obtained can then be readily generalized to the case of extended scattering centers, or atoms, by superimposing these centers on each and every point. Similar unified formalisms have been presented elsewhere.[12,25,35,54]

The neutron scattering cross section $\frac{d\sigma}{d\Omega}$ after the data reduction, for a multicomponent system in terms of the partial structure factors, $S_{ij}(Q)$ is given by Ramesh Rao et al.[58]

$$\frac{d\sigma}{d\Omega}=\sum_{i,j}\left(c_i c_j\right)^{1/2}\overline{b_i b_j} S_{ij}(Q)+\sum_i c_i \frac{\sigma_i^{\text{inc}}}{4\pi}, \tag{4.1}$$

where c_i, b_i, and σ_i^{inc} are the concentration, coherent scattering length, and incoherent scattering cross section of the ith component, respectively. The total structure factor, S(Q), can be expressed in terms of the partial structure factors $S_{ij}(Q)$ as

$$S(Q)=1+\frac{1}{\overline{b}^2}\sum_{i,j}\left(c_i c_j\right)^{1/2}\overline{b_i b_j}\left[S_{ij}(Q)-\delta_{ij}\right], \tag{4.2}$$

where $\langle\overline{b}\rangle=\left|\sum_i \overline{b_i}\right|$.

The Fourier transformation of $S(Q)$ gives the pair distribution function, $g(r)$,

$$g(r)=1+\frac{1}{2\pi^2\rho}\int_0^\infty Q\left(S(Q)-1\right)\sin Qr\, dQ, \tag{4.3}$$

where ρ is the total number density. In literature, three other correlation functions are generally used. The differential correlation function, $D(r)$ is given by

$$D(r) = (2/\pi)\int Q(S(Q) - 1)\sin Qr\, dQ \tag{4.4a}$$

$$D(r) = 4\pi\rho[g(r) - 1] \tag{4.4b}$$

The total correlation function, $T(r)$, is used for obtaining the peak positions by Soper[67] and is given by

$$T(r) = 4\pi r\rho + D(r), \tag{4.5a}$$

$$T(r) = 4\pi\rho g(r). \tag{4.5b}$$

And the radial distribution function, $N(r)$, is given by

$$N(r) = rT(r) = 4\pi r^2 \rho g(r). \tag{4.6}$$

$N(r)$ is used to derive the average number of atoms lying within a range, r to $r + dr$, of a given atom. The area of a peak in $N(r)$ gives the average coordination number $C_{e,}$[71]which in turn is related to the partial coordination numbers N_{ij} expressing the average number of j-type atoms around any i-type atom within the range of integration,

$$C_e = \frac{1}{\langle \bar{b} \rangle^2} \sum_{i,j} c_i \overline{b_i b_j} N_{ij} \tag{4.7}$$

Since $c_i N_{ij} = c_j N_{ji}$

$$C_e = c_1 \left| \frac{\overline{b_1}}{\overline{b}} \right|^2 N_{11} + 2c_1 Re\left(\frac{\overline{b_1 b_2}}{\overline{b}^2} \right) N_{12} + c_2 \left| \frac{\overline{b_2}}{\overline{b}} \right|^2 N_{22}, \tag{4.8}$$

$$C_e = W_{11}N_{11} + W_{12}N_{12} + W_{22}N_{22}, \tag{4.9}$$

where W_{ij} are the weighting factors defining the contributions of each of the partial N_{ij} to C_e. Using the above formulae one can arrive at the coordination numbers and distances between the pairs of atoms of interest. Thus, we will be able to understand the SRO as well as network connectivity in the glasses. Real space functions for a monatomic liquid or glass: (1)

pair distribution function $g(r)$, (2) density function $D(r)$, and (3) radial distribution function RDF(r) are shown in Figure 4.1.

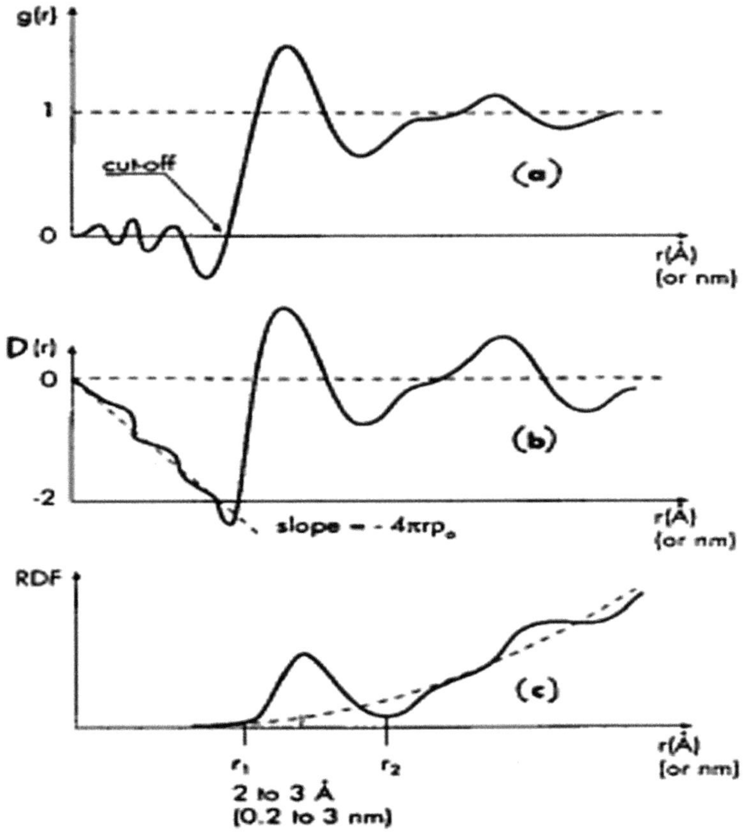

FIGURE 4.1 Real-space functions for a monatomic liquid or glass: (a) pair-distribution function $g(r)$, (b) density function $D(r)$, and (c) radial distribution function $N(r)$=RDF(r).[25]

4.2.4 ELECTRICAL SWITCHING IN CHALCOGENIDES

Several reports of non-ohmic behavior and electrical switching effects had appeared in the early 1960s.[14,29,30,68] However, it was the publication by Ovshinsky that brought enormous interest in the electrical switching behavior of amorphous semiconductors.[47] He first proposed that the internal field ionization and emission from the localized electronic states play an important role in the generation of carriers, which necessarily initiate the conduction. At a critical voltage called *threshold/switching voltage* (V_{th} or V_T), a rapid transition (in about 10^{-10} s) occurs from a high resistive (OFF) state to a very low resistive (ON) state. This breakdown is nondestructive in nature and the high-conduction state is maintained by an almost instantaneous positive feedback leading to current filamentation associated with S-shaped instability[7,61] or current controlled negative resistance (CCNR) behavior.[1] Chalcogenide glasses, which exhibit switching, are classified into memory (irreversible) or threshold (reversible) types. Threshold switching glasses revert to the OFF state upon the removal of the switching field, whereas memory switches remain locked to the ON state.

The switching phenomenon in chalcogenide glasses, described above may be reversible (threshold type) or nonreversible (memory type), as schematically depicted in Figure 4.2; the difference between the two actually lies in the manner they respond to the *removal of electric field* that induced switching. The threshold switches are monostable and return to the initial OFF state, while the bistable memory switches get latched permanently to the ON state. The onward process (OFF to ON state) is similar in both the cases, implying the commonality in the phenomenon responsible for initiating switching.[32,49] However, the process, which succeeds the field-dependent non-ohmic conductivity is different in the two cases, and is dependent on the properties of the material and on the presence or absence of suitable feedback in the system.

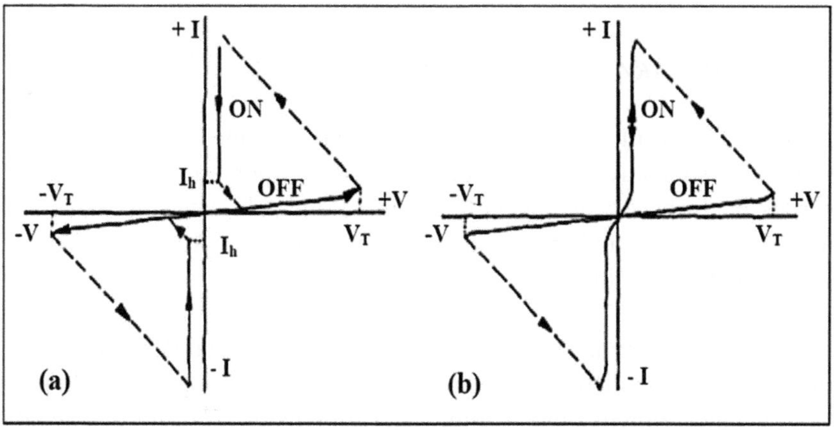

FIGURE 4.2 Schematic representations of IV characteristics of (a) threshold switch and (b) memory switch, which show symmetry between the 1st and 3rd quadrants.[19]

Switching can occur in a wide variety of materials. It is unlikely that same mechanism is responsible in all cases. The progress in the understanding of the switching behavior of glassy materials has been achieved in several stages.[1,2,5,74] In order to explain switching effects, several approaches have been adopted, which are broadly classified into three groups, namely thermal, electronic, and electro thermal theories.[10,19,20,24,26,33,51,74,75]

4.3 EXPERIMENTAL TECHNIQUES

4.3.1 INTRODUCTION

In this section, some of the experimental techniques used for both preparation and different investigations are discussed along with their working principles. The experimental set up used for glass preparation is a rocking furnace. For structural studies, techniques employed were XRD and ND. The physical properties like density/molar volume are measured using Archimedes method. Thermo mechanical investigations were carried out by thermomechanical analyzer (TMA), differential thermal analyzer (DTA), and micro hardness tester. The optical studies were carried out by UV–visible spectroscopy and FTIR and electrical properties by current–voltage (*I–V*) characteristics. The results presented in subsequent subsections are based on the experimental techniques outlined in this section.

4.3.2 PREPARATION OF GLASSES

xSb_2Se_3-$(1 - x)$CuI glasses have been prepared by melt-quenching technique in a specially designed rocking furnace operating up to 1000°C. A rocking furnace was specially designed and fabricated in Glass and Advanced Functional Materials division, Bhabha Atomic Research Centre (BARC), Mumbai. A photograph of rocking furnace is shown in Figure 4.3.

The rocking furnace is a tubular furnace, with stainless steel (SS) enclosure, having differential winding of Kanthal wire over the alumina tube of inner dimensions 65 mm, outer dimensions 75 mm and length 300 mm with uniform heating zone of 100 mm. It is mounted on a motorized rocking mechanism in which a continuously variable auto transformer is attached with a HP motor to rock the furnace for long durations while maintaining the temperature up to 1000°C. It is able to continuously rock at ± 22.5° angle. The upper end of the furnace has a removable cover for insertion and removal of the ampoule, while the lower end is kept closed.

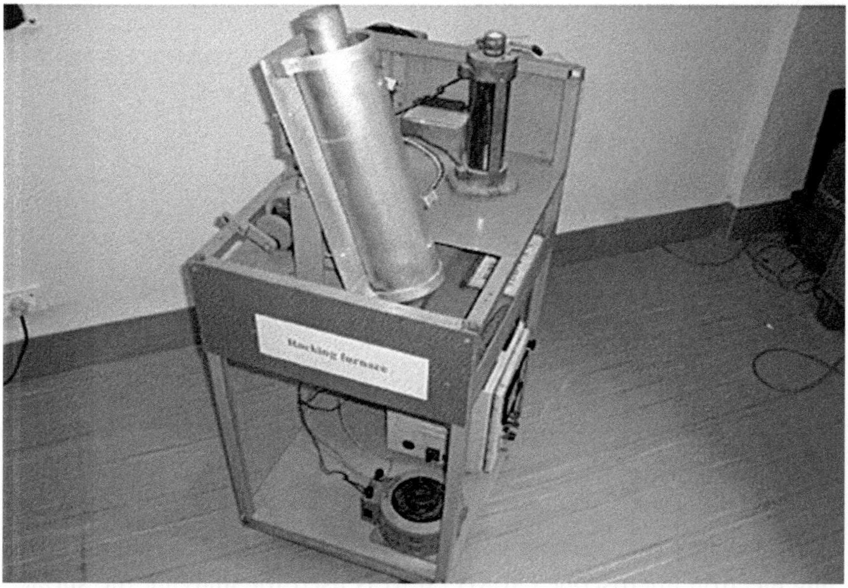

FIGURE 4.3 Schematic diagram of the rocking furnace.

A batch of about 6–8 g of sample is prepared by weighing individual constituents in appropriate proportion depending on the composition. The samples were then sealed in a specially designed quartz ampoule under a vacuum of ~10^{-5} Torr. The quartz ampoule and sealing mechanism are shown in Figure 4.4. The evacuated quartz ampoule is then securely placed in the rocking furnace. It is heated in a programmed manner to a temperature of 850–900°C for a period of 20–30 h followed by quenching in chilled salty water by holding the furnace in a vertical position. The quenching conditions were determined to obtain crack free discs. This is then annealed at a suitable temperature for 4–5 h followed by cooling to room temperature at the rate of 10°C/h.

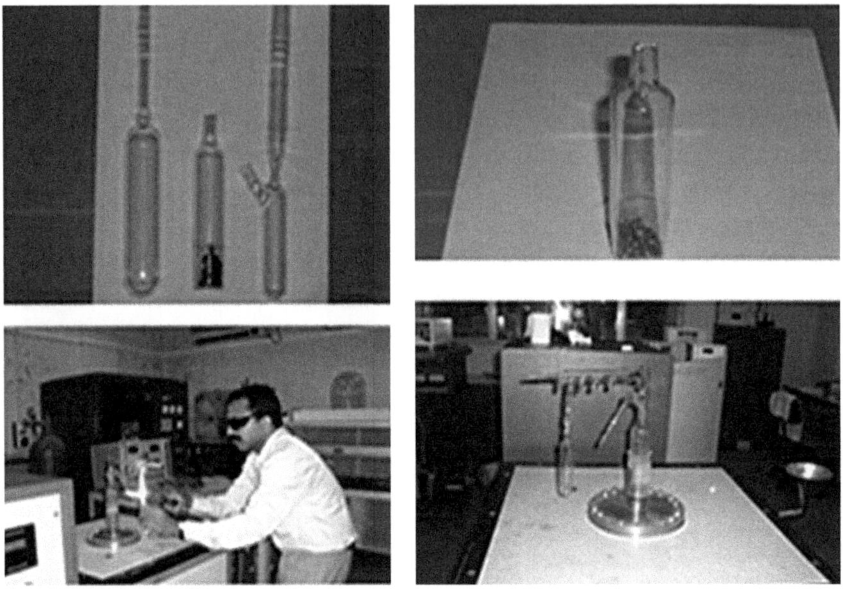

FIGURE 4.4 The quartz ampoule and sealing mechanism.

4.3.3 MICROSTRUCTURE

Microstructure and surface morphology studies can be done on scanning electron microscope (SEM). SEM is used for studying the surface topography, microstructure, and chemistry of metallic and nonmetallic specimens at magnifications from 50 up to ~100,000× with a resolution limit of <10 nm (down to ~1 nm) and a depth of focus up to several μm (at magnifications ~10,000×). In SEM, a specimen is irradiated by an electron beam and data on the specimen are delivered by secondary electrons coming from the surface layer of thickness ~5 nm and by backscattered electrons emitted from the volume of linear size ~0.5 μm. Due to its high depth of focus, SEM is frequently used for studying fracture surfaces. Sensibility of backscattered electrons to the atomic number is used for the detection of phases of different chemistry. Electron channeling in SEM makes it possible to find the orientation of single crystals by electron channeling pattern (ECP) or of grains by selected area channeling pattern (SACP).

FESEM-S4800-Hitachi was used for the investigation of surface morphology of chalcohalide glass samples of this investigation.

4.3.4 NEUTRON DIFFRACTION

Neutrons tell us "where the atoms are and what the atoms do." ND is a complementary technique to XRD and electron diffraction (ED) research reactors are typical sources of the neutrons. Neutrons in reactor possess too high energies, which are thermalized with a moderator consisting of heavy water. The thermal neutrons have kinetic energies extending over a considerable range (continuous Maxwellian distribution), but a monochromatic beam of neutrons with a single energy can be obtained by diffraction from a single crystal and this diffracted beam can be used in diffraction experiments.[77] We used High-Q Neutron diffractometer at Dhruva (100 MW), BARC, India.[34] The typical parameters of neutron High Q diffractometer at Dhruva are summarized in Table 4.1 and shown in Figure 4.5. The samples are typically held in Vanadium containers.

TABLE 4.1 Instrument Parameters of Neutron High-Q Diffractometer of Dhruva, BARC, Mumbai, India.

Neutron high-Q diffractometer (instrument parameters) at Dhruva	
Beam hole no. Monochromator Cu (220) or Cu (111)	Hs 1019
Monochromator	Cu (220) or Cu (111)
Incident wavelength (λ)	1.278 or 0.783 Å with Q_{max} = 15 Å
Range of scattering angle (2θ)	$3° < 2\theta < 140°$
Flux at sample	2×10^6 n/cm²/s for 1.278 Å
	3×10^5 n/cm²/s for 0.783 Å
	Q range 0.3–15 Å^{-1}
Sample size	5–10 mm dia, 40 mm high
Detector (1D-PSD)	10 (overlapping) at 5 positions
Resolution ($\Delta Q/Q$)	−2.5%

FIGURE 4.5 High Q Neutron diffractometer at Dhruva Reactor, BARC, Mumbai.

4.3.5 ELECTRICAL SWITCHING STUDIES

Electrical switching behaviors of the glasses have been undertaken on $0.4Sb_2Se_3$-0.6CuI and 0.6 Sb_2Se_3-0.4CuI bulk glass. I–V characteristics can be generally studied either by sourcing current and measuring voltage or vice versa. In this thesis, for studying the I–V characteristics, a constant current is passed and the resulting voltage is measured.[46] For Ohmic samples, any of the two options will produce the same result. However, in amorphous semiconductors, current increases ohmically up to a threshold voltage of V_t. At V_t, the resistance drops drastically or gradually indicating a distinct negative resistance region. If a voltage source is used directly across the sample, the reduction in resistance leads to a very high current. Therefore, a series resistance is required to limit the current flowing through the sample. On the other hand, no such precaution is needed if a constant current source is used and also, the complete information about the curve is obtained. The schematic of the sample cell used for electrical switching studies is shown in Figure 4.6. In the cell, samples polished to the required thickness are mounted in a brass holder, in between a flat bottom electrode and a point contact top electrode using a spring loading mechanism. Further, provisions are made in the cell, to study the temperature dependence of switching voltages by placing a heater in contact with the bottom electrode; a Chromel–Alumel (K-type) thermocouple is placed adjacent to the sample for measuring the sample temperature. A Keithley source measure unit (Model 2410 c), which can deliver a maximum current of 20 mA at a compliance voltage of 1100 V, is used for measuring the I–V characteristics. The current sourcing is controlled by LabVIEW 7 (NI) through a PC. Since switching voltage (V_t) is known to depend on thickness samples. It is polished to the required thickness, in the range 0.25–0.3 mm. The thickness of the samples is determined using a screw gauge of least count, 0.01 mm. We have measured I–V characteristic of chalcohalide glasses at IISC Bangalore, it is presented in the subsequent section.

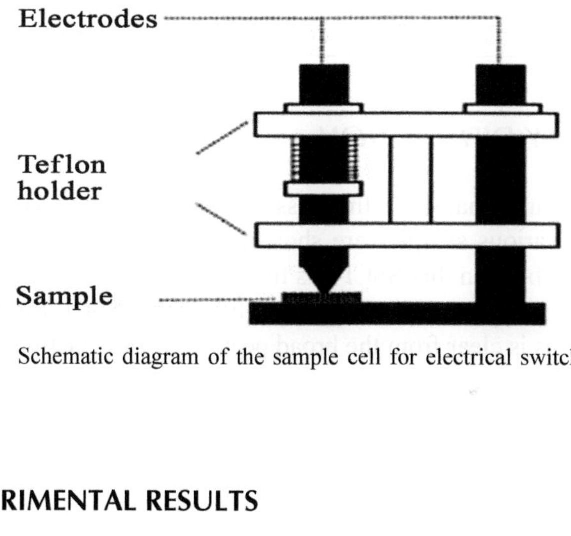

FIGURE 4.6 Schematic diagram of the sample cell for electrical switching studies of bulk glasses.

4.4 EXPERIMENTAL RESULTS

4.4.1 GLASS PREPARATION

Samples of the Sb_2Se_3-(CuI) pseudo-ternary glass were prepared by well-established melt-quenching technique. The preparation was carried out in a specially designed quartz ampoule by taking predetermined quantities of high purity raw materials of antimony (Sb), selenium (Se), and copper iodide (CuI). The initial charge is then sealed in quartz ampoules under vacuum of 10^{-5} Torr. The ampoule along with charge is placed in the quartz tube in a rocking furnace and heated in a programmed manner to a temperature of 850–900°C for a period of 16–20 h. The furnace is then held in vertical position without rocking and maintained at this temperature for about 6–8 h. Then the quartz ampoule along with the melt is taken out of the furnace and quenched in salty ice water; the quenching conditions were optimized to obtain the crack free glass discs. The ampoule is then broken open to remove the glass samples. As prepared glass with batch composition and code of glass are given in Table 4.2.

TABLE 4.2 Batch Compositions and Appearance of Melted Samples.

Sl. no.	Name of samples	Chemical composition (mol%)	Melt temperature (°C)	Nature of sample (XRD)
1	SSCI-1	50%Sb_2Se_3-50%CuI	850	Crystalline
2	SSCI-2	60%Sb_2Se_3-40%CuI	850	Partially crystalline
3	SSCI-3	30%Sb_2Se_3-70%CuI	850	Partially crystalline
4	SSCI-4	40%Sb_2Se_3-60%CuI	850	Glass

4.4.2 X-RAY DIFFRACTION

We made structural analysis of the glass samples employing XRD. The XRD plots for the various samples are shown in Figure 4.7. From these XRD patterns, it can be seen that SSCI-1 is fully crystalline, whereas SSCI-2 and SSCI-3 are partially crystalline. On the other hand, SSCI-4 is found to be purely glassy as is clear from the broad peaks in the diffraction pattern.

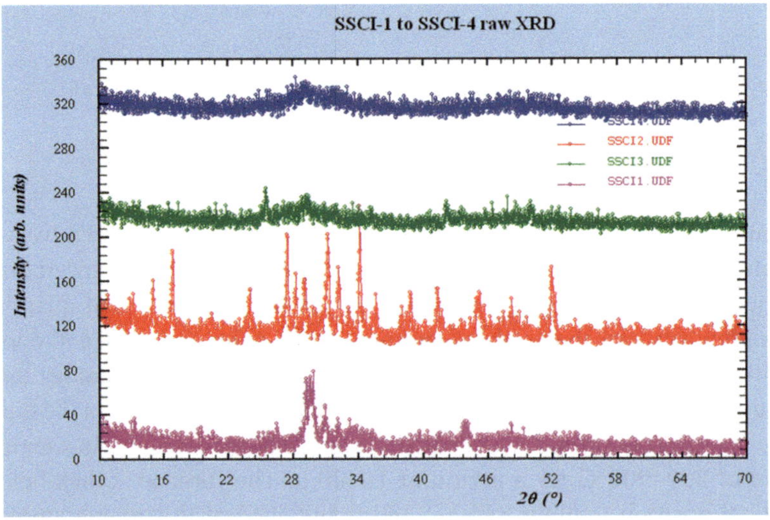

FIGURE 4.7 XRD patterns for samples SSCI-1, SSCI-3, SSCI-2, and SSCI-4 from bottom to top.

4.4.3 DENSITY MEASUREMENTS

The density of the glass sample is determined by the Archimedes method as mentioned earlier using xylene/water as an immersion liquid. The values are given in Table 4.2. The density measured at 20°C by Archimedes method was found to be 5.59 g/cc with an accuracy of +/−0.5%.

4.4.4 MICROHARDNESS MEASUREMENTS

The microhardness was determined by indentation technique using Vickers indenter on a microhardness tester. Indentation loads between 10 and 50 g were used with the dwell of 5 s. A typical indentation on the SSCI-4 sample is shown in Figure 4.8. The microhardness values are given in Table 4.2 The microhardness of glass is found to be 108.8 kg/mm^2 with a standard deviation of 4.8 kg/mm^2.

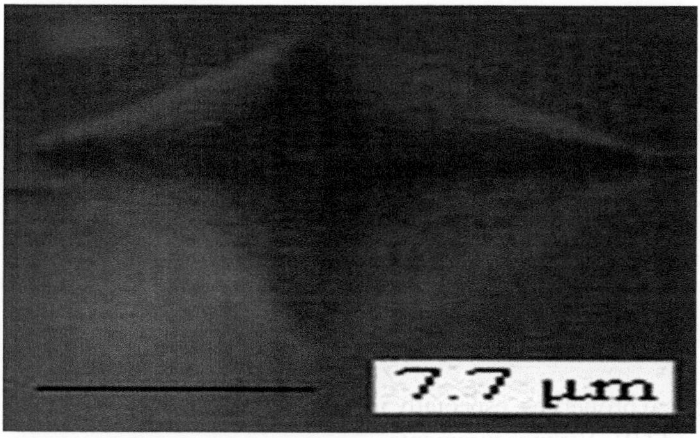

FIGURE 4.8 A typical indentation on the SSCI-\4 glass.

4.4.5 DIFFERENTIAL THERMAL ANALYSIS (DTA)

The glass transition temperature T_g was measured by using differential thermal analysis (DTA) using platinum crucible. DTA thermogram of one of chalcohalide glasses SSCI-4 is as shown in Figure 4.9. The glass transition temperature values are given in Table 4.2. From Figure 4.9, it can be seen that the glass transition temperature *Tg*, is around 167°C and crystallization temperature is around 220°C.

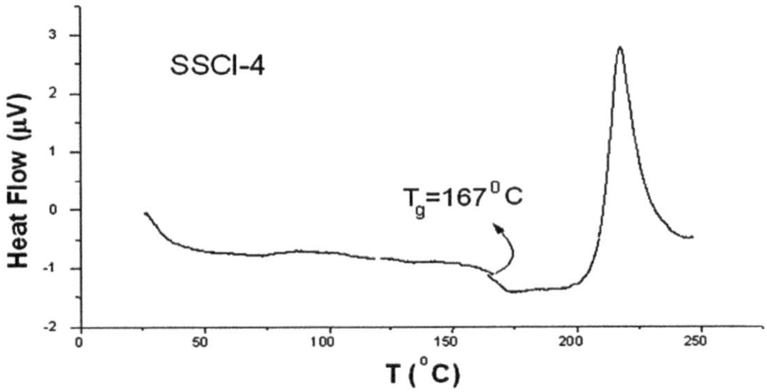

FIGURE 4.9 DTA thermogram of the chalcohalide glass.

TABLE 4.3 Values of Thermal and Mechanical Parameters for xSb_2Se_3-(100 – x) CuI Glass.

Sr. no.	Sample code	X (mol %)	Density (g/cc)	Mol. vol. (cc/mol)	DTA		Microhardness Kg/mm²	E_g eV
					T_g (°C)	T_c (°C)		
1	SSCI-2	60	5.50	66.26	156	264	116.2	–
2	SSCI-4	40	5.592	54.796	167	217	108.8	0.905

4.4.6 OPTICAL PROPERTIES

The optical properties, especially in the infrared region, were measured by FTIR and UV–Vis measurements.

4.4.6.1 FOURIER-TRANSFORM INFRARED (FTIR) SPECTROSCOPY

The infrared transmittance spectra were recorded on a FTIR in the frequency range of 200–4000 cm^{-1}. FTIR spectra of SSCI-4 glass were as shown in Figure 4.10. This composition has high transmission of 75% in the range 10–30 μm which is in the acceptable range for an infrared application.

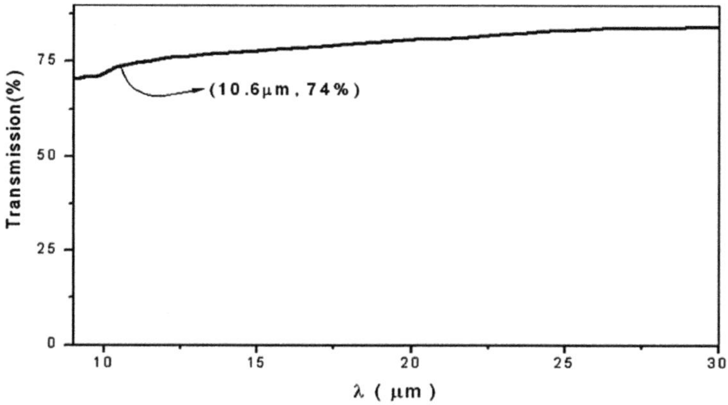

FIGURE 4.10 FTIR spectra of $0.4Sb_2Se_3$-0.6CuI glass.

4.4.6.2 UV–VIS NIR MEASUREMENTS

UV–Vis transmission of the polished glass disc was measured on JASCO spectrophotometer in the range of 200–2700 nm. The graph of absorbance (α) v/s λ is as shown in Figure 4.11. The optical gap E_g is determined by extrapolation of the straight line portions of the $(\alpha h\nu)^{1/2}$ v/s ($h\nu$) graph as shown in Figure 4.12. The values of E_g are given in Table 4.3. All the measurements have been carried out at room temperature.

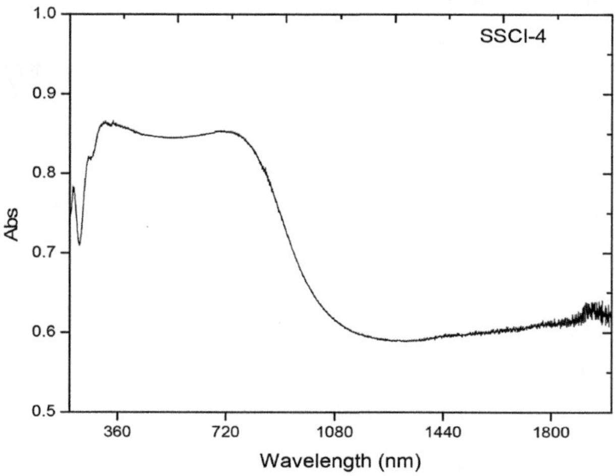

FIGURE 4.11 Variation of absorbance vs. wavelength for SSCI-4.

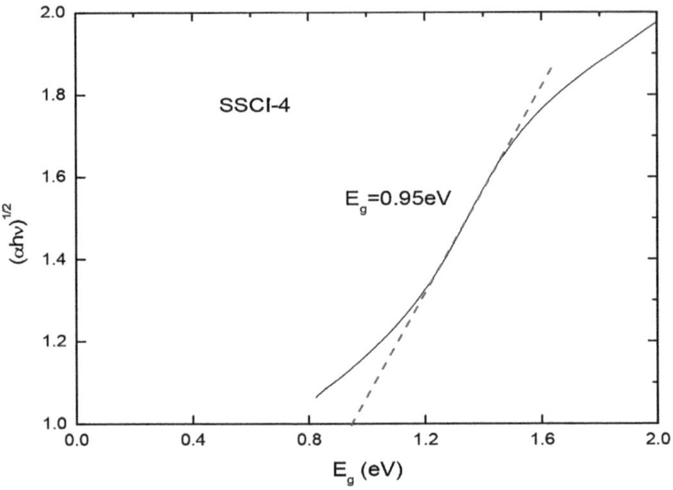

FIGURE 4.12 Plot of $(\alpha h\nu)^{1/2}$ vs. $h\nu$ for SSCI-4.

4.4.7 FIELD EMISSION SCANNING ELECTRON MICROSCOPY

FESEM was used for the investigation of surface morphology of the glass. The FESEM micrographs of the SSCI-2 and SSCI-4 glass are presented in Figures 4.13 and 4.14. The FESEM images show how the nanocrystals were formed in the form of strips of length 100–150 nm in the glass matrix (Fig. 4.13) and nanoparticles of 50–100 nm (Fig. 4.14).

FIGURE 4.13 FESEM image of SSCI-2.

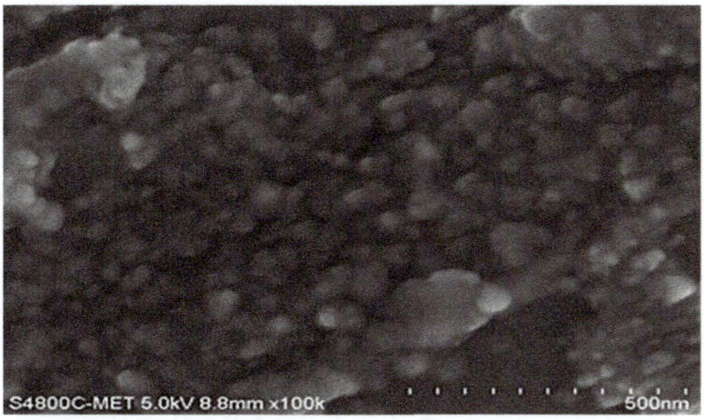

FIGURE 4.14 FESEM image of SSCI-4.

4.4.8 NEUTRON DIFFRACTION

We have done neutron diffraction (ND) studies on the sample $0.4Sb_2Se_3$-0.6CuI on the High-Q diffractometer at the Dhruva reactor, BARC, Mumbai India, to understand the SRO and network connectivity. The measured scattering intensities were divided by monitor spectra, normalized on the scattering from standard vanadium bar, and corrected for absorption, multiple scattering, self-shielding, and scattering from the container.[16] The data was transformed into real space to obtain the total correlation function, T(r) of this sample using Monte Carlo G(r) method as the S(Q) data is finite in Q.[58] The structure factor S(Q) vs. scattering wave vector Q is as shown in Figure 4.15 and total pair correlation function $T(r) = 4\pi r\rho g(r)$ vs. r in Figure 4.16.

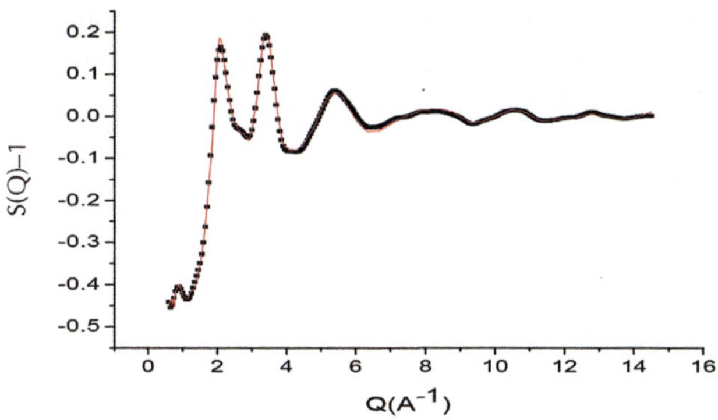

FIGURE 4.15 Total structure factor $S(Q)$ v/s scattering wave vector Q. Line is Monte Carlo $g(r)$ fit and dots are experimental data.

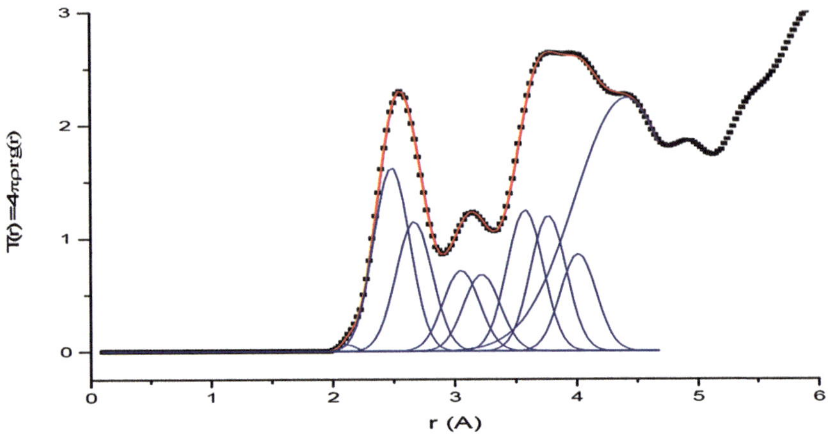

FIGURE 4.16 Total correlation function $T(r)$ vs. r. Line is fit to $T(r)$ from individual gaussian fits of various correlations and dots are the experimentally obtained $T(r)$ values.

4.4.9 ELECTRICAL SWITCHING STUDIES

I–V characteristics of these samples were measured by using Keithley source meter (Model 2410) controlled by LabVIEW6i (National Instruments). The source meter is capable of sourcing a constant current in the range 0–20 mA at a compliance voltage of 1100 V (max). The glass samples polished to about 0.35 mm thickness were mounted in a special holder, between a point contact top electrode and a flat plate bottom. A constant current is passed through the sample and the voltage developed across it, is measured. The sample exhibited switching behavior. The electrical switching characteristics of SSCI-2 and SSCI-4 samples are shown in Figure 4.17, which indicates the compositional dependent switching behavior. The composition with low as well as high CuI content shows switching behavior. It may be noted that the threshold switching is reversible whereas memory switching is irreversible in both SSCI-2 and SSCI-4.

4.5 DISCUSSION

The thermophysical and mechanical properties of chalcogenide and chalcohalide glasses depend on the structural and energetic factors. The formation of structural units at the atomic scale is correlated with the bond energy and the preparation condition. The density measured at 20°C by Archimedes method was found to be 5.59 g/cc with an accuracy of +/−0.5%. The DTA thermogram for SSCI-4 sample is shown in Figure 4.9. We observe from this graph that the glass transition temperature T_g is at 167°C and the crystallization peak T_c, is at 220°C. The thermal stability of the glass samples has been described in terms of temperature difference between glass and crystalline state (i.e., $T_c - T_g$). The higher value of about 50 reflects a good thermal stability for SSCI-4 glass. The glass transition temperature (T_g) appears to depend on the bond energy between the constituents. The T_g also depends on the structure of glasses; it tends to increase with increasing dimensionality of glass structures. The connectivity of their atomic networks can be controlled since the glasses comprise electronegative chemical species with notably different binding characters. Hence, when an ionic halogen is added to a covalently bonded chalcogen-based network, it can form non-bridging structural motifs that reduce the network connectivity and glass transition.[44] Further, the high atomic mass chalcogens and halogens bond with different kinds of metal or semimetal to form structures characterized by low phonon like energies. The chalcohalide glasses, therefore, compete with pure chalcogen and pure halogen-based glasses in terms of their excellent infrared transmission and opportunity for optical devices operating in this wavelength range.[37] The recent theoretical and experimental investigations indicate that in several chalcogenide glassy systems, the rigidity percolation spans over a range of compositions.[22,41,42,64] These systems are characterized by three different phases namely, the floppy, the intermediate, and the rigid phases. There are two distinct transitions in such glasses, namely from the floppy to the isostatically rigid state and from the isostatically rigid to the stressed rigid state. Further, the compositions between the onset and completion of an extended stiffness transition are now commonly known as the thermally reversing window as the non-reversing enthalpy is found to nearly vanish for these compositions. In chalcohalide glasses, the thermally reversing window is generally found to be very sharp, restricted to a narrow composition range.[8,57] The differential thermal analysis studies reveal that there is a sharp thermally reversing window for SSCI-4. Halogens in covalent systems are monovalent, and possess a coordination number of one. Consequently, the addition of halogens to the chalcogenide glasses results in a disruption of their three-dimensional network, leading to interesting properties. This composition has high transmission of 75% in the range 10–30 μm which is in the acceptable range for an infrared application. The FESEM micrograph of the SSCI-2 and SSCI-4 glass is presented in Figures 4.13 and 4.14. It shows nano crystals in the form of strips of length 100–150 nm size in the glass matrix; some crystalline growth of nano phase has also been observed (Fig. 4.13). In SSCI-4 glass the nanoparticles of 50–100 nm size were observed (Fig. 4.14). All these nano particles are X-ray amorphous as evidenced by X-ray diffraction data.

We have further carried out neutron diffraction studies to understand the SRO and the network connectivity in this glass.[60] The S(Q) v/s Q is as shown in Figure 4.15 and T(r) v/s r is as shown in Figure 4.16. In the total correlation graph, the peaks give us the positions of various interactions present in the system. From these peaks, we can understand the SRO and network connectivity in this glass. The SRO mainly consists of Sb—Se bonds as well as Cu—Se bonds. Se—Se bonds are ruled out as the first 2 peaks in T(r) that are larger and match with Sb—Se and Cu—Se (also Cu—I) bonds. The distances are 2.48(1) and 2.66(1) Å. From the bond energy considerations, these are more preferable bonds. The coordinations obtained are 2.9(2) and 3.7(2). The Cu—I bonds also are preferred and come around the same distances. If we assume one Cu—I bond, then there are 3.0(2) Cu—Se bonds. Similarly, Sb—I bonds also come around the identified Sb—Se distance. So, if we assume that there is one Sb—I bond, then we will have 2.2(2) Sb—Se bonds. From these results, we can say that Sb—(Se, I) pyramids are connected to Cu—(Se, I) bonds with coordinations of 3 and 4. We have further investigated the network connectivity for SSCI-4. Se—Se (non-bonding), Sb—Se (2nd neighbor), and Cu—Cu distances are identified to be in the r-range of 3.5–4.2 Å. Our results are in good agreement with the known distances in the chalcohalide glasses.[63] Cu seems to be actively participating in the network formation. Usually, chalcogenide glasses are understood using COCRN model. In this model, stability of bonds is the guiding factor in determining the closest interactions. We found that SRO mainly consists of Sb—Se bonds as well as Cu—Se bonds. From the bond energy calculations, these are the more preferred bonds. Thus, Cu seems to be actively participating in the network formation showing strong covalent nature. Since the composition SSCI-2 has higher amount of Sb_2Se_3 and it is prone to devitrification, the growth of nano-size strips/crystallites seems to be quite reasonable. The Cu supports nucleation of nanosized particles in glass matrix in this composition. The crystallites could be having a composition similar to Sb_2Se_3.

Figure 4.17 shows that these samples (SSCI-2 and SSCI-4) exhibit a compositional dependent switching behavior. The samples with high as well as low CuI content show memory switching behavior but seems to have dependence on composition as well as microstructure. This can be understood on the basis of the sharp increase in the thermal diffusivity. Memory switching in chalcogenides is mostly a thermal process, which involves phase transformation from amorphous to crystalline state. During the process of switching Se atoms can easily move because of lesser atomic radii and also their bond angles can be easily deformed due to higher compressibility. Hence, in Se-based glasses, the tendency toward regaining its initial state is less after deformation and as a result they always exhibit memory. In addition, FESEM studies exhibited the formation of nano-size strips/crystallites in $0.6Sb_2Se_3$-0.4CuI. We find the switching behavior in this sample is different (somewhat diffused) from sharper switching for the sample having lower amount of Sb_2Se_3. We propose that role of Cu in network formation combined with the formation of nano-crystalline structure might be the cause of this type of switching in $0.6Sb_2Se_3$-0.4CuI.

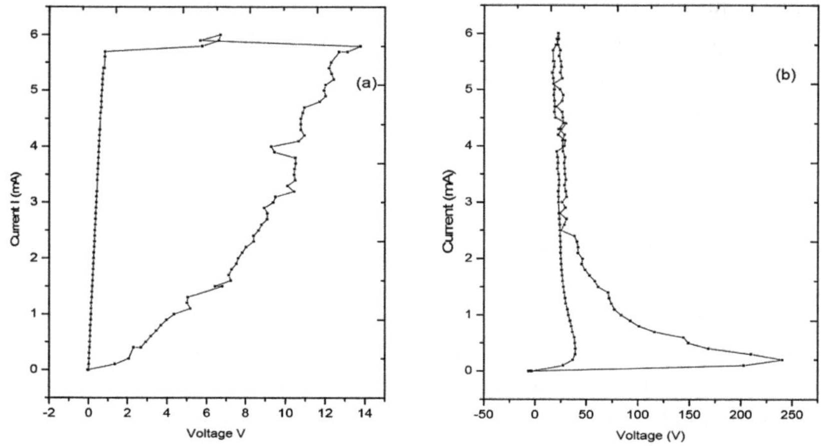

FIGURE 4.17 Variation of current *I* vs. voltage *V* for sample (a) SSCI-2 and (b) SSCI-4.

4.6 CONCLUSIONS

We obtained one of the compositions in the series ($xSb_2Se_3(1-x)CuI$ ($x = 1, 0.6, 0.5, 0.4$, and 0.3) with $x = 0.4$ in glassy form by melt quenching under vacuum showing about 75% transmission in infrared range of 10–30 μm. This shows its potential as glass for infrared application. It may be mentioned that glasses containing Cu have better resistance to cracks propagation and a higher hardness but are less stable against crystallization. We observed from FESEM measurements of chalcohalide glass nanocrystals in the form of strips of length 100–150 nm in SSCI-2 sample.

Further from neutron diffraction technique, it was clear that SRO in SSCI-4 mainly consists of Sb—Se bonds as well as Cu—Se bonds. Se—Se bonds ruled out as the first 2 peaks in T(r) were larger and matched with Sb—Se and Cu—Se (also Cu—I) bonds. The distances of these correlations were 2.48(1) and 2.66(1) Å. We can say that Sb—(Se, I) pyramids are connected to Cu—(Se,I) bonds with coordinations of 3 and 4. Cu seems to be actively participating in the network formation. The bulk, melt-quenched $xSb_2Se_3(1-x)CuI$ glasses have been found to exhibit memory switching behavior with low as well as high CuI content but seems to have a dependence on composition as well as microstructure. During the process of switching Se atoms easily move because of lesser atomic radii (1.17 Å) and also their bond angles can be easily deformed due to the higher compressibility (11×10^{-12} cm^2/dyn). Hence, in Se-based glasses, the tendency toward regaining its initial state is less after deformation and as a result, they exhibit memory switching behavior.

ACKNOWLEDGMENTS

The authors express their gratitude to Dr. B. B. Kale, C-MET Pune, for FESEM experiments. Shri A. B. Shinde and Rakesh Kumar, BARC for all other experiments. One of the authors, MSJ, indebted to Prof. H. M. Maheshwaraiah, VC, Central University, Karnataka and Dr. Meena R Chandawarkar, Ex VC, KSWU, Vijayapura for their support.

KEYWORDS

- **chalcogenide**
- **chalcohalide**
- **neutron diffraction of glass**
- **glass formation**
- **FTIR**

REFERENCES

1. Adler, D. *Amorphous Semiconductors;* Butterworth and Co. Ltd.: London, 1971.
2. Adler, D.; Henisch, H. K.; Mott, N. F. *Rev. Mod. Phy*. **1978,** *50,* 209.
3. Anderson, P. W. *Science* **1995,** *267,* 1615–1616.
4. Aravinda Narayanan, R.; Asokan, S.; Kumar, A. *Phys. Rev. B.* **1996,** *54,* 4413.
5. Arun, M.; Melvin, P. S. *The Physics and Applications of Amorphous Semiconductors;* Academic Press Inc.: London, 1988.
6. Baldwin, C. M.; Almeida, R. M.; Mackenzie, J. D. *J. Non-Cryst. Solids*. **1981,** *43,* 309.
7. Boer, K. W.; Jahne, E.; Neubauer, E. *Phys. Status Solid* **1961,** *1,* 231.
8. Boolchand, P.; Georgiev, D. G.; Goodman, B. *J. Optoelectron. Adv. Mater.* **2001,** *3,* 703.
9. Boolchand, P. Series on Directions in Condensed Matter Physics. In *Insulating and Semiconducting Glasses;* World Scientific: Singapore, 2000; Vol. 17.
10. Buckley, W. D.; Holmberg, S. H. *Phys. Rev. Lett.* **1974,** *32,* 1429.
11. Cable, M. In *Materials Science and Technology, Glasses and Amorphous Materials*. Zarzycki, J., Ed.; VCH: Weinheim, Germany, 1991; Vol. 9, pp 1–89.
12. Champeney, D. C. *Fourier Transforms and Their Physical Applications;* Academic: London, 1973.
13. Cornet, J. In *Structure and Properties of Non-crystalline Semiconductors;* Kolomiets, B. T., Ed.; Nauka: Leningrad, USSR, 1975; p 72.
14. Eaton. D. L. *J. Amer. Ceram. Soc.* **1964,** *47,* 554.
15. Ediger, M. D.; Angell, C. A.; Nagel, S. R. *J. Phys. Chem.* **1996,** *100,* 13200–13212.
16. Egelstaff, P. A. *Methods of Experimental Physics* (Chapter 14). *Neutron Scattering;* Price, D. L., Skold, Ed.; Academic Press: London, New York, 1987; Vol. 23, pp 405–70.
17. Elliot, S. R. *Physics of Amorphous Materials,* 1st ed.; Longman: London, 1984.
18. Elliott, S. R. In *Material Science and Technology: A Comprehensive Treatment, 9*; VCH Publishers Inc.: New York, NY, 1991; Chapter 7.
19. Fritzsche, H. *J. Phys. Chem. Solids.* **2007,** *68,* 878.
20. Fritzsche, H.; Ovshinsky, S. R. *J. Non-Cryst. Solids* **1970,** *2,* 393.
21. Gaskell, P. H. *Materials Science and Technology 9;* Zrzycky, J., Ed.; VCH: Weinheim, Germany, 1991; Chapter 4.

22. Georgiev, D.; Boolchand, P.; Micoulaut, M. *Phys. Rev. B.* **2000,** *62,* R9228.
23. Halpern, O.; Johnson, M. H. *Phys. Rev.* **1939,** *55,* 898.
24. Henisch, H. K.; Fagen, E. A.; Ovshinsky, S. R. *J. Non-Cryst. Solids.* **1970,** *4,* 538.
25. Henry Fischer, E.; Adrian Barnes, C.; Philip Salmon, S. *Rep. Prog. Phys.* **2006,** *69,* 233–299.
26. Jackson, J. L.; Shaw, M. P. *Appl. Phys. Lett.* **1974,** *25,* 666.
27. Kingery, W. D.; Bowen, H. K.; Uhlmann, D. R. *Introduction to Ceramics,* 2nd ed.; John Wiley and Sons: New York, NY, 1976.
28. Kingery, W. D. *Introduction to Ceramics;* McGraw-Hill: New York, NY, 1972; Vol. 1, p 351.
29. Kolomiets, B. T.; Nazarova, T. F. *Sov. Phys. Solid State.* **1960,** *2,* 369.
30. Kolomiets, B. Y.; Lebedev, E. A. *Radiotechnika i Elektronika* (russ), **1963,** *7,* 2087.
31. Kothiyal, G. P.; Rakesh Kumar.; Goswami, M.; Shrikande, V. K.; Bhattacharya, D.; Roy, M. *J. Non. Cryst. Solids* **2007,** *353,* 1337.
32. Kotz, J.; Shaw, M. P. *Physica B C.* **1983,** *117–118,* 986.
33. Kroll, D. M. *Phys. Rev. B.* **1974,** *9,* 1669.
34. Krishna, P. S. R.; Shinde, A. B. *Solid State Phys.* **2002,** *45,* 121.
35. Leadbetter, A. J.; Wright, A. C. *J. Non-Cryst. Solids* **1972,** *7,* 23.
36. Lovesey, S. W. Theory of Neutron Scattering from Condensed Matter (International Series of Monographs on Physics Vol. 72) Vol. 1 and 2, Oxford: Clarendon, 1984.
37. Lucas, J. In *Insulating and Semiconducting Glasses;* Boolchand, P., Ed.; World Scientific: Singapore, 2000; p 691.
38. Lucovsky, G.; Hayes, T. M. *Amorphous Semiconductors;* Brodsky, M. H., Ed.; Springer-Verlag: Berlin, 1979.
39. Mahantappa Jogad, S.; Madhumita, G.; Arjun, S.; Tahir, M.; Lingappa Udachan, A.; Govind Kothiyal, P. *Mater. Lett.* **2002,** *57,* 619–627.
40. Minaev, V. S. *Vitreous Semiconductor Alloys;* Moscow, Metallurgia, 1991.
41. Micoulaut, M. *Phys. Rev. B.* **2006,** *74,* 184208.
42. Micoulaut, M.; Phillips, J. C. *Phys. Rev. B.* **2003,** *67,* 104204.
43. Minaev, V. S.; Timoshenkov, S. P.; Oblozkho, S.; Rodionov, P. V. *J. Optoelec. Adv. Mater.* **2004,** *6,* 791.
44. Mitkova, M.; Boolchand, P. *J. Non-Cryst. Solids* **1998,** *240,* 1.
45. Morey, G. W. *The Properties of Glasses,* 2nd ed.; Reinhold: New York, NY, 1954.
46. Murugavel, S.; Asokan, S. *Phys. Rev. B.* 1998; *55,* 3022.
47. Ovshinsky, S. R. *Phys. Rev. Lett.* **1968,** *21,* 1450.
48. Ovshinsky, S. R.; Fritzsche, H. *IEEE Trans. Elec. Dev.* **1973,** *20,* 91.
49. Owen, A. E.; Robertson, J. *IEEE Trans. Electron Devices* **1973,** *20,* 105.
50. Phillips, J. C. *J. Non-Cryst. Solids.* **1979,** *34,* 153.
51. Popescu, M. *Solid-St. Electron.* **1975,** *18,* 671.
52. Popescu, M. *J. Optoelectron. Adv. Mat.* **2005,** *7* (4), 2211.
53. Poulain, M. *J. Non-Cryst. Solids* **1983,** *56,* 1.
54. Price, D. L. Saboungi, M. L.; Bermejo, F. J. *Rep. Prog. Phys.* **2003,** *66,* 407–80.
55. Prakash, S.; Asokan, S.; Ghare, D. B. *J. Phys. D: Appl. Phys.* **1998,** *29,* 2004.
56. Pattanayak, P.; Asokan, S. *Solid State Commun.* **2008,** *148,* 378–381.
57. Pattanayak, P. N.; Manikandan, N.; Paulraj, M.; Asokan, S. *J. Phys: Condens Matter.* **2007,** *19,* 036224.

58. Ramesh Rao, N.; Krishna, P. S. R.; Saibal, B.; Dasannacharya, B. A.; Sangunni, K. S.; Gopal, E. S. R. *J. Non-Cryst. Solids* **1998,** *240,* 221.
59. Rawson, H. *Inorganic Glass-Forming System;* Academic Press: London, New York, NY, 1967.
60. Rashmi, M. J.; Krishna, P. S. R.; Jogad, M. S.; Rakesh, K.; Mathad, R. D.; Kothiyal, G. P. *Int. J. Sci. Res.* **2012,** *01–02,* 47–48.
61. Ridley, B. K. *Proc. Phys. Soc.* **1963,** *81,* 996.
62. Salman, P. S. *Nature* **2005,** *435,* 75–78.
63. Salman, P. S.; Xin, S. *Phys. Rev. B.* **2002,** *65,* 064202.
64. Selvanathan, D.; Bresser, W. J.; Boolchand, P. *Phys. Rev. B.* **2000,** *61,* 15061.
65. Shelby, J. E. Introduction to Glass Science and Technology, 2nd ed.; CRC Press: Boca Raton, FL, 2005.
66. Soga, N.; Ota, R. *XIV Int. Congress on Glass,* 1986, Collected Papers 1, 74.
67. Soper, A. K. *Chem. Phys.* **2000,** *258,* 121–37.
68. Southworth, M. P. *Central. Engin.* **1964,** *11,* 69.
69. Stoch, L. *Glass Phys. Chem.* **2001,** *27,* 167.
70. Sun, K. H. *J. Am. Ceram. Soc.* **1947,** *30,* 277.
71. Susman, S.; Volin, K. J. Montague, D. G. Price, D. L. *J. Non-Cryst. Solids.* **1990,** *125,* 168.
72. Tanaka, K. *Phys. Rev. B.* **1989,** *39,* 1270.
73. Thorpe, M. F. *J. Non-Cryst. Solids.* **1983,** *57,* 355.
74. Vezzoli, G. C.; Peter, J.; Walsh.; William Doremus, L. *J. Non-Cryst. Solids.* **1975,** *18,* 333.
75. Warren, A. C. *IEEE Trans. Elec. Dev.* **1973,** *20,* 123.
76. Warren, B. E. *J. Appl. Phys.* **1937,** *8,* 645.
77. Warren, B. E. *X-Ray Diffraction;* Dover Publications Inc.: New York, NY, 1990.
78. Williams, W. G. *Polarized Neutrons;* Oxford University Press: Oxford, UK, 1988.
79. Xiujian, Z.; Liangying, X.; Hongbing, Y.; Sumio, S. *J. Non-Cryst. Solids.* **1994,** *167,* 70–73.
80. Zachariasen, W. H. *J. Am. Chem. Soc.* **1932,** *54,* 3841.
81. Zallen, R. *The Physics of Amorphous Solids;* John Wiley and Sons Inc.: Hoboken, NJ, 1983.
82. Zarzycki, J. *Glasses and the Vitreous State;* Cambridge University Press: Cambridge, England, 1991.
83. Ziman, J. M. *Models of Disorder;* Cambridge Univ. Press: Cambridge, England, 1979.

CHAPTER 5

BLOCK COPOLYMER MICELLES AND NANOPARTICLES HYBRID ASSEMBLIES: SYNTHESIS AND CHARACTERIZATION

LAVNAYA TANDON and POONAM KHULLAR*

Department of Chemistry, BBK DAV College for Women, Amritsar 143005, Punjab, India

**Corresponding author. E-mail:virgo16sep2005@gmail.com*

CONTENTS

ABSTRACT

Block copolymers are a versatile category of usually non-toxic polymers with enormous applications in cosmetics, food, and pharmaceutical industries. This chapter demonstrates its promising applications in materials chemistry in terms of synthesis, characterization, and applications of nanomaterials. Its micellar form is highly effective in the synthesis of gold nanoparticles (Au NPs) by using the surface cavities in the form of nanoreactors. The core–shell configuration of such micelles helps in loading and unloading of the NPs simply by altering the nature of solvents. A change in the hydrophilicity and hydrophobicity parameters induce changes in the shape and size of the micelles and hence, control the overall shape and size of the NPs. Both template and seed growth methods have been employed. Potential of hydrogen (Ph) plays an important role for the synthesis of Au NPs. Thus, applications of block polymers in materials chemistry open up a new direction for materials synthesis in different directions. The block copolymer micelles are used as drug delivery carriers for the various hydrophobic anti-cancer drugs like paclitaxel (PTX), doxorubicin, etc.

Nanoscience is one of the emerging sciences of nanoscale materials. For the synthesis of nanomaterials and fabrication of nanostructures, two approaches are used, that is, "Top down" and "Bottom up." "Top down" approach begins with the bulk piece of material which is step by step removed to form nanometer size object. "Bottom up" approach begins with the atoms and molecules that are arranged and assembled into large nanostructures. Most of the nanomaterials are prepared by using the bottom-up routes as they possess the ability to generate uniform size, shape, and distribution. In the synthesis of nanomaterials, micelle forming block polymers and surfactants are extensively used. This chapter accounts for the recent advances in the synthesis of Au NPs of various morphologies using template as well as seed growth method and future scope of these nanomaterials in biomedical applications.

Different methods, such as seed-mediated growth processes,[1] template-directed patterning,[2,3] biomineralization, two-phase reactions, and inverse micelles[4] have been used to synthesize nanomaterial of different morphologies. Block copolymers are defined as the large blocks that are made up of chemically distinct monomers that are covalently linked together. The interactions among the building blocks (hydrophobic, electrostatic, etc.) are what drives these block copolymers to self-assemble into different sites in solution. Triblock polymers (TBPs) are of considerable fundamental and industrial importance. TBPs are made up of different combination of polyethylene oxide (PEO) and polypropylene oxide (PPO)[5–7] (Fig. 5.1). Due to the presence of both hydrophilic (i.e., PEO) and hydrophobic predominant (i.e., PPO) moieties in the same polymer macromolecule, these have tremendous advantages than conventional neutral polymers. These are commercially available in a range of molecular weights and PEO–PPO composition.[8] The overall nature gets shifted from predominant hydrophilic to predominant hydrophobic by appropriate variation in the number of PEO and PPO repeating units. These polymers possess a versatile flexibility and are, therefore, useful in various industrial applications like emulsification, detergency, dispersion, lubrication, etc.[9,10] The amphiphilic character of block copolymers leads to self-assembled behavior which resembles that of classical nonionic surfactant.[11–14] As these polymers also act as non-ionic surfactants, new insights into the aggregation behavior has been gained through various techniques like light scattering,[15,16] fluorescence spectroscopy,[17,18] small-angle neutron scattering (SANS),[19,20] specific volume,[21,22] and nuclear magnetic resonance (NMR).[23]

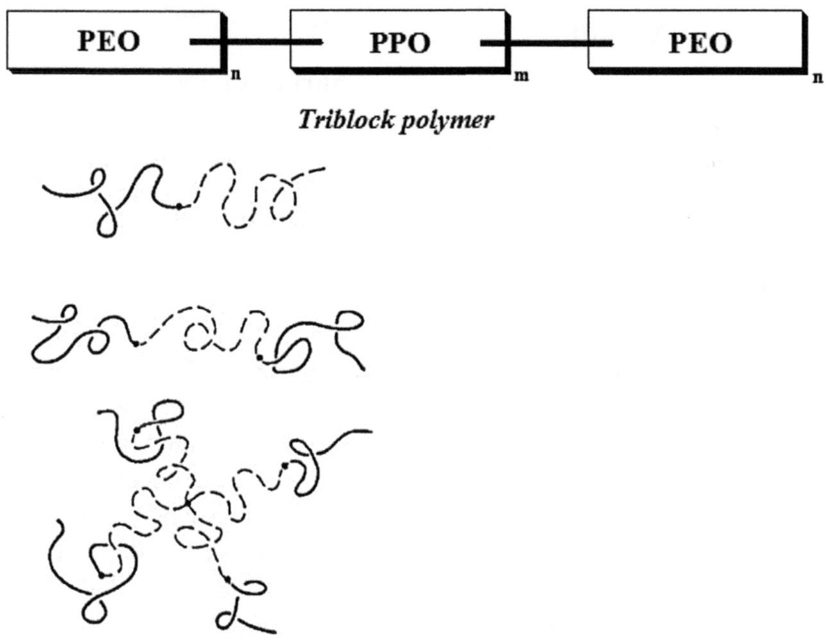

FIGURE 5.1 Schematic representation of the block copolymer macromolecules.

Micellization in TBPs (Fig. 5.2) is understood to arise due to the following reasons.[24–26] The PPO block loses its hydration sphere as the temperature is raised, and this result in greater interaction between the PPO blocks. But the PEO blocks do not lose its interaction with the water and, therefore, interaction with the water is retained. It is common for all amphiphilic molecules, that the differing phase preferences of the block drive the copolymers to form the micelles. A TBP becomes predominantly hydrophobic when the number of PEO units exceeds the PPO unit, and on the other hand, acquires a predominant hydrophobic nature when the reverse happens.[27,28]

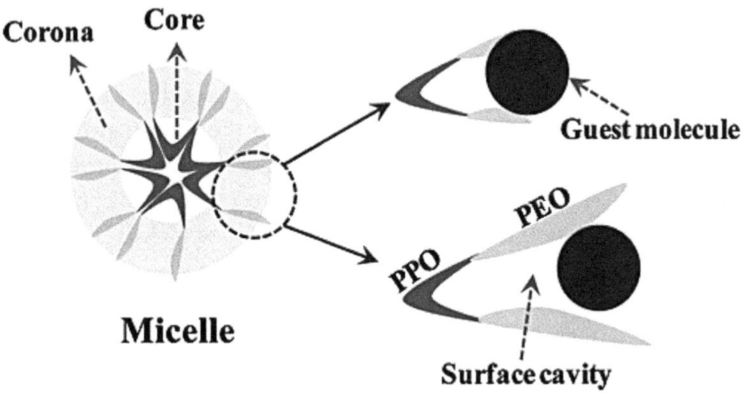

FIGURE 5.2 Schematic representation of a TBP micelle showing the core and corona. The core is mainly made up of PPO and the corona by PEO units. Arrangement of PEO units leads to the formation of surface cavities to accommodate guest molecules. (Reprinted with permission from Khullar, P.; Singh, V.; Mahal, A.; Kaur, H.; Singh, V.; Banipal, T. S. Tuning the Shape and Size of Gold Nanoparticles with Triblock Polymer Micelle Structure Transitions and Environments. *J. Phys. Chem. C.* **2011**, *115*, 10442–10454. © American Chemical Society.)

It has been shown from the structural studies[29–35] that the micelles form a hydrophobic core which consists of weakly hydrated PPO blocks, are surrounded by outer shell known as corona of fully hydrated PEO blocks. A broad temperature range exists above cmt, where the micelles coexist with unimers in solution. Most of the block copolymer molecules form micelles above the transition region. Cloud point (Cp) is the temperature above the cmt where the phase separation of micellar solution into two phases occurs.[36] The most important parameter is *cmc*, that is, the copolymer concentration at which the micelle starts forming.[37] There is some inherent complexity in the micellization of the block copolymers as compared with those of conventional low molecular weight surfactants. The micelle demonstrates interesting shape transitions which are closely associated with a number of factors, such as molar masses of PEO and PPO unit, nature of solvent, salt additives, concentration, and temperature.[38–40] Since the TBP micelles are usually very large (a few to hundreds of nm) in comparison to the monomeric surfactants, therefore, they undergo several structure transitions (i.e., spherical micelle →thread like micelle→vesicles) with concentration as well as temperature variation.[38–40] The reducing ability of the cavity which is formed by the PPO and PEO block is directly connected to its site. Since one polymer is contributing only one surface cavity, therefore, it accepts only one guest ion (i.e., oxidizing agent). A large cavity is more capable to fit the guest ion rather than a small cavity. In addition to the cavity size, another important factor is the micelle environment as fully hydrated surface cavities with low aggregation number at low temperature may not accept as many as guest ions or oxidizing agents as accepted by the partially hydrated or dehydrated cavities with high aggregation number at high temperature.

Micellization of the block copolymers depends strongly on their composition.[41–50] No sharp *cmc* and cmt have been observed for block polymers as the blocks are not completely monodispersed for a polymer having narrow distribution of molecular weight. In general, span of *cmc* is much over a larger concentration interval than with the conventional surfactant. The *cmc* is sensitive to temperature which extends the concentration range over which the *cmc* occurs[50] (Fig. 5.3).

Triblock copolymers have been used for the synthesis of Au NPs and these Au NPs find potential applications in many frontier areas, such as biology, medicine, catalysis, sensors, and surface-enhanced spectroscopy.[51–55] But the most important requirement is the presence of micellar phase along with the gold salt. The micelle formation of block copolymers is similar to that of the conventional surfactants. There exists a thermodynamic equilibrium between the more predominant micellar phase and the less predominant monomeric phase. Pluronics and tetronics are water-soluble block copolymers and these generate well-defined micelles lined with surface cavities at water solution interface. Electrostatic interactions occur when micelles come in contact with the gold ions. These interactions occur through the surface cavities to initiate the reduction reaction. The block copolymer-mediated synthesis method provides an advantage in the

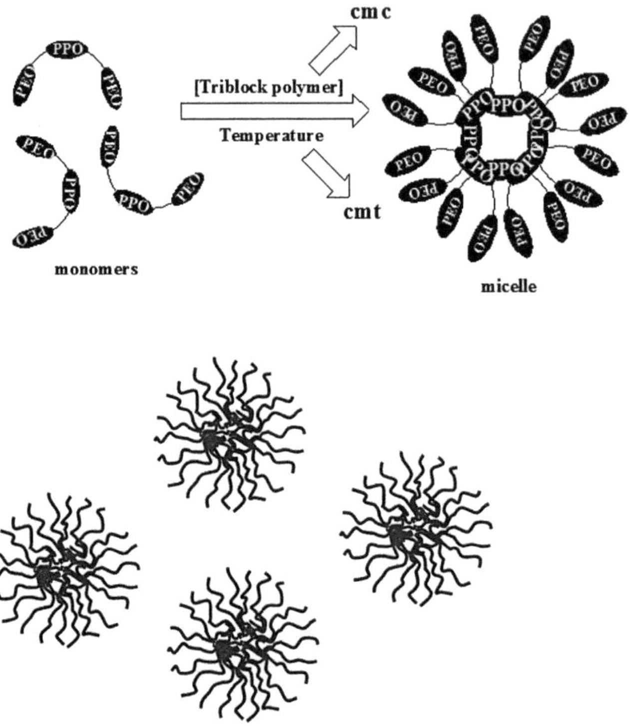

FIGURE 5.3 Schematic representation of the micelle formation of the block copolymers under the effect of concentration and temperature (upper). Micelles with core and corona model (lower).

environmental and economic aspect since the synthesized method requires a simple procedure which is the mixing of a metal salt with a block copolymer solution. Ternary combination of block copolymer + gold salt + water in the absence of any other reducing agent[56–58] is used to study the synthesis of Au NPs. In the reduction reaction, the monomeric form of block copolymer is not effectively involved due to lack of surface cavities and, therefore, the reduction is primarily exhibited by the micelles only. Reduction potential depends upon the number of surface cavities and concentration of the micelles. This type of reducing agent is known as structural reducing agent in which reduction is directed by the structural factors. The monomeric form cannot create surface cavities and hence, cannot participate in reduction reaction, therefore, the monomeric form is not *structured reducing agent.*[57,58] For the synthesis of Au NPs, the formation of structured reducing agent is an important step to develop efficient nanoreactors. Micelles of copolymers like polystyrene block poly (4-vinyl pyridine), poly[tert-butyl styrene-b-sodium(sulfamate/carboxylate)], and poly(*n*-isopropyl acryl amide)-b-poly(1-(30-amino-propyl)-4 acryl amine-1,2,3 triazole hydrochloride[59–62] are being employed as nanoreactors for the synthesis of Au NPs. These systems do not generate well-defined micelle Au NPs-hybrid morphologies. The micelle formation is an aqueous phase process where well-defined morphologies are derived from hydrophobic interactions and these are rarely generated in the non-aqueous phase. However, in biomedical applications, the stability in the aqueous phase is required which is used as efficient drug release vehicle. Various techniques have been employed for the determination of micelle formation of block copolymer micelle, and the most common techniques are surface tension, fluorescence, and ultraviolet (UV) measurement and these are highly sensitive toward the aggregation behavior of block copolymers. In literature, a limited work is available on the synthesis of anisotropic gold nanostructures with tunable optical properties via green chemistry (Fig. 5.4).

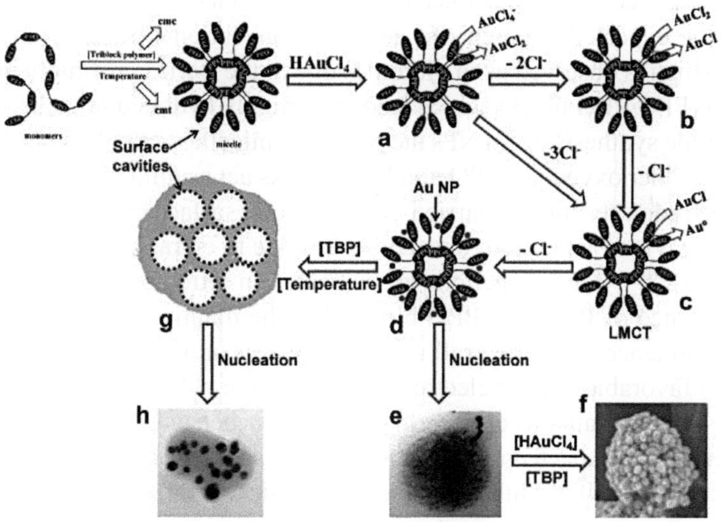

FIGURE 5.4 Demonstration of the overall redox process taking place in the surface cavities at the micelle–solution interface of block copolymer micelles. (Reprinted with permission from Khullar, P.; Mahal, A.; Singh, V.; Banipal, T. S.; Kaur, G.; Bakshi, M. S. How PEO–PPO–PEO Triblock Polymer Micelles Control the Synthesis of Gold Nanoparticles: Temperature and Hydrophobic Effects. *Langmuir* **2010**, *26*, 11363–11371. © American Chemical Society.)

Tetronics (Fig. 5.5) belong to the category of star polymer and these have unique physiochemical properties.[63–68] These possess self-assembled behavior and are highly temperature sensitive. A slight increase in the temperature causes dehydration of PPO and PEO blocks and that results in micelle-like assemblies. Tetronics are highly pH sensitive due to the presence of diamine core and, therefore, their physiochemical properties are dependent on the pH variation.[63] Amine moiety is converted into ammonium group at low pH and the aggregation is restricted as the macromolecule acquires a net positive charge. Neutral form is attained at high pH which produces micelle-like aggregation under the effect of concentration and temperature. The micelle formation in tetronics is different from the conventional micelle formation of pluronics. Pluronics consists of long-chain block polymers and are devoid of amine functionality. The morphology of long-chain pluronic micelle is easily predicted in which the hydrophobic PEO groups are present in the shell. It is not easy to predict the arrangement of star-shaped polymer like tetronic in which PPO and PEO exist in the form of four arms connected to the diamine moiety which reside in the center of the molecule to produce star-shaped geometry. This is not an expected geometry to generate conventional micellar arrangement which is observed as in pluronics due to the steric constraints in four arms of PPO/PEO blocks which leads to a highly hydrated micelle.

For the synthesis of Au NPs the pluronics micelles act as fine nanoreactors and ether oxygen of PPO and PEO blocks act as mild reducing agents and the redox reaction is carried out in micelle surface cavities and these surface cavities act as active reaction sites due to entrapment of gold ions (Fig. 5.5).[56,59,69] Same reactions are expected in tetronics micelles due to the presence of PEO and PPO blocks.[70,71] The diamine moiety has additional influence on the overall redox process as amino groups at low pH act as a favorable site for electrostatic interaction with negatively charged $AuCl_4^-$ ion. A dramatic change in aggregation behavior of tetronic with temperature is expected and thus the redox process results in the synthesis of Au NPs. Au NPs synthesis is the good indicator for the onset of aggregation process and subsequent structure transition.

The tetronic micelles are highly responsive to pH and temperature variations. pH-responsive biological behavior is due to the diamine functional group of tetronic macromolecule which in turn produced the pH-sensitive micelle. The water molecule induces the micelle hydration at low pH (Fig. 5.6). This is due to the presence of ether oxygen in PEO and PPO block along with the quaternary amine functional group. Micelles formed at low pH are more temperature sensitive than the micelle produced at high pH. In the synthesis of Au NPs, the micelles are used as nanoreactors and the pH and temperature responsive behavior are reflected in the synthesis of Au NPs. At high pH, NPs are synthesized and the amine moiety readily reduces Au (III) to Au (0) in comparison to low pH. To initiate the reduction reaction at low pH the protonation of amine moieties is required and also high temperature is required to initiate the reduction reaction. The synthesized Au NPs are adsorbed on the micelle surface so as to attain the colloidal stabilization and tetronic micelles have proved to be excellent nanoreactors to synthesize nanomaterials under different reaction condition.

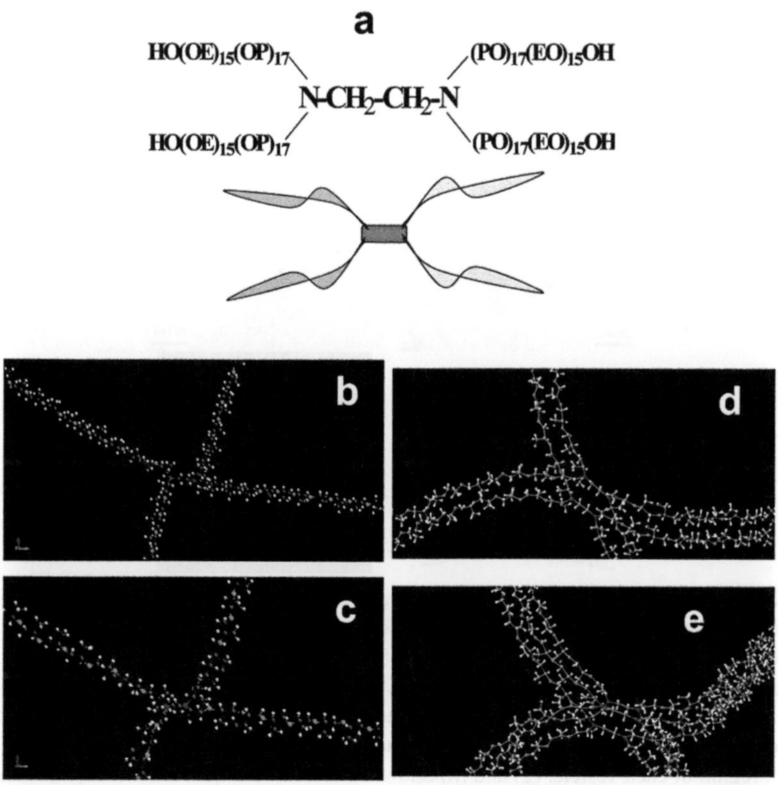

FIGURE 5.5 (a) Structural formula of star-shaped T904 and its schematic representation H (white), C (grey), O (red), and N (blue). (b) A graphical model of the structure and (c) its relaxed structure. (d) A graphical model of an aggregate of three molecules and (e) their relaxed structure.

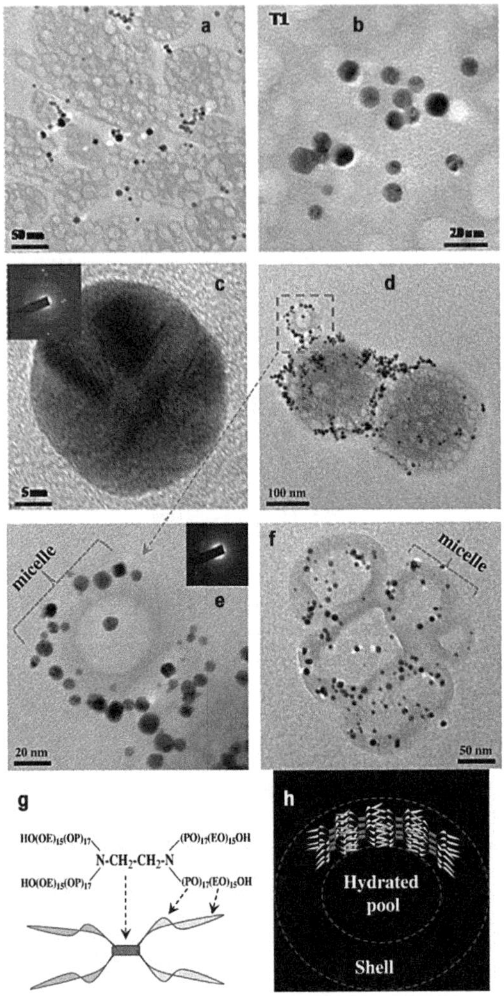

FIGURE 5.6 (a) TEM image showing large patches of vesicular assemblies of T904 along with Au NPs as dark dots. Small black dots of (a) close up in (b) and high resolution in (c) are fine spherical-shaped Au NPs of 17.5 nm. (d) Shows two sidewise fused vesicular assemblies bearing groups of small Au NPs while (e) is the high-resolution image showing the presence of Au NPs mainly on the surface of the core–shell type vesicle. (f) Shows several core–shell type vesicles fused together in a group bearing Au NPs mainly in the shell region of each vesicle. (g) and (h) Show the corresponding schematic representation of core–shell type vesicle and the alignment of P904 macromolecules based on energy minimization. (Reproduced from Singh, V.; Khullar, P.; Dave, P. N.; Kaura, A.; Bakshi, M. S.; Kaur, G. pH and Thermo Responsive Tetronic Micelles for the Synthesis of Gold Nanoparticles: Effect of Physiochemical Aspects of Tetronics. *Phys. Chem. Chem. Phys.* **2014**, *16*, 4728–4739, with permission from from the PCCP Owner Societies.)

Micelle stability is the most important factor to attain self-assembled arrangement of NPs on the micelle surface because a stable micelle has a proper arrangement of surface cavities which are the reaction sites for the conversion of Au (III) into Au (0). Predominantly, hydrophilic micelles with greater hydration are prone to self-association and form compound micelles (Figs. 5.6 and 5.7) with no clear surface arrangement of surface cavities, therefore, predominantly hydrophobic micelles are always good templates.[47,72–75] Since the reduction of Au (III) into Au (0) is facilitated in the aqueous phase, therefore, the stability of predominantly hydrophobic micelles as of L121 in aqueous phase can be achieved only by incorporating the surfactant monomers which impart required hydration to the stern layer (i.e., shell of TBP micelle).[76,77] Surfactants of different polarities in the mixed state with L121 can have dramatic influence on the synthesis of Au NPs by modifying the surface arrangement of PEO surface cavities. Addition of ionic surfactants like sodium dodecyl sulfate (SDS) and cetyl trimethyl ammonium chloride (CTAC) to a relatively less hydrophobic P123 disrupts the polymer aggregates and induces significant hydration in PPO block. Both surfactants form liquid crystalline phase with L121 dismantle the polymer aggregates with predominate effect demonstrated by cetyl trimethyl ammonium bromide (CTAB) than SDS.[78,79] Low concentrations of ionic surfactants mainly form polymer rich mixed micelles (Figs. 5.7 and 5.8) while intermediate concentrations simultaneously produce small surfactant-rich aggregates.[78] Similar interactions are observed with nonionic surfactants but with relatively of lesser magnitude than ionic surfactants based upon the regular solution approximations.[79] Therefore, choice of an appropriate surfactant is an important aspect to be taken into consideration to obtain suitable micelle templates for Au NPs.

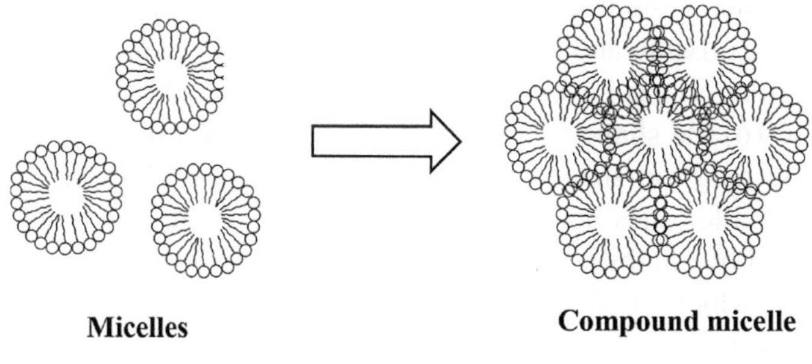

FIGURE 5.7 Formation of compound micelles from the individual micelles of block copolymer.

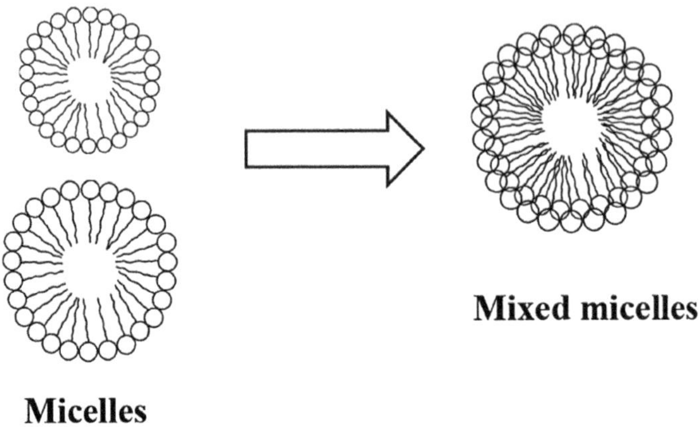

FIGURE 5.8 Formation of mixed micelles from the micelles of two individual surfactants.

Figure 5.10 depicts the variation of shape and size of micelle template with mole fraction and hydrophobicity of zwitterionic surfactants. The phase diagram has been divided into TBP and surfactants rich regions. In TBP rich region, large micelle template of different shapes and sizes loaded with self-assembled NPs are present. Micelle templates are produced with L121 with DPS (C12) over the whole mole fraction range. It systematically increases the size closer to equimolar range, therefore, L121 rich micelles break and convert into surfactant rich compound micelle in the DPS rich region. Micelle templates are loaded with the tiny NPs of 2–6 nm whether they are produced in L121 or DPS rich region in accordance with the origin from the cavities.[78–84] Similarly, L121 with TPS (C14) also results in the formation of micelle template of different shapes (Fig. 5.9) which are well defined and loaded with tiny NPs in the L121 rich region but these are deformed when mole fraction increases and finally disappears in TPS rich region of the mixture and no compound micelle is formed. But large plate-like triangular NPs is generated along with small polyhedral NPs. In case of HPS (C16) the shape transition become more prominent and in L121 rich region long rod-shaped micelles loaded with tiny NPs are produced from the smaller ones because of inter particle fusion. The latter process is facilitated by solubilizing[82–84] ability of HPS in view of its long hydrocarbon tail which confines PEO–PPO surface cavities in small packets and allow tiny NPs to undergo inter particle fusion so as to produce larger ones. However, any further increase in mole fraction results in the formation of plate-like NPs equimolar in HPS region.

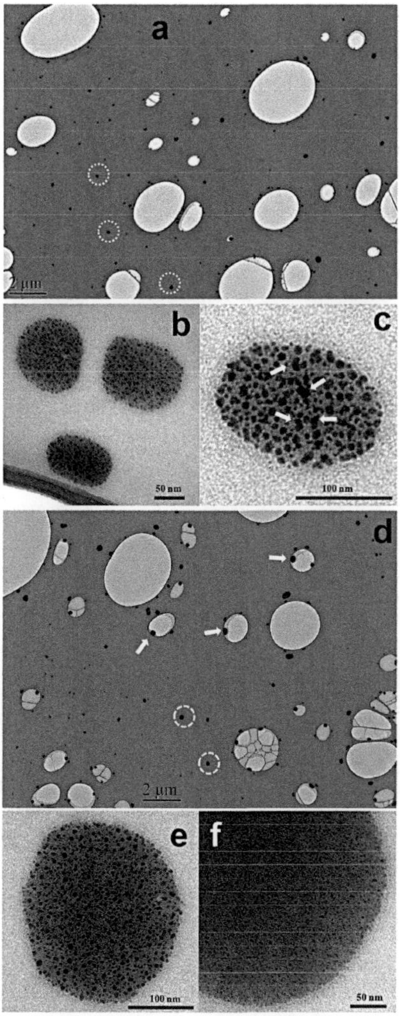

FIGURE 5.9 (a) Low-resolution TEM image of a sample prepared with DPS/L121 mole ratio = 0.5 at 70°C. Dotted white circles enclose single micelles. (b,c) High-resolution images of different roughly spherical micelles with self-assembled Au NPs of 2–6 nm in size. White arrows in (c) indicate some fused NPs. (d) Low-resolution TEM image of a sample prepared with DPS/L121 mole ratio = 0.8 at 70°C. Again, dotted white circles enclose single micelles. (b,c) High-resolution images of a single micelle with self-assembled NPs. (Reprinted with permission from Khullar, P.; Singh, V.; Mahal, A.; Kumar, H.; Kaur, G.; Bakshi, M. S. Block Copolymer Micelles as Nanoreactors for Self-assembled Morphologies of Gold Nanoparticles. *J. Phys. Chem. B.* **2013**, *117*, 3028–3039. © 2013, American Chemical Society.)

Figure 5.10 indicates that the hydrophobicity and concentration are important parameters of surfactants which determine the self-assembled arrangement of NPs on the templates of micelles. The stability of L121 micelle in aqueous phase is maintained by the low hydrophobicity and low concentration without disturbing the surface cavities. High hydrophobicity requires space in the core so as to arrange the long hydrocarbon tail.

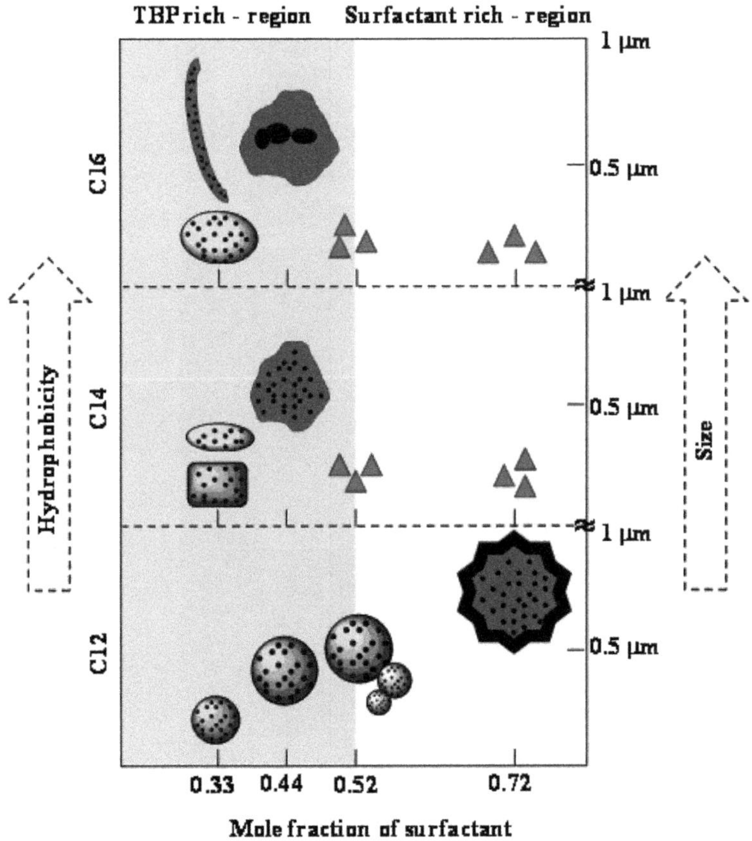

FIGURE 5.10 Proposed phase diagram showing the relationship among hydrophobicity, size of the micelles and NPs, and mole fraction of different surfactants. C12, C14, and C16 represent DPS, TPS, and HPS, respectively. Phase diagram has been divided into TBP-rich and surfactant-rich regions. (Reprinted with permission from Khullar, P.; Singh, V.; Mahal, A.; Kumar, H.; Kaur, G.; Bakshi, M. S. Block Copolymer Micelles as Nanoreactors for Self-assembled Morphologies of Gold Nanoparticles. *J. Phys. Chem. B.* **2013**, *117*, 3028–3039. © 2013, American Chemical Society.)

To adjust in the shell the hydrophobicity needs more zwitterionic head groups. In the former, the surfactant provides the solubilization of PPO blocks in the hydrocarbon environment and the L121 micelle are dismantled and the latter disturbs the arrangement of surface cavities and cause a reduction in the reduction potential. To facilitate the solubilization of otherwise insoluble L121 micelles in the aqueous phase so that they can be used as templates is the primary purpose of the surfactant. In case of DPS, it is achieved by the concentration effect due to its low hydrophobicity. But TPS and HPS cannot be used over a similar concentration range due to their enhanced hydrophobicity and thus to produce well-defined micelle template low concentration is required.

Another important factor is the presence of surface cavities of the micelle template as without surface cavity it is difficult to accommodate NPs of micelle template. On the micelle template, all the synthesized NPs remain self-assembled and none exist independently in the L121 rich region. As long as the surface cavities exist, self-assembled arrangement of NPs survives on the micelle template. To achieve self-assembled arrangement of Au NPs, the presence of surface cavities or reducing sites on micelle templates is required for the direct visualization through imaging studies. This observation can be used for other hydrophobic copolymers as well.

Metallic NPs can also be produced by using the polymeric micelles or nanoreactors within the core, hybrid polymeric micelles (HPMs) (Fig. 5.10).[85,86] First step is the synthesis of polymeric micelles which are further used to prepare HPMs. In the core of these micelles precursor of the desired NPs are loaded by the electrostatic interaction between the core and precursor.[87,88] HPMs are then formed due to the reduction of precursor. The size distribution of NPs can be easily controlled[89,90] as the reduction reaction is localized in the nanospace of the core. The formation of polymeric micelle and loading of precursor takes place simultaneously.[91–93] The common solvent, a block copolymer is used to dissolve the block copolymer and then inorganic precursor is added that forms complex with the interacting block. Micellization is induced by the complexation and the polymeric micelles are formed which consists of non-interacting block as shells and complexes act as cores. HPMs are formed by the subsequent reduction of the inorganic precursor. HPMs have various applications, such as catalyst[94] and biological probes.[95] These are based on the catalytic and photoactivities of the NPs that are encapsulated[96] (Fig. 5.11).

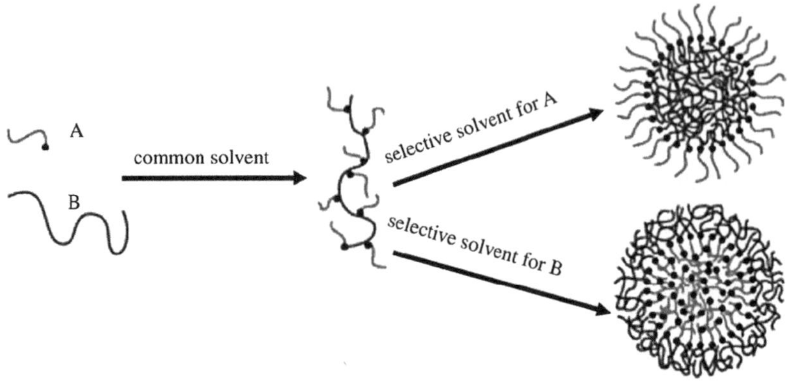

FIGURE 5.11 Schematic representation of the formation of hydrogen bonding graft polymers and the corresponding NCCMs. Reproduced from Ref. [90] with permission.

Core–shell inversions of the corresponding HPMs is a convenient way to prepare reversed HPMs; by using core as nanoreactors the facile control over the size distribution is maintained—the core and shell turn-by-turn. It is usually achieved by switching the selectivity of solvent for the core and the shell turn-by-turn producing reverse HPMs. The strong interaction between the interacting block and NPs play an important role in such process. Figure 5.12 clearly demonstrates the preparation and the stabilization of HPMs containing Au NPs with a narrow size distribution in the core. HPMs are obtained by adding the tetrachloroauric acid to PSB–P4VP solution in chloroform, a common solvent of both blocks so as to form PS shell and P4VP /HAuCl4 core micelle and subsequent reduction of HAuCl4. The protonation of core forming P4VP block chains having the HPMs are being stabilized as protonated P4VP are insoluble in chloroform and bind sufficiently strongly to Au NPs. The core–shell reversion of Au NPs[83,97] HPMs takes place by the continuous addition of methanol into the chloroform solutions that produces vesicles like RHPMs with Au NPs anchored to the shell forming protonated P4VP chain. For the stabilization of Au NPs in the shell of reversed HPMs the protonation of P4VP chain is necessary. The Au NPs can be released from vesicles like aggregates once the P4VP block chains are deprotonated.

Self-assembly of the diblock copolymer poly[tert-butylstyrene-b-sodium (sulfate) carboxylate isoprene] in water can carry Au NPs[98] by micelles having a hydrophobic core of poly (tert-butyl styrene) and a hydrophilic corona of poly(sodium sulfanate/carboxylate-isoprene), anionic polystyrene in corona block Au NPs are subsequently formed by the reduction of Au (III) to Au (0) without introducing any reducing agent. UV–Visible spectroscopy monitors the kinetics of Au reduction. This is possible by direct observation of the exact position and the intensity of the surface plasmon resonance (SPR) band of Au created NPs. Optical properties are changed due to the structural changes and are observed via shifts in the SPR band of the system. Agglomeration of Au NPs is promoted by increase in ionic strength of the solution. The interaction with the model globular protein can be investigated by considering the polyelectrolyte nature and biocompatibility of the nano assembly.

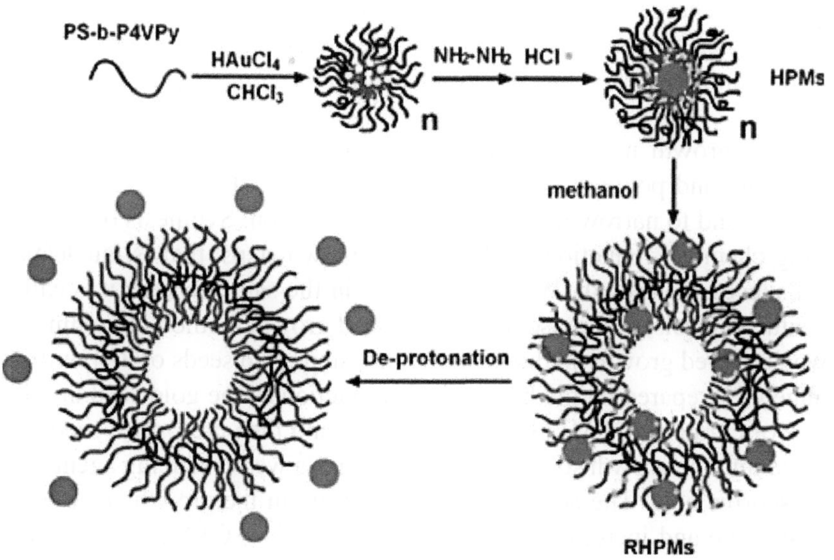

FIGURE 5.12 Schematic description of the preparation of HPMs, vesicle-like reversed HPMs. Reproduced from Ref. [58] with permission.

The kinetics of reduction of Au (III) to Au (0) is studied by mixing the tetrachloroaurate with the hybrid micellar solution at certain molar N/Au ratios. By heating the resulting solution at 50°C or at room temperature the bands are observed by UV spectroscopy every 30 min the UV absorption spectra is collected from mixed solution by naked eye, onset of the reaction is observed on the development of light pink color in the solution after 30 min under heating or after 10 h at room temperature.

At neutral pH, the positively charged lysozyme interacts with the negatively charged hybrid micelle and a complex material is formed. It consists of a dense core which is hydrophobic and pH sensitive, hydrophilic shell that carries the Au NPs that are complexed with protein globules. Initially, a decrease is observed in the hydrodynamic radius of the complex and later increase is observed due to the excess of negative charges at low lysozyme concentration. A complete coordination of the protein around the hybrid micelle is observed, due to such complexation. Due to complexation, the compact shape with small size is observed and the conformational rearrangement of the components of the complex occurs. The hydrodynamic radius of the complex increases as the lysozyme concentration increases. The stretched conformation allows the positively charged protein molecule to interact with the remaining negative charges on the corona. The dependence of the absorption intensity of the complex paves the way to utilize it as a component of biosensor system.

Seed growth method is the soft template approach which is a very important and popular technique that has been used for the synthesis of Au NPs and to narrow the particle size distribution. Synthesis of Au NPs using chemical reduction method involves two steps, that is, nucleation and successive growth. When both occur in the same process it is called in situ synthesis; otherwise, it is called seed growth method or commonly two-step seed growth methods. In the first step, gold seeds of 3–4 nm size range are prepared by the borohydride reduction of the gold seed in presence of citrate or CTAB as a capping agent and then preformed seeds are added into the growth solution which contains a weak reducing agent such as ascorbic acid. The addition of silver nitrate in the growth stage helps to improve and increase the yield of nanorods.[99–104] CTAB usually binds strongly to the surface and restricts subsequent functionalization. Triblock polymer like F127 has been employed as a cosurfactant in the modified seed growth method.[105] The nanorods thus prepared in the presence of block copolymer not only prevented any reshaping upon storage but also led to the enhanced efficiency of the seed-mediated process, which resulted in the higher yield of nanorods. The NPs obtained by the normal seed growth method could have potential application in medicine due to tunable plasmonic properties but it is retained due to the toxicity of CTAB or citrate.

The stabilization of gold nanorods is supposed to be due to the formation of stable complex between CTAB and block copolymer, hydrophobic interaction between the PPO block of block copolymer and hydrophobic surfactant.[105] The CTAB pluronic complex thus get adsorbed over the synthesized gold nanorods and hence present any further physical and chemical change. It was observed that the reducing nature of PEO block of pluronic is responsible for the enhanced yield of nanorods.[106] Ascorbic acids play a vital role in the synthesis of gold nanorods and pluronics alone is not sufficient to reduce the metal salt in the presence of CTAB. The addition of block copolymer without ascorbic acid results in a simple color change from bright orange to colorless Au^{3+} and Au^{+} but no seed was formed. It is due to the poor reducing power of PEO blocks. At high concentration, PEO can reduce the gold salt but PPO blocks get adsorbed into the gold clusters and lead to the stabilization of Au NPs that hinders the growth of NPs into nanorods. Seed growth method is noteworthy due to the following aspects: (1) monodispersed NPs are produced as compared to the Frens method,[107] (2) smaller particles grow into larger particles of predetermined size, and (3) its applicability to the surface-confined Au NPs.

pH is a crucial factor in the synthesis of Au NPs using block copolymers. In case of tetronics, the morphology of the NPs can be easily controlled by varying the pH values because of the presence of four reducing NH_2 amino group in tetronics.[108] At low pH, the addition of protons strengthens the hydrogen bond between the block copolymer and water molecules and due to the attachment of protonated water molecules with the PEO block and hence PEO can hold more positive charge. However, when gold NPs are produced, H^+ ions are also produced at the same time and the presence of H^+ interrupts the binding of metal ions to PEO blocks and the reduction will be slowed down. It results in an increase in the size of micelle[109] and hence less number of Au NPs will be produced. At low pH, because of the hydration of the micelle, the Au NPs cannot be entrapped by the micelle; hence, they find their way into the bulk where they get aggregated.

But as we increase the pH, according to Le Chatelier's principle, more and more number of H^+ ions will be neutralized and hence the reduction of gold salt is increased resulting in a large number of Au NPs. The stability of the Au NPs at the basic conditions can be attributed to the hydrophobic interaction because of the coating of NPs with the block copolymer, therefore, this coating makes the NPs hydrophilic, and easily dispersible in water without further surface modification, that is, desirable for biomedical applications. Increasing the pH[110] value enhances both the reaction rates and the coordination ability of the block copolymer as well as results in a change of the morphology of Au NPs and also causes Ostwald ripening process. Also, the pluronic polymers are adsorbed on the gold surfaces to form a cage shell structure via hydrophobic interactions.

Block copolymers form nanosized mixed micelle with core–shell morphology composed of hydrophobic PPO core of a hydrophilic PEO shell layer.[111] The anti-cancer drug such as PTX and doxorubicin has poor aqueous solubility due to its hydrophobic nature, however, block copolymer micelles can be used as drug delivery carrier for such kind of drugs. This application of block copolymer is limited due to the less stability of micelle which is attributed to the poor hydrophobic interaction of the PPO block. However, such a limitation can be improved by using thin film hydration method[112] that results in the formation of thiol-terminated block copolymer-PTX micelle, and then Au NPs can be covalently attached to the thiolated surface of such micelle. Keeping the concentration of block copolymer constant and varying the amount of drug PTX, it is observed that formulation of 4/270 (drug/carrier) has an appropriate drug loading capacity suitable for practical applications.[113]

5.1 FUTURE PERSPECTIVES

This chapter introduces the applications and uses of generally non-toxic and environment-friendly block copolymers for the synthesis, characterization, and applications of nanomaterials. The micellar assemblies of block copolymers act as nano-reactors and by changing the micellar environment, one can easily control the shape and size of the nanomaterials. A unique property of block copolymer micelles to exist in the core–shell configuration further helps in loading and unloading of the nanomaterials according to their use as drug delivery vehicles. An appropriate choice of block copolymer can achieve a desired synthesis of the nanomaterials and hence, explores a new direction in materials synthesis, characterization, and applications.

KEYWORDS

- **block copolymers**
- **micelles**
- **gold nanoparticles**
- **synthesis**
- **characterization**
- **seed-growth**
- **anti-cancer drugs**

REFERENCES

1. Iqbal, M. I.; Chung, Y.; Tae, G. An Enhanced Synthesis of Gold Nanorods by the Addition of Pluronic (F-127) Via a Seed Mediated Growth Process. *J. Mater. Chem.* **2007,** *17,* 335–342.
2. Liang, H. P.; Wan, L. J.; Bai, C. L.; Jiang, L. Gold Hollow Nanospheres: Tunable Surface Plasmon Resonance Controlled by Interior-cavity Sizes. *J. Phys. Chem. B.* **2005**, *109,* 7795–7780.
3. Johnson, C. J.; Dujardin, E.; Davis, S. A.; Murphy, C. J.; Mann, S. Growth and Form of Gold Nanorods Prepared by Seed-mediated, Surfactant-directed Synthesis. *J. Mater. Chem.* **2002,** *12,* 1765–1770.
4. Pileni, M. P. The Role of Soft Colloidal Templates in Controlling the Size and Shape of Inorganic Nanocrystals. *Nat. Mater.* **2003,** *2,* 145–150.
5. Janert, P. K.; Schick, M. Phase Behavior of Ternary Homopolymer/Diblock Blends: Influence of Relative Chain Lengths. *Macromolecules.* **1997,** *30,* 137–144.
6. Janert, P. K.; Schick, M. Phase Behavior of Binary Homopolymer/Diblock Blends: Temperature and Chain Length Dependence. *Macromolecules* **1998,** *31,* 1109–1113.
7. Alexandridis, P. Structural Polymorphism of Poly(Ethylene Oxide)–Poly(Propylene Oxide) Block Copolymers in Nonaqueous Polar Solvents. *Macromolecules* **1998,** *31,* 6935–6942.
8. Parsipany, N. J. *Pluronic and Tetronic Surfactants, Technical Brochure*; BASF-Corporation: Florham Park, NJ, 1989.
9. *Edens, M. W.* Application of Polyoxyalkylene Block Copolymer Surfactants. *In Nonionic Surfactants*; Nace, V. M., Ed.; Marcel Dekker: New York, *1996;* Vol. *60, p 185.*
10. Alexandridis, P. Amphiphilic Copolymers and their Applications. *Curr. Opin. J. Colliod Interface Sci.* **1996,** *1,* 490–501.
11. Schillén, K.; Jansson, J.; Löf, D.; Costa, T.; Mixed Micelles of a PEO–PPO–PEO Triblock Copolymer (P123) and a Nonionic Surfactant (C12EO6) in Water. A Dynamic and Static Light Scattering Study. *J. Phys. Chem. B.* 2008, *112,* 5551–5562.

12. Yoncheva, K.; Petrov, P.; Pencheva, I.; Konstantinov, S. Triblock Polymeric Micelles as Carriers for Anti-Inflammatory Drug Delivery. *J. Microencapsul.* 2015, *32,* 224–230.
13. Zhang, Y.; Lam, Y. M.; Study of Mixed Micelles and Interaction Parameters for Polymeric Nonionic and Normal Surfactants. *J. Nanosci. Nanotechnol.* **2006,** *6,* 1–5.
14. Messaoud, T.; Duplâtre, G.; Waton, G.; Michels, B. Behavior of Pluronic P84 Block Copolymer Micelles above the Gelification Temperature as Probed by Positron Annihilation Lifetime Spectroscopy. *Phys. Chem. Chem. Phys.* 2005, *7,* 3839–3844.
15. Thurn, T.; Couderc, S.; Sidhu, J.; Bloor, D. M.; Penfold, J.; Holzwarth, J. F.; Wyn-Jones, E.; Study of Mixed Micelles and Interaction Parameters for ABA Triblock Copolymers of the Type EO_m–PO_n–EO_m and Ionic Surfactants: Equilibrium and Structure. *Langmuir* 2002, *18,* 9267–9275.
16. Nolan, S. L.; Phillips, R. J.; Cotts, P. M.; Dungan, S. R.; Light Scattering Study on the Effect of Polymer Composition on the Structural Properties of PEO–PPO–PEO Micelles. *J. Colloid Interface Sci.* **1997,** *191,* 291–302.
17. Holland, R. J.; Parker, E. J.; Guiney, K.; Zeld, F. R. Fluorescence Probe Studies of Ethylene Oxide/Propylene Oxide Block Copolymers in Aqueous Solutions. *J. Phys. Chem.* 1995, *99,* 11981–11988.
18. Nakashima, K.; Anzai, T.; Fujimoto, Y. Fluorescence Studies on the Properties of a Pluronic F68 Micelle. *Langmuir* **1994,** *10,* 658–661.
19. Almgren, M.; Bahadur, P.; Jannson, M, Li, P.; Brown, W.; Bahadur, A. Static and Dynamic Properties of a (PEO–PPO–PEO) Block Copolymer in Aqueous Solution. *J. Colloid Interface Sci.* 1992, *151,* 157–165.
20. Williams, R. K.; Simard, M. A.; Jolicoeur, C. Volume Changes for Thermally Induced Transitions of Block Copolymers of Propylene Oxide and Ethylene Oxide in Aqueous Solution as Model Systems for Hydrophobic Interaction. *J. Phys. Chem.* 1985, *89,* 178–182.
21. Wen, X. G.; Verrall, R. E. Temperature Study of Sound Velocity and Volume-related Specific Thermodynamic Properties of Aqueous Solutions of Poly (Ethylene Oxide)-Poly(Propylene Oxide)-Poly (Ethylene Oxide) Triblock Copolymers. *J. Colloid Interface Sci.* 1997, *196,* 215–223.
22. Goldmints, I.; von Gottberg, F. K.; Smith, K. A.; Hatton, T. A. Small-angle Neutron Scattering Study of PEO–PPO–PEO Micelle Structure in the Unimer-to-micelle Transition Region. *Langmuir* **1997,** *13,* 3659–3664.
23. Goldmints, I.; Holzwarth, J. F.; Smith, K. A.; Hatton, T. A. Micellar Dynamics in Aqueous Solutions of PEO–PPO–PEO Block Copolymers. *Langmuir* **1997,** *13,* 6130–6134.
24. Alexandridis, P.; Holzwarth, J. F.; Hatton, T. A. Micellization of Poly (Ethylene Oxide)-Poly (Propylene Oxide)-Poly (Ethylene Oxide) Triblock Copolymers in Aqueous Solutions: Thermodynamics of Copolymer Association. *Macromolecules* **1994,** *27,* 2414–2425.
25. Alexandridis, P.; Holzwarth, J. F. Block Copolymers. *Curr. Opin. Colloid Interface Sci.* **2000,** *5,* 312–314.
26. Yang, L.; Alexandridis, P.; Steytler, D. C.; Kositza, M. J.; Holzwarth, J. F. SANS Investigation of the Temperature Dependent Aggregation Behavior of the Block Copolymer Pluronic L64 in Aqueous Solution. *Langmuir* **2000,** *16,* 8555–8561.

27. Kozlov, M. Y; Melik-Nubarov, N. S; Batrakova, E. V; Kabanov, A. V. Relationship Between Pluronic Block Copolymer Structure, Critical Micellization Concentration and Partitioning Coefficients of Low Molecular Mass Solutes. *Macromolecules.* **2000,** *33,* 3305–3313.
28. Su, Y. l; Wang, J; Liu, H. Z. Formation of a Hydrophobic Microenvironment in Aqueous PEO–PPO–PEO Block Copolymer Solutions Investigated by Fourier Transform Infrared Spectroscopy. *J. Phys. Chem. B.* **2002,** *106,* 11823–11828.
29. Almgren, M.; Brown, W.; Hvidt, S. Self-aggregation and Phase Behavior of Poly (Ethylene Oxide)-Poly(Propylene Oxide)-Poly(Ethylene Oxide) Block Copolymers in Aqueous Solution. *Colloid Polym. Sci.* 1995, *273,* 2–15.
30. Mortensen, K.; Pedersen, J. S. Structural Study on the Micelle Formation of Poly(Ethylene Oxide)-Poly(Propylene Oxide)-Poly(Ethylene Oxide) Triblock Copolymer in Aqueous Solution. *Macromolecules* 1993, *26,* 805–812.
31. Hecht, E.; Hoffmann, H. Kinetic and Calorimetric Investigations on Micelle Formation of Block Copolymers of the Poloxamer Type. *Colloids Surf. A.* **1995,** *96,* 181–197.
32. Goldmints, I.; Yu, G. E.; Booth, C.; Smith, K. A.; Hatton, T. A. Structure of (Deuterated PEO) (PPO) (Deuterated PEO) Block Copolymer Micelles as Determined by Small Angle Neutron Scattering. *Langmuir* **1999,** *15,* 1651–1656.
33. Genz, A.; Holzwarth, J. F. Dynamic Fluorescence Measurements on the Main Phase Transition of Dipalmitoylphosphatidylcholine Vesicles. *Eur. Biophys. J.* **1986,** *13,* 323–330.
34. Kositza, M. J.; Bohne, C.; Alexandridis, P.; Hatton, T. A.; Holzwarth, J. F. Dynamics of Micro- and Macrophase Separation of Amphiphilic Block-copolymers in Aqueous Solution. *Macromolecules.* **1999,** *32,* 5539–5551.
35. Bakshi, M.; Kaur, N.; Mahajan, R.; Singh, J.; Singh, N. Estimation of Degree of Counterion Binding and Related Parameters of Monomeric and Dimeric Cationic Surfactants from Cloud Point Measurements by Using Triblock Polymer as Probe. *Colloid Polym. Sci.* 2006, *284,* 879–885.
36. Hunter, R. J. *Foundations of Colloid Science,* 1st ed.; Oxford University Press: New York, NY, 1987; pp 1–565.
37. Thurn, T.; Couderc, S.; Sidhu, J.; Bloor, D. M.; Penfold, J.; Holzwarth, J. F.; Wyn-Jones, E. Study of Mixed Micelles and Interaction Parameters for ABA Triblock Copolymers of the Type EO_m-PO_n-EO_m and Ionic Surfactants: Equilibrium and Structure. *Langmuir* **2002,** *18,* 9267–9275.
38. Kadam, Y.; Ganguly, R.; Kumbhakar, M.; Aswal, V. K.; Hassan, P. A.; Bahadur, P. Time Dependent Sphere-to-rod Growth of the Pluronic Micelles: Investigating the Role of Core and Corona Solvation in Determining the Micellar Growth Rate. *J. Phys. Chem. B.* **2009,** *113,* 16296–16302.
39. Denkova, A. G.; Mendes, E.; Coppens, M. O. Kinetics and Mechanism of the Sphere-to-rod Transition of Triblock Copolymer Micelles in Aqueous Solutions. *J. Phys. Chem. B.* **2009,** *113,* 989–996.
40. Khimani, M.; Ganguly, R.; Aswal, V. K.; Nath, S.; Bahadur, P. Solubilization of Parabens in Aqueous Pluronic Solutions: Investigating the Micellar Growth and Interaction as a Function of Paraben Composition. *J. Phys. Chem. B.* **2012,** *116,* 14943–14950.

41. Bakshi, M. S.; Kaura, A.; Kaur, G. Effect of Temperature on the Unfavorable Mixing Between Tetraethylene Glycol Dodecyl Ether and Pluronic P103. *J. Colloid Interface Sci.* 2006, *296,* 370–373.
42. Bakshi, M. S.; Kaura, A.; Mahajan, R. K. Effect of Temperature on the Micellar Properties of Polyoxyethylene Glycol Ethers and Twin Tail Alkylammonium Surfactants. *Colloids Surf. A.* **2005,** *262,* 168–174.
43. Bakshi, M. S.; Sharma, P.; Kaur, G.; Sachar, S.; Banipal, T. S. Synergistic Mixing of L64 with Various Surfactants of Identical Hydrophobicity Under the Effect of Temperature. *Colloids Surf. A.* 2006, *278,* 218–228.
44. Bakshi, M. S.; Sachar, S. Influence of Hydrophobicity on the Mixed Micelles of Pluronic F127 and P103 Plus Cationic Surfactant Mixtures. *Colloids Surf. A.* 2006, *276,* 146–154.
45. Bakshi, M. S.; Sachar, S. Influence of Temperature on the Mixed Micelles of Pluronic F127 and P103 with Dimethylene-*bis*-(Dodecyldimethylammonium Bromide). *J. Colloid Interface Sci.* 2006, *296,* 309–315.
46. Mahajan, R. K.; Kaur, N.; Bakshi, M. S. Cyclic Voltammetry Investigation of the Mixed Micelles of Cationic Surfactants with Pluronic F68 and Triton X-100. *Colloids Surf. A.* **2005,** *255,* 33–39.
47. Bakshi, M. S.; Singh, J.; Kaur, G. Antagonistic Mixing Behavior of Cationic Gemini Surfactant and Triblock Polymers in Mixed Micelle. *J. Colloid Interface Sci.* **2005,** *285,* 403.
48. Bakshi, M. S.; Kaur, N; Mahajan, R. K. A Comparative Behavior of Photophysical Properties of Pluronic F127 and Triton X-100 with Conventional Zwitterionic and Anionic Surfactants. *Photochem. Photobiol.* **2006,** *183,* 146–153.
49. Mahajan, R. K.; Kaur, N.; Bakshi, M. S. Cyclic Voltammetry Investigation of the Mixed Micelles of Conventional Surfactants with L64 and F127. *Colloids Surf. A.* **2006,** *276,* 221–227.
50. Khullar, P.; Singh, V.; Mahal, A.; Kumar, H.; Kaur, G.; Bakshi, M. S. Block Copolymer Micelles as Nanoreactors for Self-assembled Morphologies of Gold Nanoparticles. *J. Phys. Chem. B.* **2013,** *117,* 3028–3039.
51. Daniel, M. C.; Astruc, D. Gold Nanoparticles: Assembly, Supramolecular Chemistry, Quantum-size-related Properties, and Applications Toward Biology, Catalysis, and Nanotechnology. *Chem. Rev.* 2004, *104,* 293–346.
52. Haruta, M.; Date, M. Advances in the Catalysis of Au Nanoparticles. *Appl. Catal. A.* 2001, *222,* 427–437.
53. Brust, M.; Kiely, C. J. Some Recent Advances in Nanostructure Preparation from Gold and Silver Particles: A Short topical Review. *Colloids Surf. A.* 2002, *202,* 175–186.
54. Sepulveda, B.; Angelome, P. C.; Lechuga, L. M.; Liz-Marzan, L. M. LSPR-based Nanobiosensors. *Nano Today* **2009,** *4,* 244–251.
55. Patra, C. R.; Bhattacharya, R.; Mukhopadhyay, D.; Mukherjee, P. Fabrication of Gold Nanoparticles for Targeted Therapy in Pancreatic Cancer. *Adv. Drug Deliv. Rev.* **2010,** *62,* 346–361.
56. Khullar, P.; Singh, V.; Mahal, A.; Kaur, H.; Singh, V.; Banipal, T. S. Tuning the Shape and Size of Gold Nanoparticles with Triblock Polymer Micelle Structure Transitions and Environments. *J. Phys. Chem. C.* **2011,** *115,* 10442–10454.

57. Khullar, P.; Mahal, A.; Singh, V.; Banipal, T. S.; Kaur, G.; Bakshi, M. S. How PEO–PPO–PEO Triblock Polymer Micelles Control the Synthesis of Gold Nanoparticles: Temperature and Hydrophobic Effects. *Langmuir* **2010,** *26,* 11363–11371.
58. Cheng, F.; Yang, X.; Peng, H.; Chen, D.; Jiang, M. Well-controlled Formation of Polymeric Micelles with a Nanosized Aqueous Core and Their Applications as Nanoreactors. *Macromolecules* **2007,** *40,* 8007–8014.
59. Hou, G.; Zhu, L.; Chen, D.; Jiang, M. Core–Shell Reversion of Hybrid Polymeric Micelles Containing Gold Nanoparticles in the Core. *Macromolecules* **2007,** *40,* 2134–2140.
60. Meristoudi, A; Pispas, S. Polymer Mediated Formation of Corona-embedded Gold Nanoparticles in Block Polyelectrolyte Micelles. *Polymer* **2009,** *50,* 2743–2751.
61. Zhou, Y.; Jiang, K.; Chen, Y.; Liu, S. Gold Nanoparticle-incorporated Core and Shell Crosslinked Micelles Fabricated from Thermoresponsive Block Copolymer of N-isopropylacrylamide and a Novel Primary-amine Containing Monomer. *J. Polym. Sci. A. Polym. Chem.* **2008,** *46,* 6518–6531.
62. Nivaggioli, T.; Tsao, B.; Alexandridis, P.; Hatton, T. A. Microviscosity in Pluronic and Tetronic Poly(Ethylene Oxide)-Poly(Propylene Oxide) Block Copolymer Micelles. *Langmuir* **1995,** *11,* 119–126.
63. Ganguly, R.; Kadam, Y.; Choudhury, N.; Aswal, V. K.; Bahadur, P. Growth and Interaction of the Tetronic 904 Micelles in Aqueous Alkaline Solutions. *J. Phys. Chem. B.* **2011,** *115,* 3425–3433.
64. Gonzalez-Lopez, J.; Alvarez-Lorenzo, C.; Taboada, P.; Sosnik, A.; Sandez-Macho, I.; Concheiro, A. Self-associative Behavior and Drug-Solubilizing Ability of Poloxamine (Tetronic) Block Copolymers. *Langmuir* **2008,** *24,* 10688–10697.
65. Seo, J. W.; Green, A. A.; Antaris, A. L.; Hersam, M. C. High-concentration Aqueous Dispersions of Graphene Using Nonionic, Biocompatible Block Copolymers. *J. Phys. Chem. Lett.* **2011,** *2,* 1004–1008.
66. Liu, T. G.; Xu, H.; Pang, G. J.; He, F. Effect of Alcohols on Aggregation Behaviors of Branched Block Polyether Tetronic at an Air/Liquid Surface. *Langmuir* **2011,** *27,* 9253–9260.
67. Juau¨rez, J. U.; Goy-Lou¨pez, S.; Cambou¨n, A.; Valdez, M. A.; Taboada, P.; Mosquera, V. U. Surface Properties of Monolayers of Amphiphilic Poly(Ethylene Oxide)-Poly(Styrene Oxide) Block Copolymers. *J. Phys. Chem. C.* **2010,** *114,* 15703–15712.
68. Yusa, S. I.; Fukuda, K.; Yamamoto, T.; Ishihara, K.; Morishima, Y. Synthesis of Well-defined Amphiphilic Block Copolymers Having Phospholipid Polymer Sequences as a Novel Biocompatible Polymer Micelle Reagent. *Biomacromolecules* **2005,** *6,* 663–670.
69. Goy-Lou¨pez, S.; Taboada, P.; Cambou¨n, A.; Juau¨rez, J. U¨.; Alvarez-Lorenzo, C.; Concheiro, A.; Mosquera, V. U¨. Modulation of Size and Shape of Au Nanoparticles Using Amino-X-Shaped Poly(Ethylene Oxide)Poly (Propylene Oxide) Block Copolymers. *J. Phys. Chem. B.* **2009,** *114,* 66–76.
70. Singh, V.; Khullar, P.; Dave, P. N.; Kaura, A.; Bakshi, M. S.; Kaur, G. pH and Thermo Responsive Tetronic Micelles for the Synthesis of Gold Nanoparticles: Effect of Physiochemical Aspects of Tetronics. *Phys. Chem. Chem. Phys.* **2014,** *16,* 4728–4739.

71. Antonietti, M.; Wenz, E.; Bronstein, L.; Seregina, M. Synthesis and Characterization of Noble Metal Colloids in Block Copolymer Micelles. *Adv. Mater.* **1995,** *7,* 1000–1005.
72. Moffitt, M.; McMahon, L.; Pessel, V.; Eisenberg, A. Size Control of Nanoparticles in Semiconductor–Polymer Composites. Control Via Sizes of Spherical Ionic Microdomains in Styrene-based Diblock Ionomers. *Chem. Mater.* **1995,** *7,* 1185–1192.
73. Zhao, M.; Crooks, R. M. Dendrimer-encapsulated Pt Nanoparticles: Synthesis, Characterization, and Applications to Catalysis. *Adv. Mater*. **1999,** *11,* 217–220.
74. Bronstein, L. M. Nanoparticles Made in Mesoporous Solids. *Top. Curr. Chem*. **2003,** *226,* 55–89.
75. Tanori, J.; Duxin, N.; Petit, C.; Lisiecki, I.; Veillet, P.; Pileni, M. P. Synthesis of Nanosize Metallic and Alloyed Particles in Ordered Phases. *Colloid Polym. Sci*. **1995,** *273,* 886–892.
76. Bakshi, M. S.; Bhandari, P. Fluorescence Studies on the Nonideal Mixing in Triblock Copolymer Binary Mixtures Under the Effect of Temperature: A Block Hydration Effect. *J. Photochem. Photobiol. A.* **2007,** *186,* 166–172.
77. Cardoso da Silva, R.; Olofsson, G.; Schillén, K.; Loh, W. Influence of Ionic Surfactants on the Aggregation of Poly(Ethylene Oxide)–Poly(Propylene Oxide)–Poly(Ethylene Oxide) Block Copolymers Studied by Differential Scanning and Isothermal Titration Calorimetry. *J. Phys. Chem. B*. **2002,** *106,* 1239–1246.
78. Lof, D.; Schillen, K.; Torres, M. F.; Muller, A. J. Rheological Study of the Shape Transition of Block Copolymer–Nonionic Surfactant Mixed Micelles. *Langmuir* **2007,** *23,* 11000–11006.
79. Thurn, T.; Couderc, S.; Sidhu, J.; Bloor, D. M.; Penfold, J.; Holzwarth, J. F.; Wyn-Jones, E. Study of Mixed Micelles and Interaction Parameters for ABA Triblock Copolymers of the Type EOm–POn–EOm and Ionic Surfactants: Equilibrium and Structure. *Langmuir* **2002,** *18,* 9267–9275.
80. Sakai, T.; Alexandridis, P. Mechanism of Gold Metal Ion Reduction, Nanoparticle Growth and Size Control in Aqueous Amphiphilic Block Copolymer Solutions at Ambient Conditions. *J. Phys. Chem. B.* **2005,** *109,* 7766–7777.
81. Sakai, T.; Alexandridis, P. Spontaneous Formation of Gold Nanoparticles in Poly (Ethylene Oxide)–poly (Propylene Oxide) Solutions: Solvent Quality and Polymer Structure Effects. *Langmuir* **2005,** *21,* 8019–8025.
82. Sakai, T.; Alexandridis, P. Ag and Au Monometallic and Bimetallic Colloids: Morphogenesis in Amphiphilic Block Copolymer Solutions. *Chem. Mater*. **2006,** *18,* 2577–2583.
83. Alexandridis, P.; Andersson, K. Reverse Micelle Formation and Water Solubilization by Polyoxyalkylene Block Copolymers in Organic Solvent. *J. Phys. Chem. B.* **1997,** *101,* 8103–8111.
84. Todorov, P. D.; Marinov, G. S.; Kralchevsky, P. A.; Denkov, N. D.; Durbut, P.; Broze, G.; Mehreteab, A. Kinetics of Triglyceride Solubilization by Micellar Solutions of Nonionic Surfactant and Triblock Copolymer. *Langmuir* **2002,** *18,* 7896–7905.
85. Choucair, A.; Eisenberg, A. Interfacial Solubilization of Model Amphiphilic Molecules in Block Copolymer Micelles. *J. Am. Chem. Soc*. **2003,** *125,* 11993–12000.
86. Jiang, M; Eisenberg, A.; Liu, G. J.; Zhang, X. 2006 *Macromolecular Self-assembly* (in Chinese); Science Press: Beijing, PR China, 2006.

87. Dong, J. H.; Zhang, X. *Frontiers and Forecast of Polymer Science* (in Chinese). Science Press: Beijing, PR China, 2011.
88. Yao, J. N., et al. *Rapid Developing Chinese Chemistry, from 1982–2012* (in Chinese). Science Press: Beijing, PR China, 2012.
89. Lazzari, M.; Liu, G. J. Lecommandoux S. *Block Copolymers in Nanoscience*. Wiley-VCH: Weinheim, Germany, 2006.
90. Chen, D. Y.; Jiang, M. Strategies for Constructing Polymeric Micelles and Hollow Spheres in Solution Via Specific Intermolecular Interactions. *Acc. Chem. Res.* **2005,** *38,* 494–502.
91. Guo, M. Y.; Jiang, M. Non-covalently Connected Micelles (NCCMs): The Origins and Development of a New Concept. *Soft Matter.* **2009,** *5,* 495–500.
92. Wang, J.; Jiang, M. Polymeric Self-assembly into Micelles and Hollow Spheres with Multiscale Cavities Driven by Inclusion Complexation. *J. Am. Chem. Soc.* **2006,** *128,* 3703–3708.
93. Chen, G. S.; Jiang, M. Cyclodextrin-based Inclusion Complexation Bridging Supramolecular Chemistry and Macromolecular Self-assembly. *Chem. Soc. Rev.* **2011,** *40,* 2254–2266.
94. Wang, C.; Wang, Z. Q.; Zhang, X. Amphiphilic Building Blocks for Self-assembly: From Amphiphiles to Supra-amphiphiles. *Acc. Chem. Res.* **2012,** *45,* 608–618.
95. Zhang, X.; Wang, C. Supramolecular Amphiphiles. *Chem. Soc. Rev.* **2011,** *40,* 94–101.
96. Wang, C.; Zhang, X. Superamphiphiles as Building Blocks for Supramolecular Engineering: Towards Functional Materials and Surfaces. *Small* **2011,** *7,* 1379–1383.
97. Cheng, W. L.; Wang, E. Polymer Mediated Formation of Corona-embedded Gold Nanoparticles in Block Polyelectrolyte Micelles *J. Phys. Chem. B.* **2004,** *108,* 24–26.
98. Meristoudi, A.; Pispas, S. Polymer Mediated Formation of Corona-embedded Gold Nanoparticles in Block Polyelectrolyte Micelles. *Polymer* **2009,** *50,* 2743–2751.
99. Brown, K. R.; Walter, D. G.; Natan, M. J. Seeding of Colloidal Au Nanoparticle Solutions Improved Control of Particle Size and Shape. *Chem. Mater*. **1999,** *12,* 306–313.
100. Jana, N. R.; Gearheart, L.; Murphy, C. J. Seeding Growth for Size Control of 5–40 nm Diameter Gold Nanoparticles. *Langmuir* **2001,** *17,* 6782–6786.
101. Rodríguez-Fernández, J.; Pérez-Juste, J.; García de, A.; Liz-Marzán, F. J. Seeded Growth of Submicron Au Colloids with Quadrupole Plasmon Resonance Modes. *Langmuir* **2006,** *22,* 7007–7010.
102. Eguchi, M.; Mitsui, D.; Wu, H. L.; Sato, R.; Teranishi, T. Simple Reductant Concentration-Dependent Shape Control of Polyhedral Gold Nanoparticles and their Plasmonic Properties. *Langmuir* **2012,** *28,* 9021–9026.
103. Njoki, P. N.; Luo, J.; Kamundi, M. M.; Lim, S.; Zhong, C. J. Aggregative Growth in the Size-Controlled Growth of Monodispersed Gold Nanoparticles. *Langmuir* **2010,** *26,* 13622–13629.
104. Bastús, N. G.; Comenge, J.; Puntes, V. C. Kinetically Controlled Seeded Growth Synthesis of Citrate-Stabilized Gold Nanoparticles of up to 200 nm: Size Focusing Versus Ostwald Ripening. *Langmuir* **2011,** *27,* 11098–11105.
105. Iqbal, M.; Chung, Y. I.; Tae, G. *An Enhanced Synthesis of Gold Nanorods by the Addition of Pluronic (F-127) Via a Seed Mediated Growth Process J. Mater. Chem.* **2007,** *17,* 335–342.

106. Ivanova, R.; Alexandridis, P.; Lindman, B. Interaction of Poloxamer Block Copolymers with Cosolvents and Surfactants. *Colloid Surf. A. Physicochem. Eng. Asp.* **2001,** *183,* 41–53.
107. Frens, G. Controlled Nucleation for the Regulation of the Particle Size in Monodisperse Gold Suspensions. *Nat. Phys. Sci.* **1973,** *241,* 20–22.
108. Goy-Lopez, S.; Taboada, P.; Cambon, A. J.; Juarez, C.; Alvarez-Lorenzo, A.; Concheiro, A.; Mosquera, V. Modulation of Size and Shape of Au Nanoparticles Using Amino-X-Shaped Poly(Ethylene Oxide)–Poly(Propylene Oxide) Block Copolymers. *J. Phys. Chem. B.* **2010,** *114,* 66–76.
109. Yang, B.; Guo, C.; Chen, S. Ma.; Wang, J. H.; Liang J.; Zheng, X. F.; Liu, L. H. Z. Effect of Acid on the Aggregation of Poly(Ethylene Oxide)-P(Propylene Oxide)-P (Ethylene Oxide) Block Copolymers. *J. Phys. Chem. B.* 2006, *110,* 23068–23074.
110. Piao, Y. Z.; Jang, Y. J.; Shokouhimehr, M.; Lee, I. S.; Hyeon, T. Facile Aqueous-phase Synthesis of Uniform Palladium Nanoparticles of Various Shapes and Sizes. *Small.* **2007,** *3,* 255–260.
111. Zhang, W.; Shi Y.; Chen, Y.; Ye J.; Sha, X.; Fang, X. Multifunctional Pluronic P123/F127 Mixed Polymeric Micelles Loaded with Paclitaxel for the Treatment of Multidrug Resistant Tumors. *Biomaterials.* 2011, *32,* 2894–2906.
112. Zhang, W.; Yuan, S.; Chen, Y. Z.; Yu, S. Y.; Hao, J. G.; Luo, J. Q.; Sha, X. Y.; Fang, X. L. Enhanced Anti-glioblastoma Efficacy by PTX-loaded PEGylated Poly (É›-caprolactone) Nanoparticles: In Vitro and In Vivo Evaluation. *Eur. J. Pharm. Biopharm.* **2010,** *75,* 341–353.
113. Youhua, T.; Jiangfeng, H.; Chunting, Y.; Tima, T.; Huanyu, D. Reduction-responsive Gold-nanoparticle-conjugated Pluronic Micelles: An Effective Anti-cancer Drug Delivery System. *J. Mater. Chem.* **2012,** *22,* 18864–18871.

CHAPTER 6

ELECTRON–HOLE BILAYER SYSTEMS IN SEMICONDUTORS: A THEORETICAL PERSPECTIVE

AYAN KHAN[1*] and B. TANATAR[2]

[1]*Department of Physics, School of Engineering and Applied Sciences, Bennett University, Greater Noida 201310, India*

[2]*Department of Physics, Bilkent University, Ankara 60032, Turkey*

[*]*Corresponding author. E-mail: ayan.khan@bennett.edu.in*

CONTENTS

ABSTRACT

In this generation of semiconductor-based miniature gadgets, it is always an intriguing issue to understand the underlying mechanism of these highly sophisticated devices. Systematic exploration uncovers the fascinating world of nano-scale physics; however, a deeper probe actually unearths the salient nature of electron–hole interaction which is the soul of semiconductor physics. In this essay, we plan to shed some light on the interplay of different interactions in electron–hole systems. This study is purely theoretical; however, we put forward recent experimental successes at suitable places. Here, we explicate the intricacies involving electron(hole)–electron(hole) and electron–hole interaction in the realm of mean-field theory. To make our description more comprehensive and complete, we elaborate the theoretical analysis by means of both path integral formalism and canonical transformation method. The discussion clearly suggests that the interplay of intra-layer and inter-layer interactions results in the formation of several exotic phases such as Sharma phase and Fulde–Ferrell–Larkin–Ovchinniov (FFLO) phase.

6.1 INTRODUCTION

The last century, widely believed as the century of physics, has witnessed several experimental and theoretical accomplishments in condensed matter physics and optics, which has paved the way for future industrial revolution in communication and computation. There was an enormous jump in terms technology when vacuum diodes were replaced by semiconductor transistors, but that was a good 60 years back. As we are exhausting the ceiling of the Moore's Law,[1] it is now evident that we need to try something different while keeping the flavor of our old prodigy (semiconductor physics).

The emergence of ultra-cold atom research in the last decade,[2–10] has widened the possibility to study different exotic phases of matter by controlling the interaction. The smooth transition of Cooper pairs from BCS superconductors to Bose superfluid of composite bosons for trapped atoms is one of such developments. This phenomenon is known as Bardeen-Cooper-Schrieffer–Bose-Einstein-Condensate (BCS–BEC) crossover. Although trapped atoms represent an ideal testing ground for a fundamental understanding of the BCS–BEC crossover, technological applications exploiting the occurrence of condensates will most probably rely on semiconductor systems. In these systems, excitons, made up of electrons and holes, play the role of composite bosons.[11] Formation of excitons between spatially separated electrons and holes and their subsequent condensation have long been predicted and arguably observed nearly 30 years later experimentally.[12]

The main objective of this chapter is to provide a comprehensive theoretical guide to deal with electron–hole (e–h) systems in the mean-field level. We start with a brief description of BCS–BEC crossover in ultra-cold atomic gases, excitonic systems, and the meaning of crossover in e–h systems. Usually, an exitonic system is modeled by taking into account their inter- and intra-species interactions. Here we first elaborate the system in the absence of intra-layer interaction and later incorporate it. However, for a better theoretical view, we employ two different mathematical techniques, namely path integral formalism and canonical transformation, in our discussion. At the fag end, we briefly comment on the possible way to include impurity in these systems. We conclude with current experimental status in this field with possible future applications.

6.1.1 BCS–BEC CROSSOVER

The experimental realization of Bose-Einstein condensate in trapped ultra-cold alkali gases[13,14] appeared as a new ray of light to the science community. For its broad appeal, people from diverse communities of physics, such as atomic and molecular physics, condensed matter physics, and nuclear physics, came under a single umbrella. The excitement led to the study of both bosonic and fermionic gases. Further, the application of Fano-Feshbach resonance (which we will discuss later)[15–17] in atomic gases[18–21] gave freedom to evolve a composite boson state to Cooper pairs passing the crossover.

In normal superconductors, electrons with opposite spins pair to form Cooper pairs below the superconducting critical temperature. The average pair size ξ_{pair} for these superconductors is much larger than the mean inter-particle distance k_F^{-1}. Therefore, the quantity $\xi_{pair}k_F$ is much larger than 1. So the Cooper pairs are largely overlapping and it is not appropriate to consider them as spin-zero bosons. It is better to appreciate them as correlation of two opposite spin fermions at a certain distance and BCS theory holds for them perfectly. But the advent of high-temperature superconductor, where $\xi_{pair}k_F$ is of the order 5−10, forced to think beyond the BCS theory. The coupling between the fermions in these novel superconductors suggests an intermediate state between Cooper pairs and composite bosons. This intuitive idea prompted to develop a theory which can connect both the BCS theory for Cooper pairs and BEC for composite bosons. Already at that time, there were some works[23–25] where the evolution of the fermionic pairs from Cooper pairs to composite bosons had been studied. Since the transition between these two limits occurs without an intermediate phase transition, the phenomenon has referred as BCS–BEC crossover. Figure 6.1 gives a physical feeling of the situation where densely packed Cooper pairs and sparsely distributed composite bosons are on both sides and in between there exist the strongly interacting fermionic pairs which are the main players in the crossover. Subsequent to the discovery of the high temperature superconductors, the interest in the crossover physics has surged.[26–31]

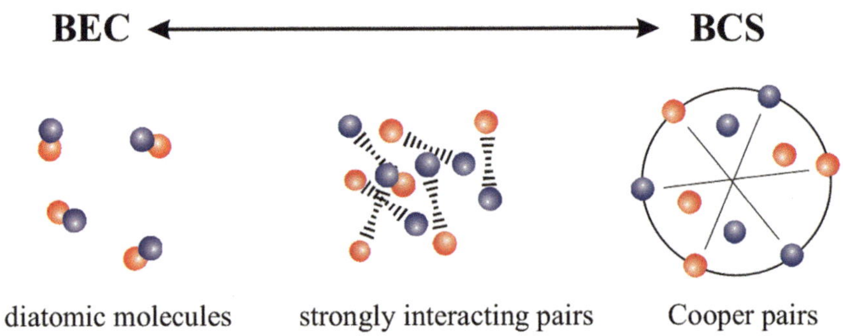

FIGURE 6.1 Representation of BCS–BEC crossover pictorially.[22]

The beauty of the BCS ground state is that it not only describes the superconductivity pretty efficiently but also contains the essence of the bosonic limit. If one starts from the BCS ground state wave function $|\tilde{\Psi}>$ and carries out the necessary algebraic rearrangements in the following manner,

$$
\begin{aligned}
|\tilde{\Psi}> &= \prod_{\mathrm{k}} (u_{\mathrm{k}} + v_{\mathrm{k}} c^{\dagger}_{k\uparrow} c^{\dagger}_{-k\downarrow}) |0> \\
&= \prod_{\mathrm{k}} u_{\mathrm{k}} (1 + \frac{v_{\mathrm{k}}}{u_{\mathrm{k}}} c^{\dagger}_{k\uparrow} c^{\dagger}_{-k\downarrow}) \\
&= \prod_{\mathrm{k}} u_{\mathrm{k}} e^{\sum_{k'} g(k') c^{\dagger}_{k\uparrow} c^{\dagger}_{-k\downarrow}} |0>,
\end{aligned}
\tag{6.1}
$$

then one can define the operator $b^\dagger = \sum_k g(k) c^\dagger_{k\uparrow} c^\dagger_{-k\downarrow}$, which contains fermionic pairs. However the defined operator can not be regarded as a true bosonic operator because,

$$[b, b] = \sum_k |g(k)|^2 (1 - n_{k\uparrow} - n_{-k\downarrow} \neq c \quad \text{number}. \tag{6.2}$$

In certain conditions when $< n_{k,\sigma} ><< 1$, $[b, b^\dagger] \simeq 1$, the b turns out truly a bosonic operator and in that situation,

$$|\tilde{\Psi}> = \exp(b^\dagger)|0> \tag{6.3}$$

represents a bosonic coherent state or a condensate.[23–25]

The simplest description of the BCS–BEC crossover can be given at the mean-field level for a homogeneous system in the zero temperature limit. In this situation, it is necessary to analyze a pair of coupled equation which reads,

$$\int \frac{d\mathrm{k}}{(2\pi)^3} \left(\frac{1}{2E_k} - \frac{m}{k^2} \right) = -\frac{m}{4\pi a_F} \tag{6.4}$$

$$\int \frac{d\mathrm{k}}{(2\pi)^3} \left(1 - \frac{\xi_\mathrm{k}}{E_\mathrm{k}} \right) = n \tag{6.5}$$

where the notations are the usual BCS notations. Precisely, $\xi_\mathrm{k} = k^2/2m - \mu$, $E_\mathrm{k} = \sqrt{\xi_\mathrm{k}^2 + \Delta_\mathrm{k}^2}$ and the bare coupling strength has been replaced by the s-wave scattering length. And eq (6.4) is known as gap equation and eq (6.5) is known as density equation. The original gap equation actually contains an ultraviolet divergence (originated from the assumption that the contact potential governs fermionic interaction); however, here the equation is suitably regularized to avoid the mentioned divergence. The coupled equations can be solved simultaneously for a given density and coupling. These solutions are plotted in Figure 6.2 where we have evaluated the pairing gap (Δ) and chemical potential (μ). In describing the chemical potential, two different normalizations are adopted depending on the sign of the chemical potential. When $\mu > 0$, it has been normalized by the Fermi energy ($\epsilon_F = k_F^2/2m$) and in the negative side we used two body binding energy ($\epsilon_0 = (m a_F^2)^{-1}$) to normalize.

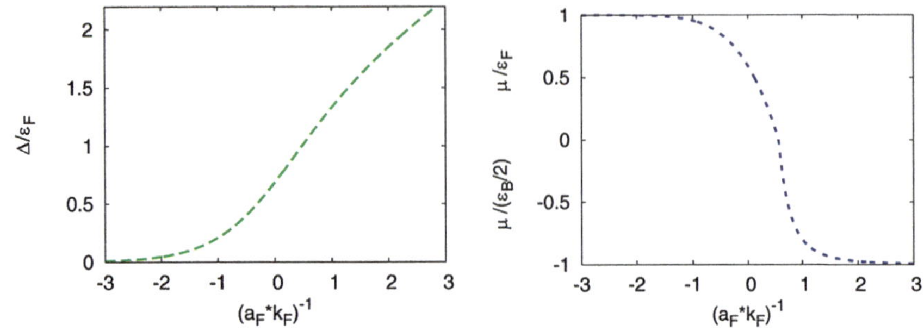

FIGURE 6.2 Variation of order parameter (pairing gap) and chemical potential with coupling strength in a homogeneous system.

6.1.2 EXCITONS

The manifestation of BEC is based on the wave nature of particles. Therefore, it is obvious that de Broglie hypothesis plays a crucial role in understanding the condensate. It can be shown that the de Broglie wavelength gets longer with decrease in temperature. When atoms are cooled to the point where the thermal de Broglie wavelength is comparable to the inter-atomic separation then BEC formation actually starts. The relation between the transition temperature and peak atomic density (n) can be estimated as $n\lambda_{dB}^3 = 2.612$ where the de Broglie wavelength (λ_{dB}) is defined as $\lambda_{dB} = h/mv = h/\sqrt{mk_BT},$ where m is the mass of the atom, k_B is the Boltzmann constant, and T is the temperature. The critical temperature for the transition works out in the range of nano-Kelvin. The above description also points out that the de Broglie wavelength is inversely proportional to the square-root of the mass of the particle. Therefore, it took a long time to develop the necessary cooling techniques to create BEC with heavy atomic mass. For solid-state systems, excitons in semiconductors have long been considered as a promising candidate for BEC because of their light mass as compared to the neutral atoms. Usually, effective mass of exciton is considered as twice the mass of electron. Now if we employ this mass in the de Broglie theory, the critical temperature for transition in two dimension turns out about $1K$. The possibility of achieving higher critical temperature in excitonic systems infuses additional interest in the research community. Unfortunately, excitons recombine quickly, too fast to allow a condensate to form. Although excitons coupled to light confined within a microcavity can form hybrid particles that do live long enough to condense, such condensates require a continuous input of light.[32] This is known as exiton-polariton condensate which has actually been observed very recently.[33] However, a very recent study based on two mono-layers of graphene separated by an insulating material poses promising platform to realize excitonic condensate.[34,35]

Excitons can be defined as a bound state of an electron and hole which are attracted to each other by the electrostatic Coulomb force. It is an electrically neutral quasiparticle that exists in insulators, semiconductors, and in some liquids. An exciton can form when a photon is absorbed by a semiconductor. This excites an electron from the valence band into the conduction band. In turn, this leaves behind a localized positively charged hole. The electron in the conduction band is then attracted to this localized hole by the Coulomb force. This attraction provides a stabilizing energy balance. The recombination of the electron and hole, that is, the decay of the exciton, is limited by resonance stabilization due to the overlap of the electron and hole wave functions, resulting in an extended lifetime for the exciton.

6.2 BCS–BEC CROSSOVER WITH EXCITONS

As mentioned earlier, electronic systems in semiconductor devices provide an alternative and technically more viable route for physical realization where BCS–BEC crossover. Electrons and holes can form bound states due to the attractive Coulomb interaction between them. These bound states are popularly known as excitons. Thus excitons are the composite bosons in this system. The interaction strength can be changed by varying the density. Here, we must note that in ultra-cold atomic systems, the controlling parameter is either *s*-wave scattering length or the density, whereas in semiconductor systems, it is only the density or the concentration of electrons and holes. Another significant difference from usual unitary Fermi gas is that the most commonly used interaction in those systems is short-range contact interaction, whereas in semiconductors, it is usually long-range Coulomb interaction. The long-range ordering due to Coulomb interaction complicates the emergence of condensate in the excitonic systems. Of late, several theoretical ideas have been floated to tackle this problem.[36–39]

However, the condensation phenomenon of excitons is a fairly old issue. It was first predicted almost 50 years before by Blatt et al.[40] In the early 1980s, Comte et al. studied the Bose condensed ground state of an electron–hole gas in a simple model semiconductor, as a function of density, using a mean field variational ansatz.[41,42] Later with the advent of new technologies and experimental observation of weakly interacting BEC, the research on excitonic condensate received a new impetus. A keen interest is paid to the bilayer quantum-well systems realized in semiconductor hetero-structures. In recent years, using electrical and optical techniques exciton condensation has been observed in several different systems. Quantum Hall experiments at half-filling investigate BEC in electron–electron and hole–hole bilayers.[11,43,44] Optically generated bilayer excitons also show evidence for condensation.[12] Recently, excitons coupled to photons to form polaritons with even smaller mass leading to higher condensation temperatures have been studied theoretically[45] and experimentally.[46,47]

6.2.1 *THEORY OF EXCITONS*

The first theoretical mean-field analysis of excitons can be found in [Refs.41 and 42] However, they involved equal electron and hole densities leading to full pairing. In recent years Pieri et al. extended the above-mentioned pioneering works for density imbalance.[48] The investigation reveals a crossover in the phase diagram from the BCS limit of overlapping pairs to the BEC limit of non-overlapping tightly bound pairs. Further, it was noted that different novel phases emerge in the crossover region when the densities of electrons and holes are varied independently. However, this analysis only takes into account the inter-layer Coulomb interaction between the electrons and holes, thereby it neglected the intra-layer electron–electron and hole–hole Coulomb interactions. Later Subasi et al. overcame this deficiency.[49] Hence a typical bilayer Hamiltonian in the mean-field level consists of a kinetic energy/hopping term, intra-layer interaction term, and inter-layer interaction as described in eq (6.6).

$$\begin{aligned} H = \sum_{\mathrm{k}} (\in_{\mathrm{k}}^{a} a_{\mathrm{k}}^{\dagger} a_{\mathrm{k}} + \in_{\mathrm{k}}^{b} b_{\mathrm{k}}^{\dagger} b_{\mathrm{k}}) + \frac{1}{2V} \sum_{\mathrm{k}_1,\mathrm{k}_2,\mathrm{q}} U_{\mathrm{q}}^{aa} a_{\mathrm{k}_1+\mathrm{q}}^{\dagger} a_{\mathrm{k}_2-\mathrm{q}}^{\dagger} a_{\mathrm{k}_2} a_{\mathrm{k}_1} + \\ \frac{1}{2V} \sum_{\mathrm{k}_1,\mathrm{k}_2,\mathrm{q}} U_{\mathrm{q}}^{bb} b_{\mathrm{k}_1+\mathrm{q}}^{\dagger} b_{\mathrm{k}_2-\mathrm{q}}^{\dagger} b_{k_2} b_{k1} + \frac{1}{V} \sum_{\mathrm{k}_1,\mathrm{k}_2,\mathrm{q}} U_{\mathrm{q}}^{ab} a_{\mathrm{k}_1+\mathrm{q}}^{\dagger} b_{\mathrm{k}_2-\mathrm{q}}^{\dagger} b_{\mathrm{k}_2} a_{\mathrm{k}_1}. \end{aligned} \tag{6.6}$$

The basis states for electrons and holes are chosen to be plane wave states. The operators $a_k / a_k^\dagger (b_k / b_k^\dagger)$ are creation and annihilation operators for electrons and holes, respectively. The single particle energies are denoted by $\in_k^a, \in_k^b$, and the matrix element U_q with respect to plane wave states becomes the Fourier transform of the corresponding two-body interaction,

$$U_q = \int e^{-iq.r} U(r) dr. \tag{6.7}$$

In eq (6.7), U^{aa}, U^{bb}, and U^{ab} denote electron–electron, holehole, and electron–hole Coulomb interactions, respectively. The explicit forms of the Coulomb potentials can be noted as,

$$U_q^{aa} = U_q^{bb} = \frac{2\pi e^2}{\in q}, \qquad U_q^{ab} = \frac{2\pi e^2}{\in q} e^{-qd}, \tag{6.8}$$

where d denotes the distance of separation between electron and hole layers.

Here, we plan to elaborate the discussion in the following manner. First we will explain the bilayer systems in the light of only the inter-layer interaction (i.e., $U^{aa} = U^{bb} = 0$).[48] Later we will move forward and include the intra-layer interaction.[49]

6.2.1.1 IN ABSENCE OF INTRA-LAYER INTERACTION

If we neglect the intra-layer interaction for the time being, eq (6.6) can be rewritten as,

$$H = \sum_k (\in_k^a a_k^\dagger a_k + \in_k^b b_k^\dagger b_k) + \sum_k (\Delta_k a_k^\dagger a_{-k}^\dagger + \Delta_k^* b_{-k} a_k). \tag{6.9}$$

Here $\Delta_k = -\sum_{k'} U_{kk'}^{ab} \langle a_{k'} b_{-k'} \rangle$ and $\in_k^a = \hbar^2 k^2 / 2m_b$, where a and b denote the electron and hole respectively.

Now eq (6.9) can be analyzed by using path integral formulation as well as canonical transformation. However, here we plan to explicate the path integral formalism at first and in the latter half (with intra-layer interaction) we will explain the canonical transformation. This will enable the readers to view the bilayer problem through two different mathematical angles, albeit both the cases are akin to each other.

6.2.1.2 PATH INTEGRAL FORMALISM

The path integral formalism deduced here is constructed in analogy with the BCS theory. However, compared to the BCS theory, the single particle energies are not equal to each other (i.e., $\epsilon_k^a \neq \epsilon_k^b$) due to the difference in electron and hole mass. Let us now introduce the chemical potential (μ) in the mean-field Hamiltonian which will ensure the density imbalance. Hence eq (6.9) reads,

$$\begin{aligned} H - \mu N &= \sum_{k,\sigma} \epsilon_k^\sigma \sigma_k^\dagger \sigma_k - \mu^\sigma \sigma_k^\dagger \sigma_k + \sum_k [\Delta_k a_k^\dagger b_{-k}^\dagger + \Delta_k^* b_{-k} a_k] \\ &= \sum_{k,\sigma} \xi_k^\sigma \sigma_k^\dagger \sigma_k + \sum_k [\Delta_k a_k^\dagger b_{-k}^\dagger + \Delta_k^* b_{-k} a_k], \end{aligned} \tag{6.10}$$

where $\sigma \in \{a, b\}$ and $\xi_k = \epsilon_k - \mu$. Hence the quantum partition function in path integral form can be written as, $Z = \int D[\Delta, \bar{\Delta}] e^{-S_{eff}}$ where,

$$S_{eff} = \int d^3x \int_0^\beta d\tau [\frac{\Delta\bar{\Delta}}{U} - \text{Tr ln } G^{-1}]. \tag{6.11}$$

G^{-1} is defined as Nambu propagator (inverse Greens function). The Greens function contains the free particle Green's function (G_0^{-1}) and the particle–particle interaction incorporated through self-energy (Σ). Thus $G^{-1} = G_0^{-1} + \Sigma$. Since the thermodynamic potential can be expressed as $\Omega = -\frac{\ln Z}{\beta}$ and density as $n = -\frac{\partial \Omega}{\partial \mu}$, therefore a careful analysis of eq (6.11) leads to

$$\begin{aligned} n_i &= \frac{1}{\beta}\frac{\partial}{\partial \mu_i} \ln \int D[\Delta, \bar{\Delta}] \left[\exp\left\{ -\int d^3x \int_0^\beta d\tau \left(\frac{\Delta\bar{\Delta}}{U_{ab}} - \text{Tr ln } G^{-1} \right) \right\} \right] \\ &= \frac{1}{\beta}\frac{\partial}{\partial \mu_i} \ln \left[\exp\left\{ \sum_{k,i\omega} \text{Tr ln } G^{-1}(k, i\omega) \int D\left[\Delta, \bar{\Delta}\right] \exp[-\frac{V\beta\Delta,\bar{\Delta}}{U}] \right\} \right] \\ &= \frac{1}{\beta} \sum_{k,i\omega} \text{Tr} \left[G \frac{\partial}{\partial \mu_i} G^{-1} \right] \end{aligned} \tag{6.12}$$

where the suffix, $i \in \{a, b\}$, the $\int D[\Delta\bar{\Delta}]$ integral reduces to 1 through Grassmann identity. The explicit definition of the Green's function can be noted as follows,

$$G^{-1} = \begin{pmatrix} i\omega + \xi_k^b & \Delta_k \\ \bar{\Delta}_k & i\omega - \xi_k^a \end{pmatrix} \text{ and } G = \begin{pmatrix} \frac{i\omega - \xi_k^a}{D} & \frac{\Delta_k}{D} \\ \frac{\bar{\Delta}_k}{D} & \frac{i\omega - \xi_k^b}{D} \end{pmatrix}$$

where, $D = -\omega^2 - i\omega(\xi_k^a - \xi_k^b) - \xi_k^a \xi_k^b - \Delta_k^2$. Hence, from eq (6.12),

$$\begin{aligned} n_a &= \frac{1}{V\beta}\sum_{k,i\omega} \mathrm{Tr}\begin{pmatrix} \frac{i\omega - \xi_k^a}{D} & \frac{\Delta_k}{D} \\ \frac{\bar{\Delta}_k}{D} & \frac{i\omega - \xi_k^b}{D} \end{pmatrix}\begin{pmatrix} 0 & 0 \\ 0 & 1 \end{pmatrix} \\ &= \frac{1}{V\beta}\sum_{k,i\omega} \frac{i\omega + \xi_k^b}{D}. \end{aligned} \tag{6.13}$$

D can be decomposed in a factorized form as follows, $D = (i\omega - \Delta\xi_k - E_k)(i\omega - \Delta\xi_k + E_k) = (i\omega - E^+)(i\omega + E^-)$. We carry out the Matsubara frequency sum from eq (6.13) as,

$$\begin{aligned} n_a &= \frac{1}{V}\frac{1}{\beta}\sum_{k,i\omega} \frac{i\omega + \xi_k^b}{i\omega - E^+)(i\omega + E^-)} f(i\omega) \\ &= \frac{1}{2V}\sum_k \left[\left(1 + \frac{\xi_k}{E_k}\right) f_k^+ + \left(1 - \frac{\xi_k}{E_k}\right)\left(1 - f_k^-\right)\right]. \end{aligned} \tag{6.14}$$

In a similar fashion we can also evaluate n_b as,

$$n_b = \frac{1}{2V}\sum_k \left[\left(1 + \frac{\xi_k}{E_k}\right) f_k^- + \left(1 - \frac{\xi_k}{E_k}\right)\left(1 - f_k^+\right)\right]. \tag{6.15}$$

Here $f(E)$ is defined as the Fermi function at zero temperature and

$$\begin{aligned} f_k^\pm &= \begin{cases} 1 & \text{if } E_k^\pm < 0 \\ 0 & \text{if } E_k^\pm > 0 \end{cases} \\ E_k^\pm &= E_k \pm \Delta\xi_k \\ \Delta\xi_k &= \frac{1}{2}\left(\epsilon_k^a - \mu_a - \epsilon_k^b - \mu_b\right). \end{aligned}$$

The other relevant equation known as gap equation can also be derived from the effective action as,

$$
\begin{aligned}
S_{eff}\left[\Delta, \bar{\Delta}\right] &= \int d^3x \int_0^\beta d\tau \left[\frac{\Delta \bar{\Delta}}{U_{ab}} - \mathrm{Tr}\ \ln\left(-G^{-1}\right)\right] \\
&= \frac{V\beta\,\Delta\bar{\Delta}}{U_{kk'}} - \ln \prod_k \left|-G^{-1}\right|.
\end{aligned}
$$

It can be shown that $|-G^{-1}| = D$, where D is same as defined earlier. One can also write, $\ln \prod_k \simeq \sum_k \ln$. Hence,

$$
S_{eff}\left[\Delta, \bar{\Delta}\right] = \beta V \frac{\Delta \bar{\Delta}}{U_{kk'}} - \sum_{k,i\omega} \ln D \tag{6.16}
$$

Applying the saddle point approximation $\frac{\delta S_{eff}}{\delta \bar{\Delta}} = 0$, we can rewrite eq (6.16) as

$$
\begin{aligned}
\frac{V\beta\Delta_k}{U_{kk'}} - \sum_{k,i\omega} \frac{\Delta}{D} &= 0 \\
\frac{V\Delta_k}{U_{kk'}} &= \frac{1}{\beta} \sum_{k',i\omega} \frac{\Delta_{k'}}{(i\omega - E^+)(i\omega + E^-)} f(i\omega) \\
&= \sum_{k'} \frac{\Delta_{k'}}{2E_{k'}} \left[f^+ - \left(1 - f^-\right)\right] \\
&= -\frac{1}{V} \sum_{k'} U_{kk'} \frac{\Delta_{k'}}{2E_{k'}} \left[1 - f_{k'}^+ - f_{k'}^-\right].
\end{aligned} \tag{6.17}
$$

One can now solve eqs (6.14) ,(6.15), and (6.17) self-consistently to study the effect of density imbalance. However, before elaborating the obtained results of Ref. [48], it is important to discuss the units of the observables. All through this discussion, the physical quantities are in Rydberg units, that is, length is measured in effective (excitonic) Bohr radius $a_B = \frac{\hbar^2 \varepsilon}{me^2}$, momentum in $1/a_B$, and energy in effective Rydberg ($\mathrm{Ryd} = \frac{\hbar^2}{2ma_B^2} = \frac{e^2}{2\varepsilon a_B}$). The reduced mass m is defined as $1/m = 1/m_a + 1/m_b$ where $m_a = m_e$ and $m_b = m_h$ are the band mass of the electron and hole, respectively. As mentioned before, the bilayer system is characterized by the electron– hole densities or by the average density parameter (r_s) and population polarization (α). α signifies the population imbalance and is defined as the ratio of density difference and total density. In other words, r_s and α can be described as,

$$n = \frac{n_a + n_b}{2} = \frac{1}{a_B^2 r_s^2} \quad \text{and} \quad \alpha = \frac{n_a - n_b}{n_a + n_b} \tag{6.18}$$

The BCS and BEC regimes are defined by means of average inter-particle spacing, that is, $r_s < 1$ implies closely packed systems thereby it is noted as BCS regime. Otherwise when $r_s > 1$, it is considered as BEC regime. By solving the coupled density and gap equation it is possible to study the smooth transition of the physical observables from one region to another region. Hence, we can borrow the BCS–BEC crossover analogy from ultra-cold atomic systems to the semiconductor systems without losing much of the generality.

Self-consistent analysis of eqs (6.14), (6.15), and (6.17) results in evaluation of the gap function. In Figure 6.3a, the wave-vector dependence of gap function is depicted. One can observe that at low r_s (BCS regime), there exists a distinct peak for the gap function; however, this peak smoother out as we increase the average inter-particle distance, that is, we move from highly overlapping BCS system to non-overlapping BEC system. Similar density-induced BCS–BEC crossover had already been studied in ultra-cold atomic systems.[53,54] Figure 6.3b describes the maximum value of the gap function for different densities with varying degree of imbalance. Here we observe that the maximum value exists at about zero average chemical potential ($\mu = (\mu_e + \mu_h)/2$). This actually implies a robust paradigm of superfluidity in the crossover region. Interestingly, this conclusion was also made from the perspective of ultra-cold clean Fermi gases[53] and dirty Fermi gases.[53] One can also notice that the magnitude of the energy gap actually reduces with increase in density imbalance.

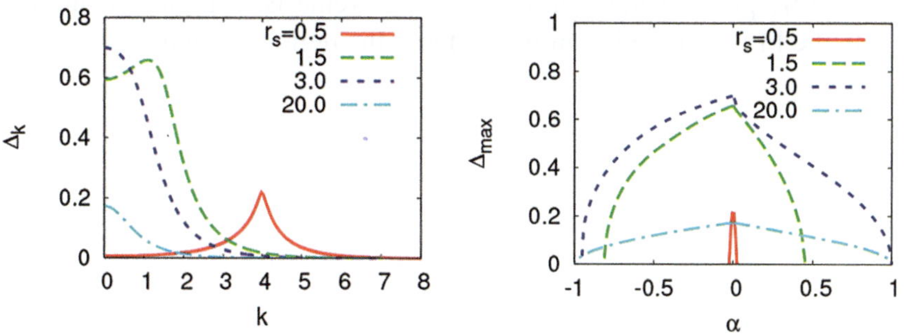

(a) (b)

FIGURE 6.3 (a) The dependence of wave vector on gap function as balanced density ($\alpha = 0$) at various values of r_s. (b) Variation of $max\{\Delta_k\}$ at different imbalances and different densities. For both the figures the inter-layer separation was considered as unity ($d = 1$). (Reprinted Ref [51] with permission from the author.)

Figure 6.4 shows the zero-temperature phase diagram for $d = 1$. We can identify various phases using Δ_k, determined from eq (6.17), and the superfluid density ρ_s. Within mean-field theory at $T = 0$ ρ_s is defined as,

$$\rho_s = m_e n_e + m_h n_h - \frac{1}{4\pi}\sum_{j,\lambda}\frac{\left(k_j^\lambda\right)^3}{\left|\frac{dE_k^\lambda}{dk}\right|_{k=k_j^\lambda}} \tag{6.19}$$

Here, k_j^λ is the j-th zero of $E_k^\lambda = 0$ with $\lambda = (+, -)$. The zeros of quasi-particle energies, E_k^+ (E_k^-), can be found only for imbalanced density scenario while no zero occurs for $\alpha = 0$.

The normal phase (N) corresponds to the trivial solution $\Delta_k = 0$. The Sarma phases corresponds to nonvanishing Δ_k when $\alpha \neq 0$ and positive superfluid density ρ_s. The S1 and S2 denote the Sarma phases for one and two Fermi surfaces, respectively. There will be one zero of E_k^λ ($j = 1$) for the S1 phase (one Fermi surface) and two zeros ($j = 1, 2$) for the S2 phase (two Fermi surfaces). A negative value of ρ_s in eq (6.19) indicates that the Sarma phase is unstable toward a phase with a spontaneously generated superfluid current, which we associate with the Fulde–Ferrell–Larkin–Ovchinniov (FFLO) phase.[56,57]

One intriguing aspect of Figure 6.4 is the dependence of the phase diagram on the sign of α. In particular, while the boundary of the normal phase does not depend appreciably on the sign of α, the region of stability of the Sarma phase with respect to the FFLO phase depends dramatically on the sign. For $\alpha < 0$, the phase diagram is dominated by the FFLO phase, with the S1 phase being confined to the extreme BEC region, while for $\alpha > 0$, the FFLO phase is compressed into the region of small r_s.

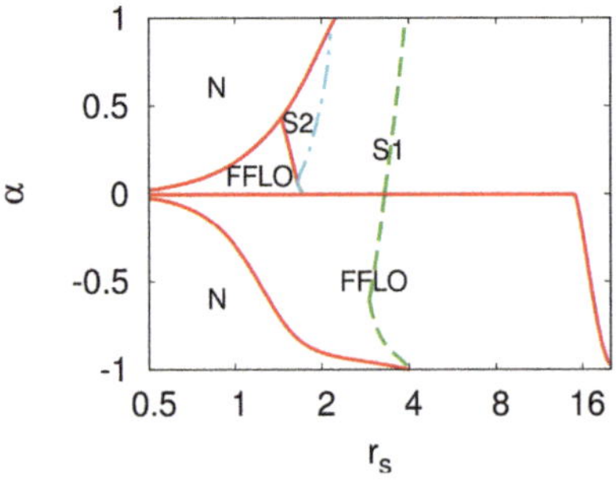

FIGURE 6.4 Zero temperature phase diagram for $d = 1$. The dashed line indicates $\mu = 0$ and the dashed dotted line indicates the separation between S1 and S2 phases.[51]

6.2.1.3 CANONICAL TRANSFORMATIONS

We can now introduce an extra complexity in the system, namely the intra-layer interaction, to explore further. However, for a transparent understanding of the mean-field formalism here we will follow the canonical transformation method to obtain the necessary mean-field equations. Our starting point is again eq (6.6) and now $U_{\mathrm{q}}^{aa} = U_{\mathrm{q}}^{bb} \neq 0$. One can now apply Bogoliubov transformation with operators $\alpha_{\mathrm{k}}^{\dagger}$ and $\beta_{-\mathrm{k}}^{\dagger}$ which are linear combinations of electron/hole creation/annihilation operators defined by

$$\begin{aligned} \alpha_{\mathrm{k}} &= u_{\mathrm{k}} a_{\mathrm{k}} - v_{\mathrm{k}} b_{-k}^{\dagger}, & \beta_{-\mathrm{k}} &= u_{\mathrm{k}} b_{-\mathrm{k}} + v_{\mathrm{k}} a_{\mathrm{k}}^{\dagger}, \\ \alpha_{\mathrm{k}}^{\dagger} &= u_{\mathrm{k}}^{*} a_{\mathrm{k}}^{\dagger} - v_{\mathrm{k}}^{*} b_{-k}, & \beta_{-\mathrm{k}}^{\dagger} &= u_{\mathrm{k}} b_{-\mathrm{k}}^{\dagger} + v_{\mathrm{k}}^{*} a_{\mathrm{k}}. \end{aligned}$$

These operators create/annihilate normalized states with excess quasi-particles orthogonal to $|\Psi\rangle_{BCS}$, which has an equal number of electrons and holes. We can also define the inverse transformations as,

$$\begin{aligned} \alpha_{\mathrm{k}} &= u_{\mathrm{k}}^{*} \alpha_{\mathrm{k}} - v_{\mathrm{k}} \beta_{-k}^{\dagger}, & b_{-\mathrm{k}} &= u_{\mathrm{k}}^{*} \beta_{-\mathrm{k}} + v_{\mathrm{k}} \alpha_{\mathrm{k}}^{\dagger}, \\ \alpha_{\mathrm{k}}^{\dagger} &= u_{\mathrm{k}} \alpha_{\mathrm{k}}^{\dagger} + v_{\mathrm{k}}^{*} \beta_{-k}, & b_{-\mathrm{k}}^{\dagger} &= u_{\mathrm{k}} \beta_{-\mathrm{k}}^{\dagger} + v_{\mathrm{k}}^{*} \alpha_{\mathrm{k}}. \end{aligned}$$

The excited states of BCS theory are states with excess (unpaired) electrons/holes or excited pairs such that

$$\begin{aligned} |\Psi\rangle &= \alpha_{\mathrm{q}}^{\dagger} |\Psi\rangle_{BCS} \\ &= a_{\mathrm{q}}^{\dagger} \prod_{\mathrm{k}\neq\mathrm{q}} \left(u_{\mathrm{k}} + v_{\mathrm{k}} a_{\mathrm{k}}^{\dagger} b_{-\mathrm{k}}^{\dagger} \right) |0\rangle. \end{aligned} \tag{6.20}$$

Here, one electron is at q state instead of the ground level. Generalizing this to a variational form we have

$$\begin{aligned} |\Psi\rangle &= \prod_{\mathrm{k}} \left(u_{\mathrm{k}}^{p} + v_{\mathrm{k}}^{+} \alpha_{\mathrm{k}}^{\dagger} + v_{\mathrm{k}}^{-} \beta_{-\mathrm{k}}^{\dagger} \right) \left(u_{\mathrm{k}} + v_{\mathrm{k}} a_{\mathrm{k}}^{\dagger} b_{-\mathrm{k}}^{\dagger} \right) |0\rangle \\ &= \prod_{\mathrm{k}} \left(u_{\mathrm{k}}^{p} + \left| v_{\mathrm{k}}^{+} \alpha_{\mathrm{k}}^{\dagger} + v_{\mathrm{k}}^{-} \beta_{-\mathrm{k}}^{\dagger} \right) \right| \Psi\rangle_{BCS.} \end{aligned} \tag{6.21}$$

The normalization can be achieved by choosing $|u_{\mathrm{k}}|^2 + |v_{\mathrm{k}}|^2 = 1$ and $|u_{\mathrm{k}}^{p}|^2 + |v_{\mathrm{k}}^{+}|^2 + |v_{\mathrm{k}}^{-}|^2 = 1$.

In a thermal state (canonical ensemble) one can write $\langle \alpha_{\mathrm{k}}^{\dagger} \alpha_{\mathrm{k}} \rangle = f(E_{\mathrm{k}}^{+}) = \left(e^{\beta E_{\mathrm{k}}^{+}} + 1 \right)^{-1}$, where β defines the inverse temperature. Thus,

$$\begin{aligned}\langle a_k^\dagger a_k\rangle &= \langle (u_k a_k^\dagger + v_k^* \beta_{-k})(u_k^* \alpha_k + v_k \beta_{-k}^\dagger)\rangle \\ &= |u_k|^2 \langle \alpha_k^\dagger \alpha_k\rangle + |v_k|^2 \langle \beta_{-k}\beta_{-k}^\dagger\rangle \\ &= (1-|v_k|^2) f_k^+ + |v_k|^2 (1-f_k^-) = f_k^+ + |v_k|^2 (1-f_k^+ - f_k^-).\end{aligned} \tag{6.22}$$

At $T = 0$, the Fermi function $f_k^\pm = 0$, therefore one can recover the usual BCS result as $\langle a_k^\dagger a_k\rangle T = 0 = |v_k|^2$. Hence, by applying the properties of canonical transformation, one can write $|u^p\, v_k|^2 = |v_k|^2\,(1-f_k^+ - f_k^-)$, $|v_k^+|^2 = f_k^+$, $|v_k^-|^2 = f_k^-$ and $|u^p\, u_k|^2 = |u_k|^2(1-f_k^+ - f_k^-)$. This signifies the probabilities of having a *pair*, type *a* particle, type *b* particle and *no* particle in the k quantum state. It must be noted here that the formalism is identical with the finite temperature BCS theory where $f_k^\pm$ are the occupation numbers of quasi particles.

To derive the mean-field energy gap equation it is now necessary to minimize the Helmholtz free energy with respect to the variational parameters,

$$\begin{aligned}\langle \hat{F}\rangle &= \langle \hat{H}\rangle - TS - \mu_a \langle \hat{N}_a\rangle - \mu_b \langle \hat{N}_b\rangle, \\ \text{where, } S &= -k_B \sum_{k\sigma} \left[f_k^\sigma \ln f_k^\sigma + (1-f_k^\sigma \ln (1-f_k^\sigma)) \right].\end{aligned} \tag{6.23}$$

Applying eqs (6.6) and (6.21) one can write,

$$\begin{aligned}\langle \Psi|\hat{H}|\Psi\rangle = \langle \hat{H}\rangle &= \sum_k (\epsilon_k^a + \epsilon_k^b)|v_k|^2 (1-f_k^+ - f_k^-) + \sum_k (\epsilon_k^a f_k^+ + \epsilon_k^b f_k^-) \\ &+ \frac{1}{V}\sum_{kk'} U_q^{ab} u_k v_k^* u_{k'}^* v_{k'} (1-f_k^+ - f_k^-)(1-f_{k'}^+ - f_{k'}^-) \\ &- \frac{1}{2V}\sum_{kk'} U_q^{aa} |v_k|^2 |v_{k'}|^2 (1-f_k^+ - f_k^-)(1-f_{k'}^+ - f_{k'}^-) \\ &- \frac{1}{2V}\sum_{kk'} U_q^{bb} |v_k|^2 |v_{k'}|^2 (1-f_k^+ - f_k^-)(1-f_{k'}^+ - f_{k'}^-) \\ &- \frac{1}{2V}\sum_{kk'} U_q^{aa} 2|v_k|^2 (1-f_k^+ - f_k^-) f_{k'}^+ - \frac{1}{2V}\sum_{kk'} U_q^{aa} f_k^+ f_{k'}^+ \\ &- \frac{1}{2V}\sum_{kk'} U_q^{bb} 2|v_k|^2 (1-f_k^+ - f_k^-) f_{k'}^+ - \frac{1}{2V}\sum_{kk'} U_q^{bb} f_k^- f_{k'}^- .\end{aligned} \tag{6.24}$$

If we assume the quasi particle wave functions as, $u_k = \cos\theta_k$ and $v_k = \sin\theta_k$ and rewrite the chemical potential in the following way, $2\xi_k^+ = \xi_k^a + \xi_k^b$,

$2\xi_k^- = \xi_k^a - \xi_k^b$, $2\mu = \mu_a + \mu_b$ and $2h = \mu_a - \mu_b$, then the free energy can be written as,

$$\begin{aligned}\langle \hat{F} \rangle = & \sum_k 2\xi_k^+ \sin^2\theta_k (1 - f_k^+ - f_k^-) + \sum_k \left(\xi_k^a f_k^+ + \xi_k^b f_k^- \right) \\ & + \frac{1}{4V} \sum_{kk'}{}' U_{kk'}^{ab} \sin 2\theta_k \sin 2\theta_{k'} (1 - f_k^+ - f_k^-)(1 - f_{k'}^+ - f_{k'}^-) \\ & - \frac{1}{V} \sum_{kk'}{}' U_{kk'}^{aa} \sin^2\theta_k \sin^2\theta_{k'} (1 - f_k^+ - f_k^-)(1 - f_{k'}^+ - f_{k'}^-) \\ & - \frac{1}{V} \sum_{kk'}{}' U_{kk'}^{aa} \sin^2\theta_k (1 - f_k^+ - f_k^-)(f_{k'}^+ + f_{k'}^-) \\ & - \frac{1}{2V} \sum_{kk'}{}' U_{kk'}^{aa} \left(f_k^+ f_{k'}^+ + f_k^- f_{k'}^- \right) \\ & + \frac{1}{\beta} \sum_k{}' \left[f_k^+ \ln f_k^+ + \left(1 - f_k^+ \ln\left(1 - f_k^+\right)\right)\right] \\ & + \frac{1}{\beta} \sum_k{}' \left[f_k^- \ln f_k^- + \left(1 - f_k^- \ln\left(1 - f_k^-\right)\right)\right]. \end{aligned} \tag{6.25}$$

In the above calculation we have considered that the magnitude of electron–electron and hole–hole interaction is same, that is, $U^{aa} = U^{bb}$. Minimizing eq (6.25) with respect to θ_k and rearranging them we obtain,

$$\tan 2\theta_k = \frac{-\frac{1}{2V}\sum_{k'}{}' U_{kk'}^{ab} \sin 2\theta_{k'} (1 - f_{k'}^+ - f_{k'}^-)}{\xi_k^+ - \frac{1}{2V}\sum_{k'}{}' U_{kk'}^{aa} \left[2\sin^2\theta_{k'} \left(1 - f_{k'}^+ - f_{k'}^-\right) + f_{k'}^+ + f_{k'}^- \right]} \equiv \frac{\Delta_k}{\xi_k}. \tag{6.26}$$

Borrowing the analogy from BCS theory we can further write,

$$\begin{aligned} \sin 2\theta_k &= \frac{\Delta_k}{\sqrt{\xi_k^2 + \Delta_k^2}} \equiv \frac{\Delta_k}{E_k} \\ \text{and, } \sin^2\theta_k &= \frac{\Delta_k^2}{2(\Delta_k^2 + \xi_k^2 + \xi_k E_k} = \frac{1}{2}\left(1 - \frac{\xi_k}{E_k}\right). \end{aligned} \tag{6.27}$$

If we now minimize the free energy with respect to f_k^+ and f_k^-, (i.e., $\partial\langle F\rangle / \partial f_k^+ = 0$, $\partial\langle F\rangle / f_k^- = 0$) and carry out necessary rearrangements, we will be able to write the mean-field gap equation. Thus, the coupled equations can be noted as

$$\Delta_{\mathrm{k}} = \frac{1}{V}\sum_{\mathrm{k}'} U_{\mathrm{kk}'}^{ab} \frac{\Delta_{\mathrm{k}'}}{2e_{\mathrm{k}'}}(1 - f_{\mathrm{k}'}^{+} - f_{\mathrm{k}'}^{-}) \tag{6.28}$$

$$\xi_{\mathrm{k}} = \xi_{\mathrm{k}}^{+} - \frac{1}{V}\sum_{\mathrm{k}'} U_{\mathrm{kk}'}^{aa}\left[\left(1 - \frac{\xi_{\mathrm{k}'}}{/} E_{\mathrm{k}'}\right) \times \left(1 - f_{\mathrm{k}'}^{+} - f_{\mathrm{k}'}^{-}\right) + f_{\mathrm{k}'}^{+} + f_{\mathrm{k}'}^{-}\right] \tag{6.29}$$

$$E_{\mathrm{k}}^{2} = \xi_{\mathrm{k}}^{2} + \Delta_{\mathrm{k}}^{2};\; f_{\mathrm{k}}^{\pm} = \begin{cases} 1 & \text{if } E_{\mathrm{k}}^{\pm} < 0 \\ 0 & \text{if } E_{\mathrm{k}}^{\pm} > 0 \end{cases}$$

where

$$E_{\mathrm{k}}^{\pm} = E_{\mathrm{k}} \pm \Delta E_{\mathrm{k}},\; \Delta E_{\mathrm{k}} = \Delta\xi_{\mathrm{k}} + \frac{1}{2A}\sum_{\mathrm{k}'} U_{\mathrm{kk}'}^{aa}\left(f_{\mathrm{k}'}^{-} - E_{\mathrm{k}}^{+}\right),\; \Delta\xi_{\mathrm{k}} = \frac{1}{2}\left(\in_{\mathrm{k}}^{a} - \mu_{a} - \in_{\mathrm{k}}^{b} + \mu_{b}\right).$$

However, the density equations remains unchanged, that is, one needs to follow eqs (6.14) and (6.15).

After solving the coupled equations self-consistently, it is possible to comment on the nature of the bilayer system and their phase separations. Figures 6.5 and 6.6 describe the variation of the energy gap and quasi particle energies with wave vector. However, the interaction energy is chosen differently for both the figures. In Figure 6.5 is calculated only with inter-layer interaction, whereas in Figure 6.6 both intra- and inter-interactions are taken into account. Moreover, in these two figures, the electron and hole numbers are considered as same that is the population balanced case ($h = 0$) thereby the density imbalance parameter $\alpha = 0$. Expectedly in the figures we do not observe any variation in occupation number for electrons and holes with varying wave vector. Nevertheless, we can definitely conclude that inclusion of intra-layer interaction suppresses the gap function as evident from the figures.

In the usual mean-field description of the electron–hole bilayer, one uses the bare Coulomb interaction as given in eq (8). In realistic systems, if taken into account the many body effects, it becomes necessary to modify the bare Coulomb interaction. The many body effects can be suitably modeled by a screening function which usually decreases the strength of the bare Coulomb interaction for electrons and holes in the normal phase. However, it is difficult to model an exact 2D screening function due to intra- and inter-layer interactions for condensed phase. Nevertheless for a qualitative understanding, one can consider the mechanism of

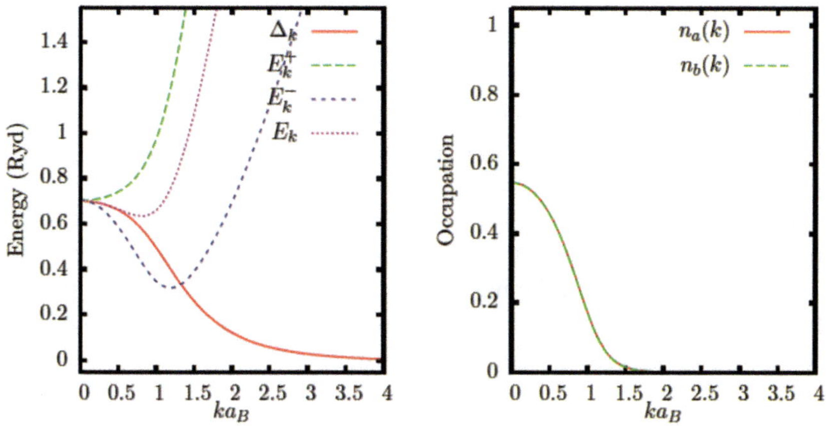

FIGURE 6.5 The left side figure depicts variation of Gap function, quasi-particle energies with wave vector for balanced density ($\alpha = 0$) and $r_s = 3$. The right side figure describes the variation of occupation number with wave vector. In this calculation, only interlayer interaction is taken into account.[58]

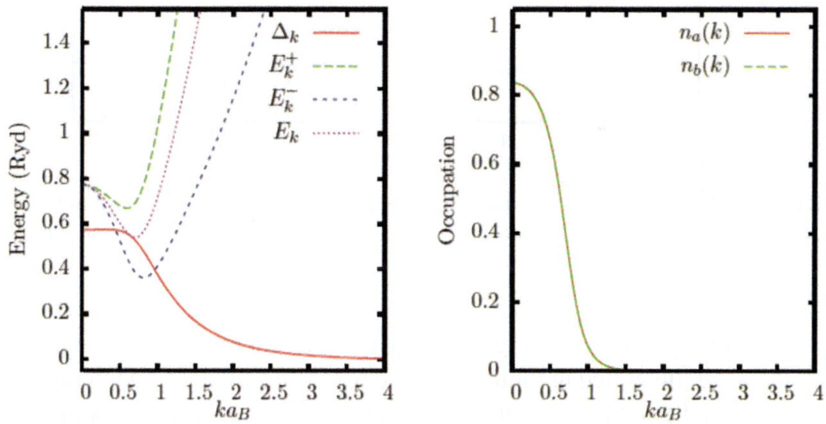

FIGURE 6.6 Both the figures are similar with Figure 6.5 in parameter values and observables; however in this figure, both inter-layer and intra-layer interactions are taken into account.[58]

gate screening. In this mechanism, the Coulomb potential of a point charge is replaced by that of a dipole consisting of the point charge and its image behind the metallic gate. An approximate description of the screening by the gate potential after taking into account the intra- and inter-layer interactions can be expressed as,[49]

$$U_q^{aa} = U_q^{bb} = \frac{2\pi e^2}{\varepsilon\sqrt{q^2 + \hat{e}^2}}, \qquad U_q^{ab} = \frac{2\pi e^2}{\varepsilon\sqrt{q^2 + \hat{e}^2}} e^{-qd}, \tag{6.30}$$

respectively, where the parameter κ is a screening wave number. In the calculation, the screening length associated with gate screening is considered as $\sim 20a_B$, that is, $\kappa = 1/20a_B$.

In Figures 6.7 and 6.8, the variation of gap function, quasi particle energy, and occupation number with wave vector is depicted for bare Coulomb interaction. In the numerical calculations, GaAs system parameters are taken into account where mass ratio turns out to be m_a/m_b = 0.07/0.30 and background dielectric constant is $\varepsilon = 12.9$.

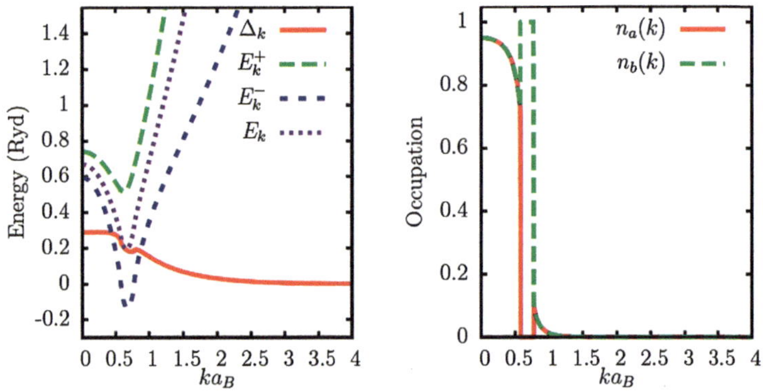

FIGURE 6.7 Gap function and quasi-particle energies associated with Sarma-2 phase at $r_s = 3$ and $\alpha = -0.3$ (excess hole) is depicted with bare Coulomb interactions for $m_e/m_h = 0.07/0.30$ and $d = a_B$.[59]

The solutions of energy gap, quasi particle energies, and occupation numbers are illustrated in Figures 6.7 and 6.8 for bare Coulomb interaction. The screened Coulomb counter part is presented in Figures 6.9 and 6.10. In all the cases the inter-layer distance is considered as one Bohr radius ($d = a_B$). The figures show the gap function (Δ_k), the quasi-particle energies ($E_k^{\pm}$) and their average (E_k) on the left panels, and the electron and hole occupation numbers $n_a(k)$, $n_b(k)$ on the right panels. At $T = 0$ in the ground state, the quasi-particle levels with negative energy are occupied, positive energy levels are empty. The two different type of excitation branches can be observed due different electron–hole mass and chemical potential values. When one of the spectra crosses the zero energy axis, population imbalance is created. If the negative energy region includes the origin at

$k = 0$, the ground state has one Fermi surface defined as S1 (as evident in Figs. 6.8 and 6.10), otherwise it has two S2. Since the quasi-particle energy branch is continuous, the system still has gapless excitations. A close investigation of the gap function Δ_k in the absence of screening (Figs. 6.7 and 6.8) shows that it has a cusp at the zero crossings of the quasi-particle energy, corresponding to a divergence in the derivative of Δ_k. This divergence leads to important consequences on the stability of the Sarma phase at $T = 0$.

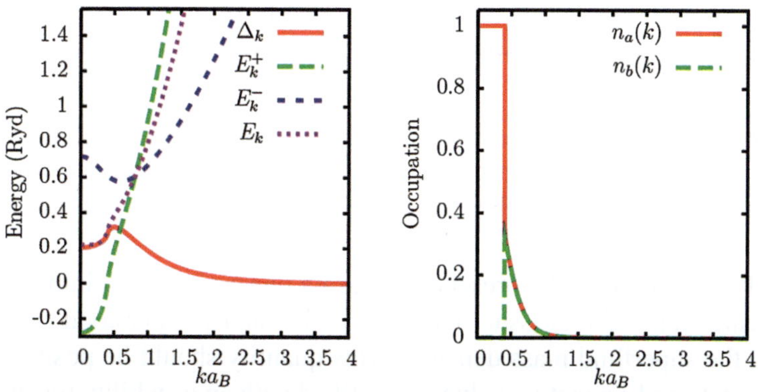

FIGURE 6.8 Gap function and quasi-particle energies associated with Sarma-1 phase at $r_s = 5$ and $\alpha = 0.5$ (excess electron) is depicted with bare Coulomb interactions for $m_e/m_h = 0.07/0.30$ and $d = a_B$.[59]

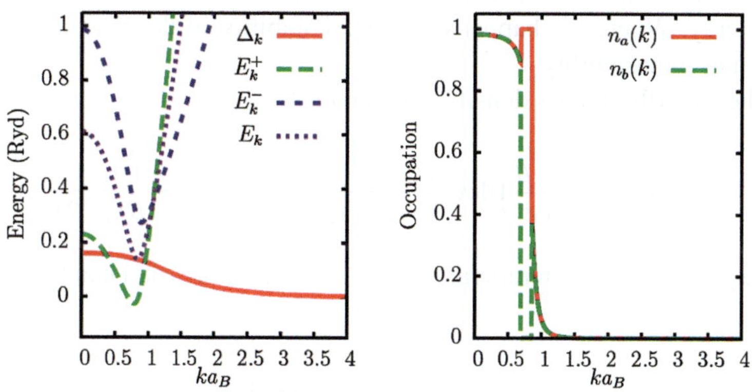

FIGURE 6.9 Gap function and quasi-particle energies associated with a Sarma-2 phase at $r_s = 2.5$ and $\alpha = 0.2$ (excess electrons) in presence of screened Coulomb interactions for $m_e/m_h = 0.07/0.30$ and $d = a_B$ is illustrated. The lower panel show a Sarma-1 phase at $r_s = 5$ and $\alpha = 0.5$ with excess electrons. Occupation numbers are shown on the right.[59]

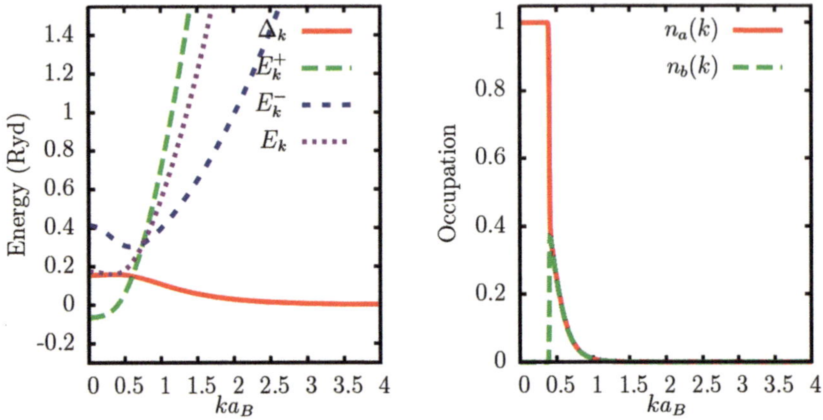

FIGURE 6.10 Gap function and quasi-particle energies associated with a Sarma-1 phase at $r_s = 5$ and $\alpha = 0.5$ (excess electrons) in presence of screened Coulomb interactions for $m_e/m_h = 0.07/0.30$ and $d = a_B$ is described.[59]

After elaborating the presence of Sarma phases it is now important to explicate their stability issue. The stability is usually understood by calculating the superfluid mass density.[48] This quantity should be positive in a stable state and a negative value is identified with an instability towards an FFLO phase.[48] The positivity of the superfluid mass density ensures that the Sarma phase is a local minimum of the energy with respect to fluctuations of the gap parameter. However, there exists another possibility of an FFLO phase with finite pair momentum leading to a global minimum of the energy. When this happens, the local stability of the Sarma phase is known as metastability.

The superfluid mass density is given by[49]

$$\rho_s = m_{\mathrm{e}} n_{\mathrm{e}} + m_{\mathrm{h}} n_{\mathrm{h}} - \frac{\hbar^2 \beta}{8\pi} \int \mathrm{d}k k^3 \frac{1}{2} \left[\frac{1}{\cosh^2(\beta E_{\mathrm{k}}^+ / 2)} + \frac{1}{\cosh^2(\beta E_{\mathrm{k}}^- / 2)} \right] \quad (6.31)$$

where β is the inverse temperature. At $T = 0$ this expression can be written as[48]

$$\rho_s = m_{\mathrm{e}} n_{\mathrm{e}} + m_{\mathrm{h}} n_{\mathrm{h}} - \frac{\hbar^2}{4\pi} \sum_{j,\lambda} \frac{\left(k_j^\lambda\right)^3}{\left|\frac{dE_k^\lambda}{dk}\right|_{k=k_j^\lambda}}. \quad (6.32)$$

where k_j^λ are the roots of E_k^λ with $\lambda = \pm$. At zero temperature the last expression involves the derivative of Δ_k at the zero crossings of $E_k^\pm$. However, it has been observed that numerical analysis turns our more efficient if the calculation is carried out for $T \rightarrow 0$ instead of $T = 0$. The numerical simulation reveals that the derivative of the gap energy diverges logarithmically as $T \rightarrow 0$. An analytical calculation demonstrates that for the bare Coulomb interaction as

$$\left|\frac{d\Delta_k}{dk}\right|_{k=k*} \approx \frac{e^2}{\pi\varepsilon 2E_{k*}}\left|\ln T\right| \quad \text{as } T \rightarrow 0 \tag{6.33}$$

where k^* is the zero crossing point at $T = 0$ as $k \rightarrow k^*$

$$\left|\frac{d\Delta_k}{dk}\right|_{T=0} \approx \frac{e^2\,\Delta_{k*}}{\pi\varepsilon 2E_{k*}}\ln\left|k - k*\right| \quad \text{as } k \rightarrow k^*, \tag{6.34}$$

This divergence is due to the presence of the long-range Coulomb interaction, which is singular at $q = 0$, and due to the discontinuity of the Fermi function at $T = 0$. Since at finite temperature, the discontinuity of the Fermi function is smeared out, the divergence is also removed. Same argument can be applied for screening potential in place of bare Coulomb interaction.

The phase diagram presented in Figure 6.11 originates from the comparison of the energies of the Sarma and normal phases and their stability scenario. As usual in the calculation, the inter-layer separation is taken as $d = a_B$. Figure 6.11 can be considered as a continuation of the phase diagram presented in Figure 6.4, where the diagram was continuously modified to take into account different realistic situations. In precise, the top-left figure is drawn for bare inter-layer interactions only; in the adjacent figure, intra-layer interaction is also added. The bottom-left figure depicts the phases for screened inter-layer interactions only, whereas the bottom right one is generated for screened inter- and intra-layer interactions.

For bare interactions, the superfluid density is always positive and the Sarma phase is stable locally as discussed earlier. Hence the top-left figure does not include the FFLO phase. However as mentioned earlier, there remains the possibility of first-order transition to FFLO phase. Therefore in the top panel two topologically distinct Sarma phases, Sarma-1 with one Fermi surface and Sarma-2 with two Fermi surfaces can be seen. The intra-layer repulsive interactions will try to delocalize the charge carriers which effectively favor the normal phase with respect to the condensed phases.

This phenomenon can be seen in the top-right figure as the r_s required to increase to draw the phase boundary between normal and condensed phases.

The bottom panel of Figure 6.11 presents the phase diagram when the gate screening is taken into account. In presence of the screening, inter-layer interactions itself induces Sarma phases to be unstable for a large portion of the phase diagram, especially with excess holes, that is, $\alpha < 0$ and one can only find S1 phase making S2 phase completely absent. However, switching on the intra-layer interactions reduces the space occupied by the FFLO phase and S2 phase can be obtained for small region in the phase diagram. Usually, the FFLO modulations of the gap function is accompanied by some density modulation in the real space. The repulsive Coulomb interaction does not favor such density modulations as an effect one cannot find FFLO states in presence of bare Coulomb interaction thereby ensuring a dominant Sarma phase. Therefore, the phase diagram becomes more intriguing when both intra-layer and screening effects are present. The presence of locally stable Sarma phases confirms that gapless superfluid states can be stable with momentum dependent interaction.

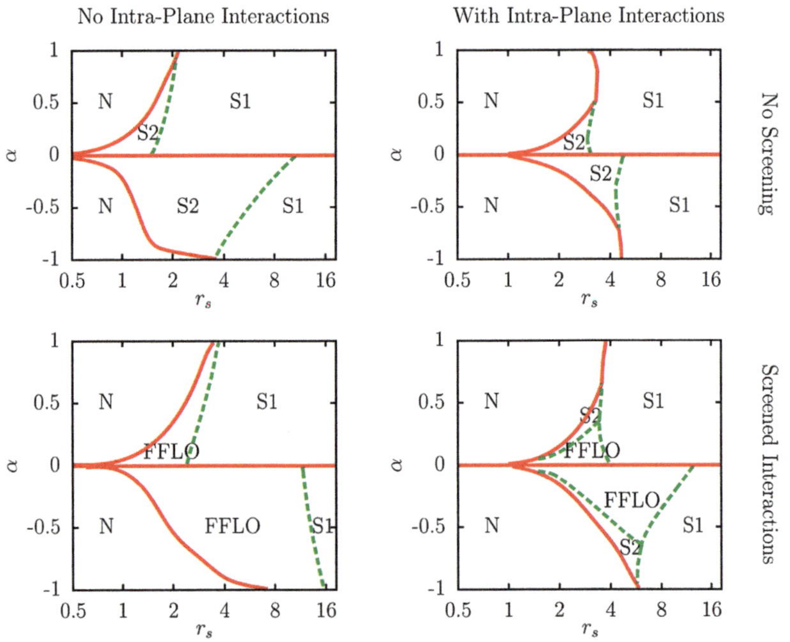

FIGURE 6.11 Zero temperature phase diagram for $d = 1$. The dashed line indicates $\mu = 0$ and the dashed dotted line indicates the separation between S1 and S2 phases.[59]

6.2.1.4 COMMENTS ON INCLUSION OF IMPURITY

From the above discussion, it is very clear that the interplay of interactions enriches the phase diagram dramatically and certain new phases emerge. We expect this situation will become more novel if we include some disorder. Here, we present a brief idea of how the disorder can be included in the mean-field formalism. It is in the same spirit as the usual atomic BCS–BEC crossover.

In atomic BCS–BEC crossover, a small amount of impurity can be embedded in the system by means Gaussian fluctuations. For that purpose, we will use the path integral formalism where we consider the Nambu propagator as

$$G^{-} = G_0^{-1} - V_d(q)\sigma_z, \tag{6.35}$$

where $V_d(q)$ is the disorder potential. We define,

$$G_0^{-1} = \begin{pmatrix} i\omega+\xi_k^b & \Delta_k^0 \\ \bar{\Delta}_k^0 & i\omega-\xi_k^b \end{pmatrix}. \tag{6.36}$$

We also assume small fluctuation about the pairing gap, that is, $\Delta_k = \Delta_k^0 + \delta\Delta_k$, where Δ_k^0 is the pairing gap in the clean system and it causes small fluctuation $\delta\Delta_k$ in presence of disorder. Hence the self-energy can be written as,

$$\sum = -V_d(q)\sigma_z + \delta\Delta_k(q)\sigma^+ + \delta\bar{\Delta}_k(-q)\sigma^-. \tag{6.37}$$

From eq (6.11), one needs to calculate the effective action appropriately. For that purpose we can rewrite ln $(1 + \Sigma G_0) = G_0\Sigma - \frac{1}{2}(G_0\Sigma' G_0'\Sigma)$ when the Dyson equation is expanded till the second order. Hence the path integral over $\frac{\Delta\bar{\Delta}}{U}$ − Tr ln G^{-1} can be expressed as sum of Bosonic action (S_B), fermionic action (S_F) and the action related to saddle point calculation S_0. In other terms, S_0 is the first order in the Dyson equation and for any extremum in the total action this term must be equated to zero. Therefore the effective action can be written as $S_{eff} = S_B + S_0 + S_F$. A detailed calculation reveals

$$S_B = \sum_k \left[\frac{\Delta_k^0}{U_{kk'}} + \sum_{k'} \frac{\Delta_{k'}^0}{2E_{k'}}\left(1 - f_{k'}^{+} - f_{k'}^{+}\right)\right]\left[\delta\Delta_k(0) + \delta\bar{\Delta}_k(0)\right]. \tag{6.38}$$

Applying the saddle point condition $\left(\frac{\delta S_{eff}}{\delta\Delta}=0\right)$, one can write

$$\Delta_k^0 = -\frac{1}{V}\sum_{k'} U_{kk'}\frac{\Delta_{k'}^0}{2E_{k'}}\left(1-f_{k'}^{+}-f_{k'}^{+}\right) \tag{6.39}$$

The fermionic action can be expressed as

$$S_F = \beta\sum_{kk'}\frac{\Delta_k^0\bar{\Delta}_k^0}{U_{kk'}} - \frac{1}{\beta}\sum_{k,i\omega}\left[Tr\ \ln\left(-G_0^{-1}(k)\right)+V_d(0)\sigma_z)\right]+\Omega_{Fd,} \tag{6.40}$$

$$\text{where, } \Omega_{Fd,} = \frac{1}{2\beta}\sum_{k,q}\mathrm{Tr}\left[G_0(k)\sigma_z G_0(k+q)\sigma_z\left\langle V_d(q)V_d(-q)\right\rangle\right].$$

The bosonic action arising through fluctuation and impurity can be written as

$$S_B = \frac{1}{2}\sum_q\left[V_d(q)\chi\lambda^\dagger + V(-q)\chi^\dagger\lambda + \lambda^\dagger M\gamma\right], \tag{6.41}$$

where

$$\chi = \begin{pmatrix} A_{k+q}\Gamma_k - B_k\Gamma_{k+q} \\ A_k\bar{\Gamma}_{k+q} - B_{k+q}\bar{\Gamma}_k \end{pmatrix}, \quad \lambda = \begin{pmatrix} \delta\Delta_k(q) \\ \delta\bar{\Delta}(-q) \end{pmatrix}.$$

Here for the convenience of calculation, we have redefined the Greens functions in the following way

$$G_0(k) = \begin{pmatrix} A_k & \Gamma_k \\ \bar{\Gamma}_k & B_k \end{pmatrix}, \quad G_0(k+q) = \begin{pmatrix} A_{k+q} & \Gamma_{k+q} \\ \bar{\Gamma}_{k+q} & B_{k+q} \end{pmatrix}. \tag{6.42}$$

However, one can always look back to eq (6.36) for exact form of the Green's function. The last term in the bosonic action turns out as

$$\lambda^\dagger M\lambda = \left(\delta\bar{\Delta}_k(-q), \delta\Delta_k(q)\right)\begin{pmatrix} \frac{1}{U_{kk'}} + A_{k+q}B_k & \Gamma_k\Gamma_{k+q} \\ \bar{\Gamma}_k\bar{\Gamma}_{k+q} & A_kB_{k+q} \end{pmatrix}\begin{pmatrix} \delta\Delta_k(q) \\ \delta\bar{\Delta}(-q) \end{pmatrix}$$

It is well known that that $\frac{1}{\beta}\frac{\partial S_F}{\partial\mu_i} = n_i$ for the clean Fermi system. However, if we consider the existence of fluctuation and insert the impurity we obtain non zero bosonic action. Hence for system described above, one needs to apply

$$\frac{1}{\beta}\frac{\partial S_{eff}}{\partial \mu_i} = n_i$$
$$\frac{1}{\beta}\left[\frac{\partial S_F}{\partial \mu_i} + \frac{\partial S_B}{\partial \mu_i}\right] = n_i \tag{6.43}$$

Using the modified density equation along with the gap equation, one can now examine the effect of impurity in the phase separation of a bilayer system in a semiconductor.

6.2.2 EXPERIMENTS WITH EXCITONS

In the above discussion, we have elaborated the consequence of interplay of different types of interaction by means of mean-field theory. This we explicate with analogy from the atomic BCS–BEC crossover. However, it is interesting to note that excitons were first to be considered for the BCS–BEC crossover.[61] On the reverse path, experiments with population imbalance in ultra-cold trapped Fermi atoms[60] have stimulated a considerable amount of theoretical work on two-component Fermi systems with density imbalance.[62,63] This upsurge in interest actually derives from the expectation of exotic phases in addition to the ordinary BCS pairing.[62,64,65] However, the presence of a trap and charge neutrality of the atoms inhibit the occurrence of many of the exotic phases. Therefore excitons are considered as a good candidate for observing such exotic phases because the Coulomb repulsion within each layer acts to suppress phase separation.[66] Additionally, it is worth noting that the different electron and hole effective masses in GaAs, m_e and m_h, and the non-local nature of the electron–hole attraction both favor the occurrence of exotic phases.[48]

In recent years, using electrical and optical techniques exciton condensation has been observed in several different systems. Quantum Hall experiments at half-filling investigate BEC in electron–electron and hole–hole bilayers. Optically generated bilayer excitons also show evidence for condensation. Off-late, excitons coupled to photons to form polaritons with even smaller mass leading to higher condensation temperatures have been experimentally detected.[58] In the last couple of years, scientist are even able to form a droplet (about five electron–hole pairs together inside semiconductor) made up of electron–hole bilayers systems.[67] This is popularly known as dropleton. The creation of dropletons was carried out in an electron–hole plasma inside a *GaAs* quantum well by ultrashort laser pulses. As the name suggested dropleton are the first quasiparticle found to behave like a liquid. This discovery was sudden and the scientist had no idea about such property a priori. This emphasizes the richness of the context of bilayer systems. Also, it reminds that there are many issues to unravel in the process to understand this system properly.

6.3 APPLICATIONS

The importance of excitonic research lies in its multifacet possible applicability. The current interests of different scientific and engineering endeavors are mostly getting converged to the field of energy and communication. Already there are different techniques of solar cells and microchips which have made substantial enrichments in these fields but it is still far beyond the goal. In the domain of information technology and computation, the idea of quantum computers and simulators are already been placed. From a theoretical point of view, quantum information processing can be considered as a well-established field by now but the key issue of the design and realization of concrete solid-state implementation protocols are subject of intense investigation at the moment.

Recently, several proposals are placed for an all optical implementation of quantum information/computation with semiconductor macroatoms (quantum dots in zero dimension). These quantum dots can be defined as a portion of matter (say semiconductor) whose excitons are confined in all three spatial dimensions. The quantum hardware consists of an array of quantum dots and the computational degrees of freedom are energy-selected inter-band optical transitions. The quantum-computing strategy exploits exciton-exciton interactions driven by ultrafast multicolor laser pulses. It allows a subpicosecond, decoherence-free, operation time scale in realistic semiconductor nanostructures.[68,69] Also, there exist proposals based on charge-plus-spin degrees of freedom in semiconductor quantum dots.[70] These propositions encourage a coherent optical control of electronic spins as well as of excitonic state. In addition, a proper tailoring of exciton-exciton Coulomb coupling (allowing for the implementation of single- as well as two-qubit gates) can introduce the full set of basic operations to implement quantum computing.

Through the above discussion, we have tried to highlight the intriguing technological issues which are mainly governed by the physics of electron–hole systems. Therefore in this chapter, we have tried to narrate a theoretical perspective to present an overview for the electron–hole bilayer systems in semiconductors.

ACKNOWLEDGMENT

This work is supported by TUBITAK (112T176) and TUBA.

KEYWORDS

- **semiconductor**
- **electron–hole system**
- **Bose-Einstein condensate**
- **resonance stabilization**
- **intra-layer interaction**

REFERENCES

1. Moore, G. E. *Electronics* **1965,** *38,* 8.
2. Regal, C. A.; Greiner, M.; Jin, D. S. *Phys. Rev. Lett.* **2004,** *92,* 040403.
3. Davis, K. B.; Mewes, M. O.; Andrews, M. R.; Van Druten, M. J.; Durfee, D.S.; Kurn, D. M.; Ketterle, W. *Phys. Rev. Lett.* **1995,** *75,* 3969.
4. Bradley, C. C.; Sackett, C. A.; Hulet, R. G. *Phys. Rev. Lett.* **1997,** *78,* 985.
5. Modugno, G.; Ferrari, G.; Roati, G.; Brecha, R. J.; Simoni, A.; Inguscio, M. *Science.* **2001,** *294,* 1320.
6. Regal, C. A.; Ticknor, C.; Bohn, J. L.; Jin, D. S. *Nature* **2003,** *424,* 47.
7. Strecker, K. E.; Partridge, G. B.; Hulet, R. G.*Phys. Rev. Lett.* **2003,** *91,* 080406.
8. Zwierlein, M. W.; Stan, C. A.; Schunck, C. H.; Raupach, S. M. F.; Gupta, S.; Hadzibabic, Z.; Ketterle, W. *Phys. Rev. Lett.* **2003,** *91,* 250401.
9. Jochim, S.; Bartenstein, M.; Altmeyer, A.; Hendl, G.; Chin, C.;Denschlag, J. H.;Grimm, R. *Phys. Rev. Lett.* **2003,** *91,* 240402.
10. Cubizolles, J.; Bourdel, T.; Kokkelmans, S. J. J. M. F.; Shlyapnikov, G. V.; Salomon, C. *Phys. Rev. Lett.* **2003,** *91,* 240401.
11. Eisenstein, J. P.; MacDonald, A. H. *Nature* **2004,** *432,* 691.

12. Butov, L. V.; Gossard, A. C.; Chemla, D. S. *Nature* **2002,** *418,* 751.
13. Anderson, M. H.; Ensher, J. R.; Matthews, M. R.; Wieman, C. E.; Cornell, E. A. *Science* **1995,** *269,* 198.
14. Davis, K. B.; Mewes, M. O.; Andrews, M. R.; vanDruten, N. J.; Durfee, D. S.; Kurn, D. M.; Ketterle, W. *Phys. Rev. Lett.* **1995,** *75,* 3969.
15. Fano, *U. Nuovo Cimento.* **1935,***12,* 156.
16. Fano, U. *Phys. Rev.* **1961,** *124,* 1866.
17. Feshbach, H. *Ann. Phys.* **1962,** *19,* 287.
18. Inouye, S.; Andrews, M. R.; Stenger, J.; Miesner, H. J.; Stamper-Kurn, D. M.; Ketterle, W. *Nature* **1998,** *392,* 151.
19. Roberts, J. L.; Claussen, N. R.; Burke, J. P., Jr.; Greene, C. H.; Cornell, E. A.; Wieman, C. E. *Phys. Rev. Lett.* **1998,** *81,* 5109.
20. Holland, M.; Kokkelmans, S. J. J. M. F.; Chiofalo, M. L.; Walser, R. *Phys. Rev. Lett.* **2001,** *87,* 120406.
21. Ohashi, Y.; Griffin, A. *Phys. Rev. Lett.* **2002,** *89,* 130402.
22. Regal, C. Advances in Atomic, Molecular, and Optical Physics. In *Experimental Realization of BCS-BEC Crossover Physics with a Fermi Gas of Atoms;* University of Colorado: Boulder, CO,2006.
23. Eagles, D. M. *Phys. Rev.* **1969,** *186,* 456.
24. Leggett, A. J. In *Lecture Notes in Physics,* Proceedings of the XVI Karpacz Winter School of Theoretical Physics, Springer, Berlin, Vol. 115, p 13, 1980.
25. Nozières, P.; Schmitt-Rink, S. *J. Low Temp. Phys.* **1985,** *59,* 195.
26. Randeria, M.; Duan, J. M.; Shieh, L. Y. *Phys. Rev. Lett.* **1989,** *62,* 981.
27. Haussmann, R. *Z. Phys. B.* **1993,** *91,* 291.
28. Pistolesi, F.; Strinati, G. C. *Phys. Rev. B.* **1994,** *49,* 6356.
29. Stintzing, S.; Zwerger, W. *Phys. Rev. B.* **1997,** *56,* 9004.
30. Jankó, B.; Maly, J.; Levin, K. *Phys. Rev. B.* **1997,** *56,* R11407.
31. Chen, Q.; Kosztin, I.; Jankó, B.; Levin, K. *Phys. Rev. Lett.* **1998,** *81,* 4708.
32. Fuhrer, M. S.; Hamilton, A. R. *Physics* **2016,** *9,* 80.
33. Byrnes, T.; Kim, N. Y.; Yamamoto, Y. *Nat. Phys.* **2014,** *10,* 803.
34. Lee, K.; Xue, J.; Dillen, D. C.; Watanabe, K.; Taniguchi, T. Tutuc, E. *Phys. Rev. Lett.* **2016,** *117,* 046803.
35. Li, J. I. A.; Taniguchi, T.; Watanabe, K.; Hone, J.; Levchenko, A.; Dean, C. R. *Phys. Rev. Lett.* **2016,** *117,* 046802.
36. Neilson, D.; Perali, A.; Hamilton, A. R. *Phys. Rev. B.* **2014,** *89,* 60502.
37. Efimkin, D. K.; Galitski, V. *Phys. Rev. Lett.* **2016,** *116,* 046801.
38. Dell'Anna, L.; Perali, A.; Covaci, L.; Neilson, D. *Phys. Rev. B.* **2015,** *92,* 220502.
39. Zarenia, M.; Perali, A.; Peeters, F. M.; Neilson, D. *Sci. Rep.* **2016,** *6,* 24860.
40. Blatt, J. M.; Böer, K. W.; Brandt, W. *Phys. Rev.* **1962,** *126,* 1691.
41. Comte, C.; Nozières, P. *J. Physique.* **1982,** *43,* 1069.
42. Comte, C.; Nozières, P. *J. Physique.* **1982,** *43,* 1083.
43. Kellogg, M.; Eisenstein, J.; Pfeiffer, L.; West, K. *Phys. Rev. Lett.* **2004,** *93,* 036801.
44. Tutuc, E.; Shayegan, M.; Huse, D. *Phys. Rev. Lett.* **2004,** *93,* 036802.
45. Littlewood, P. *Science* **2007,** *316,* 989.

46. Kasprzak, J.; Richard, M.; Kundermann, S.; Baas, A.; Jeambrun, P.; Keeling, J. M. J.; Marchetti, F. M.; Szymanska, M. H.; Andre, R.; Staehli, J. L.; Savona, V.; Littlewood, P. B.; Deveaud, B.; Dang, L. S. *Nature* **2006,** *443,* 409.
47. Balili, R.; Hartwell, V.; Snoke, D.; Pfeiffer, L.; West, K. *Science* **2007,** *316,* 1007.
48. Pieri, P.; Neilson, D.; Strinati, G. C. *Phys. Rev. B.* **2007,** *75,* 113301.
49. Subasi, A. L.; Pieri, P.; Senatore, G.; Tanatar, B. *Phys. Rev. B.* **2010,** *81,* 075436.
50. Dubi, Y.; Balatsky, A. V. *Phys. Rev. Lett.* **2010,** *104,* 166802.
51. Pieri, P.; Neilson, D.; Strinati, G.C. The Figure Is Obtained from "Effects of Density Imbalance on the BCS-BEC Crossover in Semiconductor Electron-hole Bilayers," 2006. http://lanl.arxiv.org/abs/cond-mat/0610311
52. Andrenacci, N.; Perali, A.; Pieri, P.; Strinati, G. C. *Phys. Rev. B.* **1999,** *60,* 12410.
53. Andrenacci, N.; Pieri, P.; Strinati, G. C. *Phys. Rev. B.* **2003,** *68,* 144507.
54. Khan, A.; Pieri, P. *Phys. Rev. A.* **2009,** *80,* 012303.
55. Khan, A.; Basu, S.; Kim, S. W. *J. Phys. B. At. Mol. Opt. Phys.* **2012,** *45,* 135302.
56. Wu, S. T.; Yip, S. *Phys. Rev. A.* **2003,** *67,* 053603.
57. He, L.; Jin, M.; Zhuang, P. *Phys. Rev. B.* **2006,** *73,* 214527.
58. Subasi, A. L. The Figure Is Obtained from Many-body Effects in Selected Two-Dimensional Systems. Ph.D. Thesis, Bilkent University, 2009.
59. Subasi, A. L.; Pieri, P.; Senatore, G.; Tanatar, B. The Figure Is Obtained from "Stability of Sarma Phases in Density Imbalanced Electron-hole Bilayer Systems," 2009. http://lanl.arxiv.org/abs/0912.3326
60. Zwierlein, M. W., et al. *Science.* **2006,** *311,* 492.; Partridge, G. B.; Li, W.; Kamar, R. I.; Liao, Y.; Hulet, R. G. *Science.* **2006,** *311,* 503.
61. Keldysh, L. V.; Kopaev, Y. V. *Sov. Phys. Solid State.* **1965,** *6,* 2219.; Keldysh, L. V.; Kozlov, A. N. *Sov. Phys. J.Exp.Theor.Phys.* **1968,** *27,* 521.
62. Carlson, J.; Reddy, S. *Phys. Rev. Lett.* **2005,** *95,* 060401.; Pao, C. H.; Shin-Tza, W.; Yip, S. K. *Phys. Rev. B.* **2006,** *73,* 132506.; Sheehy, D. E.; Radzihovsky, L. *Phys. Rev. Lett.* **2006,** *96,* 060401.
63. Pieri, P.; Strinati, G. C. *Phys. Rev. Lett.* **2006,** *96,* 150404.; Kinnunen, J.; Jensen, L. M.; Törmä, P. *Phys. Rev. Lett.* **2006,** *96,* 110403.; Yi, W.; Duan, L. M. *Phys. Rev. A.* **2006,** *73,* 031604.; Chevy, F. *Phys. Rev. Lett.* **2006,** *96,* 130401.
64. McNeil Forbes, M., et al. *Phys. Rev. Lett.* **2005,** *94,* 017001.
65. Son, D. T.; Stephanov, M. A. *Phys. Rev. A.* **2006,** *74,* 013614.; Bulgac, A.; McNeil Forbes, M.; Schwenk, A. *Phys. Rev. Lett.* **2006,** *97,* 020402.; Mannarelli, M.; Nardulli, G.; Ruggieri, M. *Phys. Rev. A.* **2006,** *74,* 033606.
66. Butov, L., et al. V. *J. Exp. Theor. Phys.* **2001,** *92,* 260.
67. Almand-Hunter, A. E.; Li, H.; Cundiff, S. T.; Mootz, M.; Kira, M.; Koch, S. W. *Nature* **2014,** *506,* 471.
68. Biolatti, E.; Iotti, R. C.; Zanardi, P.; Rossi, F. *Phys. Rev. Lett.* **2000,** *85,* 5647.
69. DâĂŹAmico, I.; De Rinaldis, S.; Biolatti, E.; Pazy, E.; Iotti, R. C.; Zanardi, P.; Rossi, F. *Phys. Stat. Sol. B.* **2002,** *234,* 58.
70. Pazy, E.; Biolatti, E.; Calarco, T.; D'Amico, I.; Zanardi, P.; Rossi, F.; Zoller, P. *Eur. Phys. Lett.* **2003,** *62,* 175.

PART II
Nanocomposite Properties

CHAPTER 7

POLYMER NANOCOMPOSITES: A CASE STUDY OF RUBBER TOUGHENED EPOXY/CTBN MATRIX IN THE PRESENCE OF CLAY AND CARBON NANOTUBES AS NANOFILLERS

BHAGWAN F. JOGI[1*], MADAN KULKARNI[1], P. K. BRAHMANKAR[1], D. RATNA[2], and RINUL M. DHAJEKAR[1]

[1]*Department of Mechanical Engineering, Dr. Babasaheb Ambedkar Technological University, Lonere 402103, Raigad, Maharashtra, India*

[2]*Naval Materials Research Laboratory (DRDO), Ambernath, Mumbai, Maharashtra, India*

[*]*Corresponding author. E-mail: bfjogi@dbatu.ac.in; bfjogi@gmail.com*

CONTENTS

ABSTRACT

Rubber toughened blends of epoxy and carboxyl group-terminated butadiene nitrile (CTBN) was selected as a system for this study. CTBN was used as toughening agent with different concentrations of CTBN with epoxy and 0, 5, 10, 15, and 20 parts per hundred parts resins (phr) are the combinations prepared by varying CTBN content in epoxy. Through optimization, it is found that 15 phr epoxy/CTBN shows improvement in mechanical properties; tensile properties, and dynamic mechanical thermal analysis (DMTA). Furthermore, clay and carbon nanotubes (CNTs) as nanofiller were optimized and found that 3 wt% of clay with 15 phr epoxy/CTBN blends show improvement in the mechanical properties. However, 0.5 wt% of modified CNT depicts an improvement in the properties. It may be due to good dispersion of the CNT and high exfoliation of clay in polymer nanocomposites (PNCs). Scanning electron microscopy (SEM) was carried out on few selective samples. The main aim is to optimize the system.

7.1 INTRODUCTION

The addition of nanofillers in polymer matrices creates a class of novel materials (polymer nanocomposites (PNC)) exhibiting superior mechanical, thermal, electrical, and barrier properties suitable to replace many existing materials for engineering applications. The reinforcing of polymer matrix may be carried out by much stiffer nanoparticles (NPs) of ceramics, clays, nanoclay, carbon black, CNTs, glass fibers, etc. Addition of clay with polymers has shown the improvement in mechanical properties. In the past two decades, Pötschke et al.[1] and Meincke et al.[2] developed the idea of using layered clay minerals for PNC applications. This progress was enabled by the utilization of specially designed organophilic clay as nanofillers in PNC by Potschke et al.[3,4] Furthermore, carbon nanotubes (CNTs) have been viewed as the most promising material in the area of PNC as per Ijjima in 1991.[5] In this context, use of clay and CNTs in polymer matrices has gained considerable attention in the scientific and industrial community due to the possibility to utilize the unique mechanical, thermal, and electrical properties of clay and CNTs. Several research groups have reported the encouraging results regarding the possibility of introducing clay and CNT for mechanical enhancement in polymer/CNT composites.[6–8]

Clay- and CNT-based PNCs possess high stiffness, high strength, and good electrical conductivity at relatively low concentrations of filler. A general conclusion has been drawn that nanocomposites show much improved mechanical properties over their micro-sized similar systems. Because of their small size, NPs have a high surface to volume ratio and provide high-energy surfaces. An expected result of embedding NPs into a polymer matrix is enhanced bonding between the polymer matrix and filler, resulting from the nanoparticles' high interfacial energy. Polymer composite theory predicts that improved bonding between polymer matrix and fillers leads to improved mechanical properties.[9–12]

However, the magnitude of reinforcement by clay and CNT is found to be limited due to the challenges involved, dispersion of clay, and CNT in a molecular level in polymer matrices as fillers tend to form clusters and bundles. These aggregates continue, until the physical or chemical modification of clay and CNT. A few modification methods such as vigorous mixing of the polymer damages and CNT structures affect their properties. But this problem can be overcome by chemically modifying their surfaces.[13–15,31] The functionalization led to a reduced agglomeration and evidences are given for improved interaction between the nanotubes and the epoxy resin. CNT is functionalized by half neutralized adipic acid (HNAA) and sodium salt of amino hexanoic acid (Na-AHA) for this study.

Furthermore, epoxy-based materials are extensive and include coatings, adhesives, and composite materials such as those using carbon fiber and fiberglass reinforcements. Epoxies are known for their excellent adhesion, chemical nature, heat resistance, mechanical properties, and very good electrical insulating properties. Epoxy systems are used in industrial tooling applications to produce molds, master models, laminates, castings, fixtures, and other industrial production aids. This "plastic tooling" replaces metal, wood, and other traditional materials and generally improves the efficiency. It either lowers the overall cost or shortens the lead time for many industrial processes.

Epoxy was selected for PNC as it shows low creep behavior; however, it has also high mechanical properties, high heat distortion temperature, and excellent dimensional stability.[9] To enhance the toughness of cured epoxy resin, various reactive liquid rubbers have been incorporated. Epoxy rubber toughened by carboxyl terminated butadiene nitrile (CTBN)[16,17] is selected for this study.[18] Hence, in this study epoxy/CTBN blends with varying concentration of clay and CNT is planned to optimize through various characterization techniques. Furthermore, after optimization of the rubber-toughened blends of epoxy and CTBN, in the presence of clay and CNT is mixed as a nano filler material. Clay loading varies as 2, 3, 4, and 5 wt%, while loading of CNT filler as 0.1, 0.25, 0.5, 0.75, and 1 wt%, in a 15 phr composition. These samples were planned for tensile test. Compositions depict good improvement during tensile test, and were further, characterized for dynamic mechanical thermal analysis (DMTA) and SEM analysis.

7.2 EPOXY AS THERMOSETTING POLYMER (THERMOSETS)

Thermosets are the polymers which harden permanently by heating. It has covalent crosslinks (~10–50% of mers) formed during heating which hinder bending and rotations of the polymer. Hence, thermosets are harder, more dimensionally stable, and more brittle than thermoplastics. For example, vulcanized rubber, epoxies, etc.[13,15,28,34]

Epoxy is also known as polyepoxide, a thermosetting polymer formed from the reaction of an epoxide "resin" with polyamine "hardener." Epoxy is a copolymer; that is, formed from two different chemicals. These are referred to as the "resin" or "compound" and the "hardener" or "activator." The resin consists of monomers or short chain polymers with an epoxide group at either end. Most common epoxy resins are produced from a reaction between epichlorohydrin and bisphenol-A, though the latter may be replaced by similar chemicals. When these compounds are mixed together, the amine groups react with the epoxide groups to form a covalent bond. Each NH group can react with an epoxide group, so that the resulting polymer is heavily cross-linked, and is thus rigid and strong. The structure of unmodified epoxy prepolymer resin is shown in Figure 7.1.

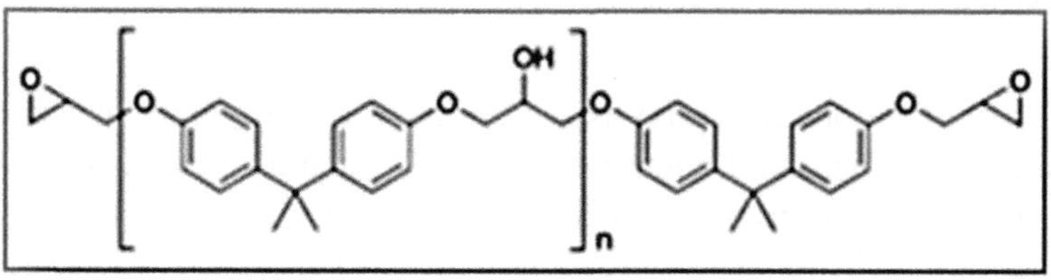

FIGURE 7.1 Structure of unmodified epoxy prepolymer resin.

The term "epoxy resin" is a generic name for compounds that have two or more oxirane rings (epoxy groups) in one molecule, and are cured three-dimensionally by a suitable curing agent. The process of polymerization is called "curing" and can be controlled through temperature, choice of resin, hardener compounds and the ratio of said compounds; the process can take minutes to hours. Some formulations benefit from heating during the cure period, whereas others simply require time and ambient temperatures. In general, epoxy adhesives cured with heat will be more heat and chemical resistant than those cured at room temperature. However, in most cases, the term refers to bisphenol-A diglycidyl ether (DGEBA), which is formed by the reaction between bisphenol A and epichlorohydrin, which currently commands a 75% share of the epoxy resin market. The products of three bond 50–60% of one part epoxy resin and more than 90% of two-part epoxy resin based on DGEBA or compounds containing DGEBA. Therefore, DGEBA is synonym for epoxy resin having specific structure shown in Figure 7.2.

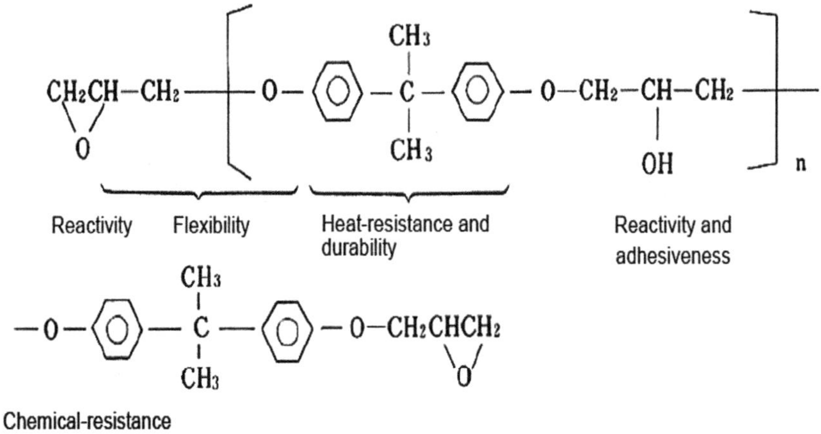

FIGURE 7.2 Structure of epoxy resin.

The applications for epoxy-based materials are extensive and include coatings, adhesives, and composite materials such as those using carbon fiber and fiberglass reinforcements. Epoxies are known for their excellent adhesion, chemical and heat resistance properties, good to excellent mechanical properties, and very good electrical insulating properties.

Epoxies typically are not used in the outer layer of a boat because they deteriorate by exposure to UV light. The strength of epoxy adhesives is degraded at temperatures above 350°F (177°C).[28,30,31,34]

7.2.1 HEALTH RISKS

The primary risk associated with epoxy use is sensitization to the hardener, which over time can induce an allergic reaction. Allergic reaction sometimes occurs at a time which is delayed several days from the exposure. Epoxy use is a main source of occupational asthma among users of plastics. Bisphenol A, which is used in epoxy resin, is a known endocrine disruptor.[35]

7.2.2 CURING AGENTS

Epoxy resins can be cured with a wide variety of curing agents. The choice of curing agents depends on the required physical and chemical properties, processing methods, and curing conditions. Osumi Y[20] studied the treatment of epoxy resins with curing agents or hardeners, which gives three-dimensional insoluble and infusible networks of PNC. Epoxy resins can be cured with either catalytic or co-reactive curing agents. Catalytic curing agents function as initiators for epoxy ring-opening homopolymerization. Epoxy resins can be catalytically cured by Lewis bases such as tertiary amines, or Lewis acids such as boron trifluoride monoethylamine. These catalytic curing agents can be used for homopolymerization, as accelerators or supplemental curing agents for other curing agents such as amines.

7.3 EPOXY TOUGHENING BY CARBOXYL TERMINATED BUTADIENE ACRYLONITRILE (CTBN)

A liquid CTBN copolymer is used predominately as a reactant with a base thermosets resin to gain product performance improvements. The important research work on rubber modified epoxy resins was pioneered by McGarry and his groups at MIT. The paper reports on the preparation, structure, and properties of ternary thermosetting blends, based on DGEBA epoxy, cured with J 300 and modified by the addition of CTBN reactive liquid rubber is shown in Figure 7.3. The toughening effect of the phase-separated rubber particles is enhanced by the presence of CNTs, through a change in the morphology. In the absence of rubber, the CNTs alone produce a minimal effect upon the thermo-mechanical characteristics of the resin as per Maria et al.[17]

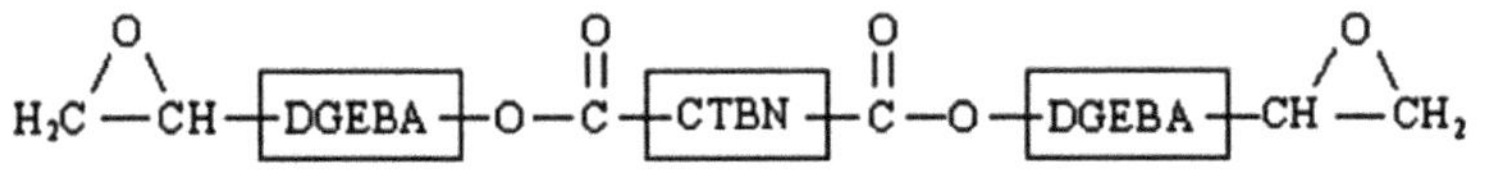

FIGURE 7.3 CTBN-epoxy adduct.[17]

The rubber system that has attracted the most attention is the family of copolymers of butadiene and acrylonitrile; these are commercially available with different acrylonitrile contents ranging from 0 to 26%. The low molecular weight (3400–4000 g/mol) butadiene–acrylonitrile rubbers are soluble in liquid DGEBA epoxy resins and can be synthesized with carboxylic acid. When a solution of rubber in epoxy is cured, rubber particles precipitate out as a second phase. With just 10 phr rubber modifier, the fracture toughness of modified epoxy resins increases dramatically with only a slight reduction in the glass transition temperature and the modulus. The purposes of adding elasticated (rubbery) agents include the following: improvements in mechanical strength, prevention of cracks due to thermal distortion, reduction of distortion, and improvements in adhesiveness, particularly improvements in peel strength by imparting elasticity to disperse stresses.

The modified epoxy resins are prepared by a simple additive method or a "pre-reacted" method. A typical "pre-reacted" preparation employs about 10 phr CTBN liquid rubber using a catalyst such as triphenylphosphine to promote the reactions between the carboxylic groups and the epoxy groups at a relatively low temperature and then further cured with other curing agents. To avoid such deteriorations in properties, a special elasticated agent, CTBN copolymer liquid rubber, may be added. CTBN has mutual solubility with epoxy resin, but does not have it with cured epoxy resin and therein forms a dispersed rubber particle phase, and serves as a cushioning material to prevent cracks as shown in Figure 7.4. Only the highly toughen epoxy resins modified with rubber toughen are known to produce an impressive toughening effect. These highly toughen ductile epoxy resins usually exhibit rather low glass transition temperatures (usually below 100°C).

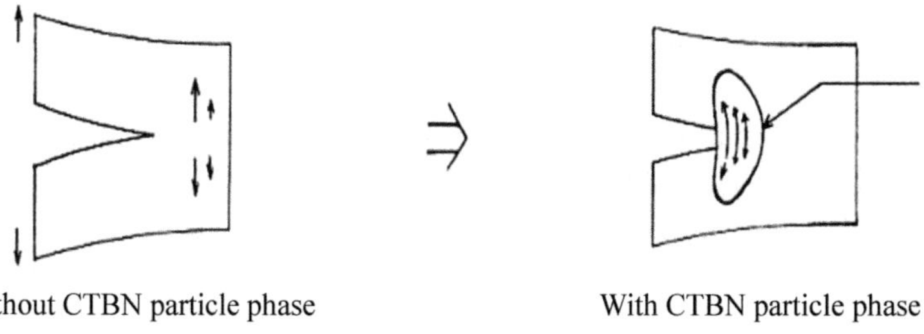

FIGURE 7.4 Effect of CTBN on polymer.[16]

7.4 POLYMER NANOCOMPOSITES FILLERS: CLAY AND CARBON NANOTUBES (CNT)

PNC is the term used for wide range of materials. The name itself suggests the polymer has been given a reinforcement of nano material to form nanocomposite. Schafer et al.[21], Sanchez[22], Ratna D[23], and others[27–31] explained in recent years, that the use of nanofiller has increased tremendously and the dramatic growth is showing every sign to continue in future. Among the various available nanofillers, clay and CNTs have been viewed as the most promising material in the area of PNC as per Meincke et al., Pötschke et al., and Ijjima.[1–5] Various types of filler materials are selected for epoxy.[17] Fillers tend to be regarded merely as bulking agents. Their roles cannot be neglected due to the fact that increasingly rigorous properties are required of epoxy resin. It is therefore necessary to select and add appropriate fillers for epoxy materials. The fillers play an important role as follows:

- improvements in mechanical strength
- reduction in thermal distortion and dimensional change
- improvements in electrical properties, particularly insulating and dielectric properties
- improvements in fire retardancy without CTBN particle phase with CTBN dispersed rubber particle phase
- improvements in heat conductivity.

For this study, clay and multiwalled carbon nanotubes (MWNTs) are selected as nanofiller materials.

7.4.1 CLAY: NANOMER I 30E

Nanomer I 30 E nanoclay is a surface modified montmorillonite mineral. It will disperse to nanoscale in epoxy resin systems. The dispersion creates a near-molecular blend commonly known as PNC. This new type of PNC exhibits enhanced strength, thermal, and barrier properties.[24] I 30E is a white powder when mixed with epoxy resin it disperses and a clear transparent PNC is prepared. Unlike conventional mineral fillers, Nanomers enhance performance at low loading also. High glass transition systems (70°C) suffer impact loss and require modifiers to achieve desired levels. Curing temperatures should be below 120°C to avoid self-polymerization.[1] Nayak[24] also studied commercial treated clay from Nanocor, USA. It contains octadecylammonium organo-ions lining the surfaces of the galleries, encouraging ingress of the monomer, and hardener. It is designed to be easily dispersed into amine cured-epoxy resins to form nanocomposites. The chemical formula of the clay used is $[CH_3\,(CH_2)_{17}\,NH_3^{(+)\,(-)}$ $[MgO.._{54}Al_3.._{38}](Si_{7.64}\,Al_{0..36})O_{20}(OH)_4]$, where $CH_3(CH_2)_{17}NH_3^{(+)}$ is the organic moiety inside the clay. It has a d-spacing of 23 Å and can contain 70–75% MMT and 25–30% of organic moiety.

7.4.2 CARBON NANOTUBES (CNTS)

Carbon—carbon covalent bond is among the strongest bonds in nature. Consequently, a material based on a suitable arrangement of these bonds can produce a strong structure by Farahani.[6] The exceptionally high aspect ratio in combination with a low density as per Li et al.[32], a high strength and stiffness as per Yu et al.[7] and Yu et al.[8] make CNT a potential candidate as reinforcement for polymeric materials. CNTs were unanticipated by product of bulk synthesis of Buckminster fullerene (C_{60}). They have intriguing electronic, magnetic, optical, and mechanical properties coupled with unusual molecular shape and size. Physicists are attracted by CNT because of its electronics properties, chemists as "nanotest tubes" material and scientists for its amazing stiffness, strength, and resilience. As graphene sheets are rolled to get CNT, to close the tips, half fullerenes or more complex structures that include pentagons are required. However, the cylindrical part of CNT, which contains only hexagons, is considered infinite in most studies. CNTs have high specific surface area.

Overall, CNTs (i.e., nanoscopically hollow fibers) are generally considered in two categories: single-walled carbon nanotubes (SWNTs) and MWNTs. SWNTs are comprised of an individual graphene sheet. As per Nalwa[16], MWNT density ranges between 1 and 2 g/cm^3. For MWNTs the measurement interlayer distance is about 0.34 nm. The structural aspects of CNTs are: tube diameters can range from 1 to 100 nm with aspect ratios (length to diameter ratio) in excess of 100 or 1000. Thermal conductivity exhibits peaking behavior as a function of temperature and strongly depends upon CNT diameter but does not or weakly depends on helicity. Compressed mats of CNT gave value of thermal conductivity: 1750–5850 Wm^{-1}°K^{-1}, as explained by Nalwa.[16]

There are various classical methods to produce CNT such as arc discharge, laser ablation, and catalytic technique, etc. Several research groups have reported the encouraging results regarding the possibility of introducing CNT for mechanical enhancement in polymer/CNT composites.[16] Figure 7.5 depicts the MWNTs.

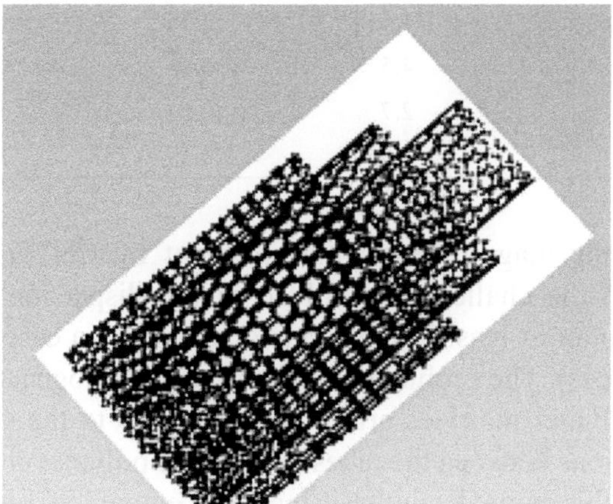

FIGURE 7.5 MWNT first observed in 1991 diameter: 4–50 nm.[5]

These materials offer improvements over conventional composites in mechanical, thermal, electrical, and barrier properties. In the case of

CNT/layered silicate (clay) nanocomposites, loading levels of 2–5 wt% result in improvement of mechanical properties similar to those found in conventional composites with 30–40 wt% reinforcing material. However, CNT-PNCs are gaining substantial attention in this decade because CNT offer opportunities to impart unique electrical and thermal properties to the polymer resin as well as enhance mechanical and physical response. The intrinsic strength of carbon—carbon sp^2 bond is expected to give CNT the highest strength and modulus among all existing whiskers. CNT have excellent elastic properties. Some of the properties of CNTs are compared with conventional fillers tabulated in Table 7.1.

In contrast to the two decades investigation of layered silicate dispersion in polymers, exploration of CNT dispersion is relatively new, hindered until recently by the limited availability of study based on CNT.

TABLE 7.1 Properties of Carbon Nanotubes Compared to Traditional Reinforcement Materials.[6]

Materials	Specific gravity (g/cm^3)	Tensile strength (GPa)	Elastic modulus (GPa)
MWNTs	2.6	1.5	850
Conventional carbon fiber	2.2	4	550
IM7 carbon composites	1.6	2.1	152
Titanium	4.5	0.9	103
Aluminum	2.7	0.5	69
Carbon steel	7.85	0.19–0.758	190–220

However, the magnitude of reinforcement by CNT is found to be limited due to the challenges involved in the dispersion of CNT in a molecular level or at least in the forms of "ropes" (in case of SWNTs in polymer matrices). The problem associated with homogeneous dispersion of CNT in polymer matrices arises primarily due to the strong Van der Waals interactions between the tubes making the polymer chain difficult to intercalate. In this context, synthesis induced "entangled" and "aggregated structure" of CNT magnifying the problem of homogeneous dispersion. Coupled with the issue of homogeneous dispersion of CNT in polymer matrices, poor interfacial adhesion between CNT and polymer is another critical parameter, which dictates the efficiency of load transfer in the CNT-based polymer composites. Hence, further modification of the CNT is a must before considering as filler materials for PNC.

7.4.2.1 MODIFICATION OF MWNT

Jogi et al.[25,27–29] and Jagtap et al.[14] explained several studies have been done to overcome the problem of poor dispersion such as ultrasonication, high sear mixing, polymer chain wrapping, surface oxidation (covalent functionalization) by acid treatments, and use of surfactant. Surface oxidation of MWNT (covalent functionalization) by acid is an effective way to improve the dispersion of MWNT by generating polar groups. These polar groups can further be used to tailor the surface using covalent grafting method. The disadvantage of this method is that it destroys the π-electron clouds of MWNT to ascertain the extent and reduces the inherent conductivity of MWNT.

As illustrated by Gojny et al.[15] MWNTs produced by arc-discharge method, were treated with oxidizing inorganic acids. The surface modification of the oxidized nanotubes was achieved by refluxing the tubes with multifunctional amines. The functionalized nanotubes were embedded in the epoxy resin and the resulting composite was investigated by transmission-electron microscopy (TEM). The functionalization led to a reduction in agglomeration and evidences are given for improved interaction between the CNT and the epoxy resin. MWNTs are also planned to incorporate in epoxy-CTBN matrix. It may enhance the mechanical properties, modulus, storage modulus, and effect on glass transition temperature, etc. To avoid agglomeration of MWNT probe ultrasonication is performed along with mechanical stirring.

7.4.2.1.1 Half Neutralized Adipic Acid (HNAA) Modified MWNT

HNAA was used as modifier[14] in 1:1 wt% ratio. MWNT and HNAA (1:1 wt%) ratio was mixed along with Millipore (distilled) water in beaker. Then the mixture is ultrasonicated for 15 min at amplitude 90%. Then it is heated on a hot plate to evaporate the water and made it completely dry. Prepared modified MWNT was preserved in air tight glass bottle.

7.4.2.1.2 Sodium Salt of Hexanoic Acid (Na-AHA) Modified MWNT

Na-AHA was used as modifier[26] in 1:1 wt% ratio. MWNT and Na-AHA (1:1 wt%) ratio was mixed along with Millipore (distilled) water in beaker.

Then mixture is ultrasonicated for 15 min at amplitude 90%. Then it is heated on hot plate to evaporate the water and made it completely dry. Prepared modified MWNT was preserved in air tight glass bottle.

7.5 PREPARATION OF POLYMER NANOCOMPOSITES (PNC)

The synthesis of PNCs is an integral aspect of polymer nanotechnology. By inserting the nanometric inorganic filler compounds, the properties of polymers improve and hence PNC has multifunctional applications depending upon the inorganic material present in the polymers. There are a few methods commonly used to incorporate CNT into polymer matrix to prepare PNC:

- film casting of suspensions of CNTs in dissolved polymers
- in situ polymerization of monomer in the presence of CNT and
- melt mixing of CNTs with polymers.

For the nanocomposites of epoxy/CTBN in presence of clay and CNT, epoxy resin was supplied by Ciba-Atul Ltd., Ahmedabad, under the trade name LY 556. It has viscosity: 10,000–20,000 MPa-s at 25°C, density: 1.15–1.20 g/cm^3 at 25°C, and equivalent weight 195 ± 5 used for further study. Epoxy resin is having 1:1 stoichiometric ratio. CTBN was supplied by CVC thermoset species, under the trade name as 1300X13 CTBN. It has Brooksfield viscosity of 360,000–640,000 at 27°C, specific gravity: 0.960 at 25°C, and molecular weight: 3150. Nanomer I 30 E, nanoclay is a surface modified montmorillonite mineral. Clay was supplied by Nanocor, USA. It has specific gravity: 1.71, density: 0.41 g/cm^3, mean dry particle size: 8–10 μ, and thickness: less than 10 Å. MWNT is available in black powder are supplied by Bayer baytubes, Germany. It has diameter of 5–100 nm, length of 10 μ, density of 1–2 g/cm^3, and 18 m^2/g surface area.

The modified epoxy resins are prepared by a simple additive method or a "pre-reacted" method.[11] Different phr of CTBN are prepared by mixing CTBN and epoxy resin and 0, 5, 10, 15, and 20 phr are the combinations planned to prepare modified epoxy, also known as rubber toughened epoxy. 20 phr sample has 100 g of epoxy; 20 g of CTBN and 0.5 g of triphenyl phosphine (TPP) to promote the reactions between the carboxylic groups and the epoxy groups at relatively low temperature are mixed together in a three neck round bottom flask. However, TPP acts as catalyst having about 0.5 wt% of epoxy. This process of modifying epoxy is also known as pre-reaction. A mechanical stirrer for three necks round bottom flask along with potentiometer is used to get the regulated speed of rotation. Temperature around 100°C is maintained during the completion of the reaction in the flask. The reaction is carried out for 3 h by maintaining the constant temperature at 100°C. The same procedure is followed to prepare all phr compositions. To check the completion of the reaction of epoxy and CTBN, acid value is calculated by following formula: Acid value = B. R. X (normality of NaOH or KOH/weight of sample in g) where, B. R. stands for burette reading. If acid value is found zero, it meant that reaction is over. Clay and MWNTs are also planned to incorporate in epoxy-CTBN matrix. To avoid agglomeration of nanofillers probe ultrasonication is performed along with mechanical stirring. HNAA and Na-AHA was used as modifier[11,14] in 1:1 wt% ratio. The mixture is ultrasonicated for 15 min at amplitude 90%. Then, it is heated on hot plate to evaporate the water and made it completely dry. Prepared modified MWNT was preserved in air tight glass bottle.

For this study, Jeffamine 300 is selected as a curing agent as it has room temperature processing capability. It also imparts improvement in the toughness and impact stress.[33] It is supplied by Sigma-Aldrich. It has molecular weight as 400 and 1:1 mol ratio with epoxy. Furthermore, the curing time for clay and MWNT was also decided as 2 and 4 days, respectively. This operation is followed by post curing at 100°C for 6 h for MWNT. The dog bone shaped tensile samples were cut as per ASTM D638 standards and tested on universal testing machine (UTM).

7.6 TESTING OF POLYMER NANOCOMPOSITES

While there has been a great amount of experimental work that has taken place in the area of PNCs, a consensus has not yet been reached on how nano-sized inclusions affect mechanical properties. Several studies have shown that reduced size improves mechanical properties, specifically elastic modulus. These studies vary in polymer nanocomposite systems and the mechanical properties characterized. The studies propose various theories to explain their results. Various characterization techniques are available, which will give the mechanical properties, the behavior of material under heating stressed condition, dispersion of filler within the mixture. The characterization techniques are described in the following sections.

7.6.1 TENSILE TESTING

Tensile testing is a destructive characterization technique. Tensile testing will be performed to determine elastic modulus, ultimate stress, and ultimate strain. In tensile testing, a "dog bone" shaped sample is placed in the grips of movable and stationary fixtures in a screw driven device, which pulls the sample until it breaks and measures applied load versus elongation of the sample. The testing process requires specific grips, load cell and extensometer for each material and sample type. The load cell is a finely calibrated transducer that provides a precise measurement of the load applied. The extensometer is calibrated to measure the smallest elongations. Output from the device is recorded in a text file including load and elongation data. Elongation is typically measured by the extensometer in volts and must be converted to millimeters. Mechanical properties are determined from a stress vs. strain plot of the load and elongation data. Tensile properties were determined by using UTM. UTM (Hounsfield Tinius Olsen), 10 kN gives the result of stress–strain curve. Test specimen dimensions and tolerances as per standard ASTM D638 Gauge length: 50 mm, width: 13 mm, and thickness: 5 mm.

7.6.2 DYNAMIC MECHANICAL ANALYSIS (DMA)

DMA, like tensile testing, is a destructive technique. Dynamic mechanical analysis (DMA) will be used to gather elastic modulus data for all six systems. DMA determines elastic modulus, loss modulus, and damping coefficient as a function of temperature, frequency, or time. The approach is often used to determine glass transition temperature, as well. Samples in DMA, depending on the equipment, can be quite small, in the range of $40 \times 5 \times 1$ mm. The sample is clamped into movable and stationary fixtures and then enclosed in a thermal chamber. The DMA applies torsional oscillation to the sample while slowly moving through the specified temperature range. Experimental inputs into the equipment include frequency and amplitude of oscillations, static initial applied load, and temperature range. Results are typically recorded as a graphical plot of elastic modulus, loss modulus, and damping coefficient versus temperature. DMA was carried out using Gabo in the bending mode and the furnace will heat at 2 K/min from −10 to 65°C, with a frequency of 1 Hz. The furnace had flushed with liquid nitrogen throughout the measurements to prevent oxidative degradation of the samples.

7.6.3 DIFFERENTIAL SCANNING CALORIMETRIC (DSC) ANALYSIS

Differential scanning calorimetric (DSC) analysis is the most widely used technique of all the thermal analysis methods (most of all in purity measurements, for example, in polymer or pharmaceutical industry). There are two different types of DSC methods, namely, power compensated DSC and heat flux DSC, in which, although the same properties are measured, differ in their instrumentation. The apparatus used in the practical course was a heat flux DSC machine only. In a heat flux DSC machine the substance which shall be measured is placed in an aluminum pan whereas an empty aluminum pan serves as the reference. These two pans are put on an electrically heated plate in order to make sure that the temperature in the sample is the same as in the reference. These plates and the pans, transferred heat to the sample and the reference with the use of a defined computer controlled heating program (the rates of heating can be adjusted). The differential heat flux to the sample and the reference as well as the sample temperature are measured. The output of a DSC measurement is called as thermogram, which is a plot of difference of heat delivered to the sample and to the reference as a function of the sample temperature.

If a physical or a chemical process which is endothermic (consuming energy as heat) is taking place, in order to maintain the same temperature of the two pans, more heat must be delivered to the sample pan than to the reference pan (where of course no transition occurs). The effect is a positive or negative peak in the thermogram (the sign of the peak is depending on the definition of the sign for the direction of the heat flow). At some point, the molecules may obtain enough freedom of motion to spontaneously arrange themselves into a crystalline form. This is known as the crystallization temperature (T_c). This transition from amorphous solid to crystalline solid is an exothermic process, and results in a peak in the DSC signal. As the temperature increases the sample eventually reaches its melting temperature (T_m). The melting process results in an endothermic peak in the DSC curve. The ability to determine transition temperatures and enthalpies makes DSC a valuable tool in producing phase diagrams for various chemical systems. TA Instrument Q100 machine was used to carry out the DSC tests at 10°C and ramp with temperature between −40 and 120°C.

7.6.4 THERMOGRAVIMETRIC ANALYSIS (TGA)

Thermogravimetric analysis (TGA) was used to determine the changes in polymer decomposition temperatures between the six samples to help determine the thickness of the polymer layer surrounding the nanoparticles. TGA continuously measures the weight of a sample as a function of temperature and time. The sample is placed in a pan held in a microbalance. The pan and sample are heated in a controlled manner and weight is measured throughout the heating cycle. Changes in weight at specific temperatures correspond to reaction or changes in the sample such as decomposition. The weight loss experienced during the decomposition experiment corresponds to the amount of polymer that was attached to the particles in the sample. All the samples were tested on TA instruments machine from 0 to 800°C maintaining 20°C ramp.

7.6.5 X-RAY DIFFRACTION (XRD) ANALYSIS

X-ray diffraction analysis (XRD) has played a central role in identifying and characterizing solids. The nature of bonding and the working criteria for distinguishing between short-range and long-range order of crystalline arrangements from the amorphous substances are largely derived from XRD analysis and thus it remains as a useful tool to obtain structural information. XRD analysis is a useful tool for determining the structure of PNSs. When there is a peak it means clay stacks present have a change in their "d" spacing or gallery spacing. Due to its easiness and its availability, XRD analysis can only detect the periodically stacked MMT layers; disordered or exfoliated layers are not detected. XRD analysis of the samples is carried out in Rigaku Ultima III diffractometer (40 KV, 40 mA) with Cu K α source ($\lambda = 1.54$ Å) to analyze the exfoliation characteristics. The scanning 2θ range is 1.5–10° with a scanning range of 1°/min.

7.6.6 SCANNING ELECTRON MICROSCOPIC (SEM) ANALYSIS

SEM analysis will be used to determine particle size and distribution of the filler in the fractured surfaces. The SEM consists of an electron gun producing a source of electrons at an energy range of 1–40 keV. Electron lenses reduce the diameter of the electron beam and place a small focused beam on the specimen. The electron beam interacts with the near-surface region of the specimen to a depth of about 1 μm and generates signals used to form an image. The smaller the beam size, the better the resolution of the image. The smaller the beam size, the less is the current available to form a clear picture. Operating the SEM requires fine-tuning to optimize picture quality with resolution. SEM is run under a vacuum to minimize beam interactions with gas molecules which would retard resolution. Non-conductive specimens, such as most polymers, often suffer from variations in surface potential which introduce astigmatism, instabilities, and false X-ray signals. Charging a condition, during which charge accumulates on the surface of a non-conducting specimen causing excessive brightness, often occurs making it difficult to obtain quality images. Sputter coating non-conductive samples with a fine gold layer is often required to avoid these issues.

7.7 RESULTS AND DISCUSSION

The mechanical behavior of blends is strongly decided by the contribution of each component (polymer and filler), the phase morphology, and the interfacial adhesion between the phases. It is well realized that elongation at break and toughness are the important tools to monitor the adhesion between phases (epoxy and CTBN) whereas the tensile strength is related to the morphology. All the compositions 0, 5, 10, 15, and 20 phr were made in presence and absence of clay.

7.7.1 TENSILE PROPERTIES OF EPOXY/CTBN/CLAY

An efficient way to toughen epoxy is to blend with rubbery phase CTBN. It is optimized as shown in Table 7.2 (0–20 phr). The toughening effect was also taken into consideration to optimize rubber concentration. To investigate the effect of clay reinforcement on the mechanical properties of the epoxy/CTBN matrix, tensile testing has been carried out. The effects of blending sequence on the mechanical behavior of the blends have also been investigated.

TABLE 7.2 Tensile Test Results of Varying CTBN Content in Epoxy as phr System.[28]

CTBN in epoxy (phr)	Modulus (MPa)	Tensile strength (MPa)	Elongation at break (%)
0	3000	28	26
5	2500	20	54
10	1301	16	56
15	1506	20	58
20	1090	15	60

(Reprinted from Jogi, B. F.; Kulkarni, M.; Brahmankar, P. K.; Ratna, D. Some Studies on Mechanical Properties of Epoxy/CTBN/Clay Based Nanocompoaites (PNC). Proceedings of International Conference on Advances in Manufacturing and Materials Engineering (AMME2014), Procedia Material Science, 2014, 5, 787–794. © 2014 with permission from Elsevier.)

Different phr compositions of epoxy and CTBN were made by varying CTBN content through prereaction. These compositions were tested on Hounsfield UTM, 10 KN. It is found that modulus was decreasing with increasing percentage of rubber CTBN. It may be due to the rubbery amorphous phase addition in the epoxy polymer matrix. At the same time, tensile strength also was found to be decreasing. As rubber percentage increases elongation is also found to be increased due to increase in the rubbery content.

Figure 7.6 depicts higher elongation at break along with modulus for the 15 phr combination and hence may be opted as a suitable combination for mechanical properties.

The stress–strain graph is shown in Figure 7.6, 0 phr is showing highest stress having a brittle nature of the material. Addition of CTBN rubber decreases the brittle nature in all compositions; however, 15 phr showed the smooth curve.

In Figure 7.6, it is also observed that 10 and 20 phr are showing good improvements in the ductility. Hence, further it is planned to add nanofiller like clay for all compositions. Through literatures[25–27] it is observed that addition of 2–3 wt% of nanofiller shows considerable improvement in mechanical properties. Hence, for this study 3 wt% of clay is planned to understand the behavior of the epoxy/CTBN materials. It is observed that enhancement in tensile properties was high in case of 15 phr material after loading 3 wt% clay in the varying content of CTBN in epoxy. It is also observed that there was an improvement in modulus as well as in tensile strength as compared to plain phr system after addition of clay as shown in Table 7.3.

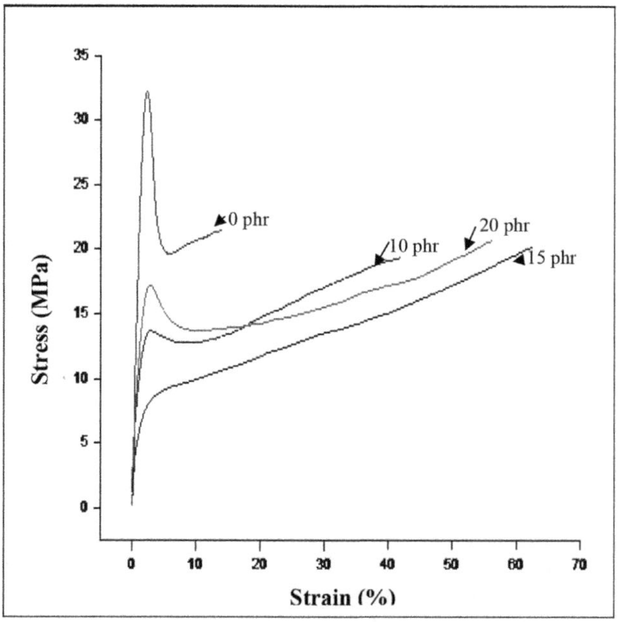

FIGURE 7.6 Stress–strain relationship of epoxy and CTBN blends with varying phr.

TABLE 7.3 Tensile Test Results of Varying CTBN Content in Epoxy System with 3 wt% of Clay.

CTBN in epoxy (parts/hundred)	Modulus (MPa)	Tensile strength (MPa)	Elongation at break (%)
0 phr + 3 wt% clay	3022	40	3
5 phr + 3 wt% clay	2805	24	3
10 phr + 3 wt% clay	2708	19	4
15 phr + 3 wt% clay	2810	28	25
20 phr + 3 wt% clay	2762	23	27

The clay filled epoxy/CTBN composites shows improved mechanical properties. Addition of rubber increases the toughness, further resulting in tensile property improvement. However, it is found that clay increases the modulus of the epoxy/CTBN material. It is also found that 15 phr composition has shown improvement in the storage modulus as well as strength as compared to other compositions as exhibited in Figure 7.7. Hence, it is planned to vary clay percentage by 2, 3, 4, and 5 wt % in the epoxy/CTBN blends.

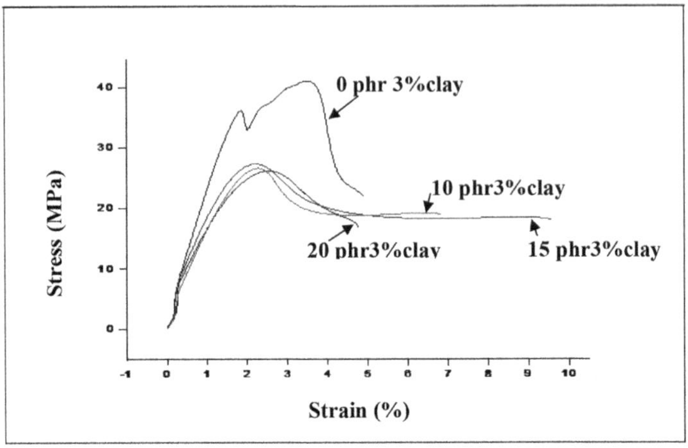

FIGURE 7.7 Stress–strain relationship of epoxy and CTBN blends with varying phr in the presence of 3 wt % of clay.

When clay percentage is varied from 2 to 5 wt%; the composition having 3 wt% clay is showing good tensile strength and modulus as compared to all other compositions as shown in Table 7.4. Tensile strength was found to be the highest in value compared to other combinations, which obtained maximum at 15 phr in presence of 3 wt% clay. For all other compositions modulus also decreases after addition of 4 and 5 wt% clay. It may be due to the agglomeration of the clay. Elongation seems to be decreasing as clay percentage is increased; however, it was found to be maximum at low percentage. It seems that dispersion of clay may be good in composition of 3 wt% clay as compared to all other compositions, hence showing improved mechanical properties.

TABLE 7.4 Tensile Test Results of Varying Clay Content in Epoxy/CTBN 15 phr System.[28]

Composition	Modulus (MPa)	Tensile strength (MPa)	Elongation at break (%)
15 phr 2 wt% clay	2583	25	32
15 phr 3 wt% clay	2810	28	25
15 phr 4 wt% clay	2533	26	13
15 phr 5 wt% clay	2515	21	12

(Reprinted from Jogi, B. F.; Kulkarni, M.; Brahmankar, P. K.; Ratna, D. Some Studies on Mechanical Properties of Epoxy/CTBN/Clay Based Nanocompoaites (PNC). Proceedings of International Conference on Advances in Manufacturing and Materials Engineering (AMME2014), Procedia Material Science, 2014, 5, 787–794. © 2014 with permission from Elsevier.)

As clay reinforcement increases beyond a certain limit, the tensile properties start decreasing, it is found after 3 wt% clay addition as shown in Figure 7.8. It is also observed that properties are improving till clay wt% reaches from smaller wt to 3 wt%. In all epoxy/CTBN phr compositions, in presence of 5 wt% clay, no improvement is found in mechanical properties.

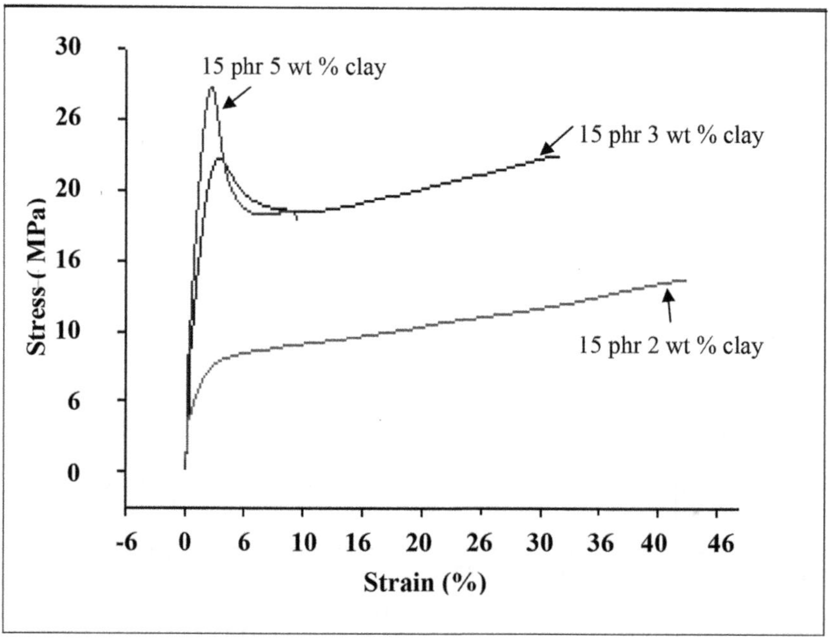

FIGURE 7.8 Stress–strain relationship between epoxy and CTBN blends with 15 phr in presence of varying clay wt% (2, 3, and 5 wt% of clay).

7.7.2 DYNAMIC MECHANICAL ANALYSIS (DMA) OF EPOXY/CTBN/CLAY

Further analysis of the epoxy/CTBN/clay PNCs was carried out by performing test on DMA. Neat, 15 phr epoxy and 15 phr epoxy samples are tested for DMA in presence of 2–5 wt% of clay. Figure 7.9 depicts the graph of effect on storage modulus with varying temperatures from −60 to 60°C. Storage modulus decreases as clay percent is increased. In case of glass transition temperature (T_g), it is found to be increasing on the addition of clay. Storage modulus of pure epoxy is higher than all other composition as it was showing brittle nature. Storage modulus of 15 phr, 3 wt% of clay is found to be more amongst all samples. Effect of toughening shows change in glass transition temperature. Also the addition of optimized wt% of clay shows better improvement as compared to pure epoxy and epoxy/CTBN rubber. Further good exfoliation of clay resulted in increment in storage modulus.

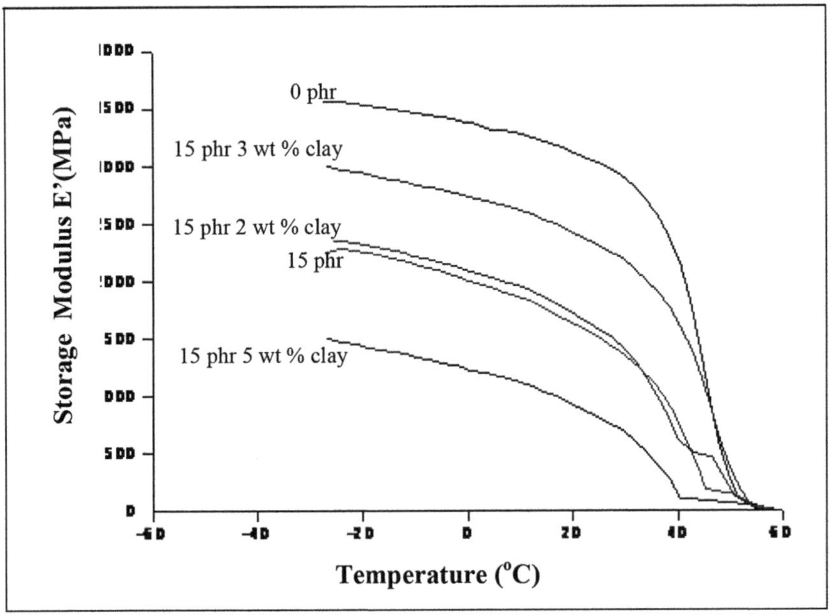

FIGURE 7.9 Graph showing change in storage modulus as wt% of clay varied. (Reprinted from Jogi, B. F.; Kulkarni, M.; Brahmankar, P. K.; Ratna, D. Some Studies on Mechanical Properties of Epoxy/CTBN/Clay based Nanocompoaites (PNC). Proceedings of International Conference on Advances in Manufacturing and Materials Engineering (AMME2014), Procedia Material Science, 2014, 5, 787–794. © 2014 with permission from Elsevier.)

It is also observed that clay samples reinforcement shows higher loss modulus as compared to neat epoxy and epoxy-CTBN samples. But 15 phr with 5 wt% clay is showing reduced loss moduli than neat epoxy. It may be due to agglomeration of clay stacks in PNC. Loss modulus is greater for 15 phr, 3 wt% clay shows good interaction of clay particles with matrix. As we observed in 15 phr, 5 wt% clay sample, it is showing reduced loss modulus due to high wt% of filler. To investigate the agglomeration of clay, XRD testing was performed.

7.7.3 X-RAY DIFFRACTION (XRD) ANALYSIS OF EPOXY/ CTBN/CLAY

XRD of pure clay nanomer I 30 E is showing a peak at 3.93 Å as depicted in Figure 7.10. Furthermore, clay samples are also investigated and found that the clay peak is at the same point or had moved forward means increased at a small extent. However, 0 and 15 phr epoxy/CTBN/clay nanocomposites samples XRD analysis depicts no peak of the clay as per Figure 7.11, which may be observed at the 3.93 Å for pure clay. It shows pure clay was giving broad diffraction peak at 3.93°. Furthermore, no peaks are found for epoxy/CTBN/clay samples. It indicated that nanoclay is highly exfoliated. It may be the indication of highly intercalated or exfoliated clay stacks in the epoxy/CTBN/clay nanocomposites. Furthermore, it may be concluded that due to the exfoliation good dispersion is achieved resulting in improved tensile properties for epoxy/CTBN/clay nanocomposites. So it can be concluded that the interaction of clay surface was good with epoxy matrix. Further, DSC and SEM may also reveal the extent of dispersion of clay in PNC. Hence, T_g and morphological behavior study of the epoxy/CTBN/clay nanocomposites is carried out.

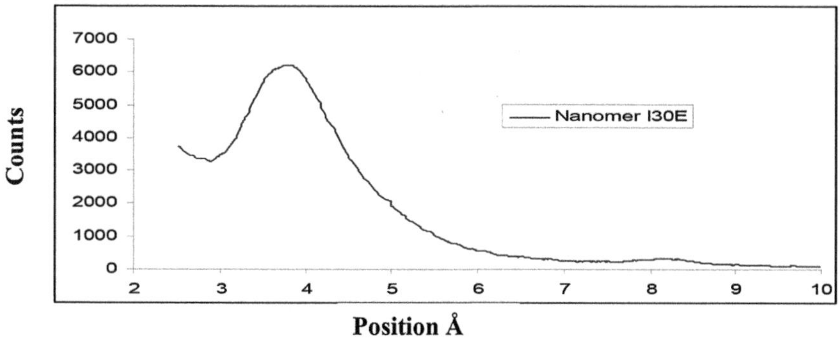

FIGURE 7.10 XRD of nanomer I 30E (clay). (Reprinted from Jogi, B. F.; Kulkarni, M.; Brahmankar, P. K.; Ratna, D. Some Studies on Mechanical Properties of Epoxy/ CTBN/Clay Based Nanocompoaites (PNC). Proceedings of International Conference on Advances in Manufacturing and Materials Engineering (AMME2014), Procedia Material Science, 2014, 5, 787–794. © 2014 with permission from Elsevier.)

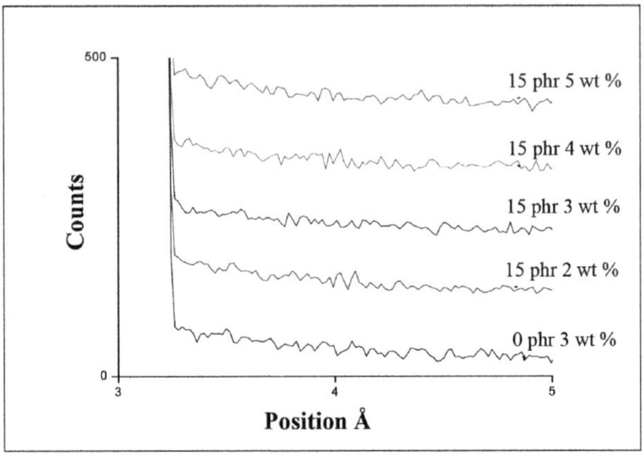

FIGURE 7.11 XRD of 0 and 15 phr epoxy/CTBN samples with varying wt% of clay.[28]

7.7.4 DIFFERENTIAL SCANNING CALORIMETRIC (DSC) ANALYSIS OF EPOXY/CTBN/CLAY

DSC analysis was carried out on selective samples of 0 and 15 phr in presence and absence of 3 wt% clay as shown in Figure 7.12. It was also found that there was an increment in T_g of the system after adding 3 wt% clay. It

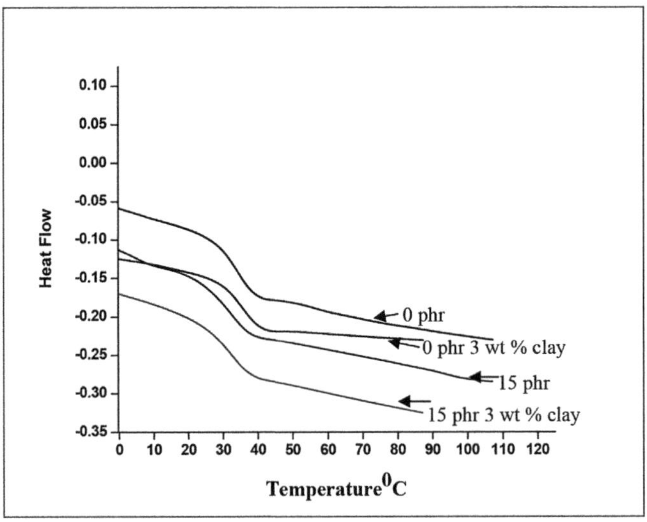

FIGURE 7.12 Comparison of T_g by DSC analysis of 0 and 15 phr with and without clay.

can be clearly seen that increased concentration of rubber decreases the T_g of the system. T_g obtained from DSC and DMA was compared as shown in Table 7.5. It is found that the trend obtained was same in both DSC and DMTA. The values may change from instrument to instrument. The structural defect may result in a decrease in T_g of the system. Hence, the maximum value is considered. It is also observed that the difference from both the testing in T_g is ranging between 10 and 20°C. This clearly reveals the selected system was a low T_g system verified by DSC and DMA.

TABLE 7.5 Comparison between T_g Obtained from DSC and DMA.[28]

Composition	T_g measured by DSC	T_g measured by DMA
0 phr	34.65	49.6
15 phr	31.93	46.8
0 phr + 3 wt% clay	35.91	53.8
15 phr + 3 wt% clay	32.60	49.9
0 phr + 5 wt% clay	34.70	58
15 phr + 5 wt% clay	32.54	49.8

7.7.5 SCANNING ELECTRON MICROSCOPIC (SEM) ANALYSIS OF EPOXY/CTBN/CLAY

SEM picture of fracture surface for pure epoxy shows smooth surface with clear and long fracture lines. Addition of CTBN rubber shows black dots in smooth epoxy surface as shown in Figure 7.13. It proved that there was a phase separation between epoxy and CTBN rubbery phase that depicts particle CTBN dispersed phase (epoxy) morphology. Rubber remains in epoxy matrix, but does not enter into epoxy layers. In presence clay, it shows the formation of interconnected cracks. It may result that adhesion between clay and polymer matrix gets improved. It is believed that clay particles restrict or hindered the crack propagation during fracture. But since clay particles are not distributed uniformly in 5 wt% clay, fracture strength of polymer nanocomposite was found to be decreased. However, 3 wt% of clay composition sample shows more uniform dispersion having better properties due to better clay distribution. It is also found that in presence of clay number of fracture lines increases, which produces a very rough surface of PNC. It seems addition of clay makes system brittle because silicate layers come between the epoxy chains and decrease interlinking. Furthermore, in presence of clay in 15 phr epoxy/CTBN matrix, black dots were seen to be filled having good surface finish, higher T_g, and improved mechanical properties as explained early.

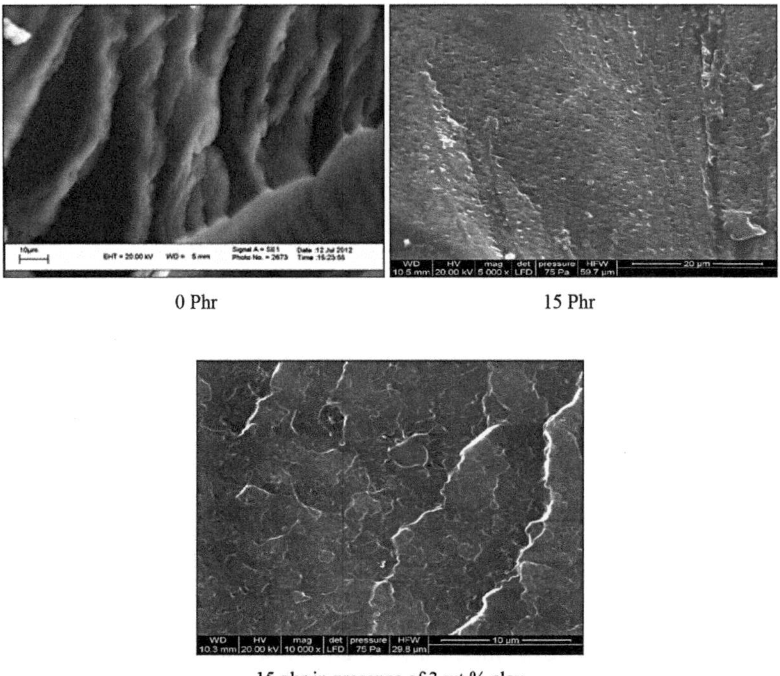

0 Phr 15 Phr

15 phr in presence of 3 wt % clay

FIGURE 7.13 SEM showing phase separation between pure epoxy and epoxy with rubber and clay. (Reprinted from Jogi, B. F.; Kulkarni, M.; Brahmankar, P. K.; Ratna, D. Some Studies on Mechanical Properties of Epoxy/CTBN/Clay Based Nanocompoaites (PNC). Proceedings of International Conference on Advances in Manufacturing and Materials Engineering (AMME2014), Procedia Material Science, 2014, 5, 787–794. © 2014 with permission from Elsevier.)

7.7.6 THERMOGRAVIMETRIC ANALYSIS (TGA) OF EPOXY/CTBN/CLAY

To evaluate the thermal stability of neat epoxy, epoxy/CTBN (rubber) and epoxy/CTBN/clay nanocomposites TGA was carried out under nitrogen environments and the results are presented in Table 7.6. Pure epoxy exhibits an onset of thermal degradation at ~328°C which is less than epoxy/CTBN composition having 15 phr. Degradation temperature of the samples decreases as the clay wt% is increased. No significant effect of the addition of rubber is found on the degradation temperature. TGA values for epoxy/CTBN/clay nanocomposites exhibits a decrease in the thermal degradation temperature due to reinforcement.

TABLE 7.6 Comparison of Degradation Temperatures.

Composition	Onset of degradation temperature (°C)	Maximum degradation temperature (°C)	Complete TGA weight loss at 500 °C (%)
0 phr	328	450	8
15 phr	332	480	10
15 phr 3% clay	306	628	12
15 phr 5% clay	295	560	16

It concluded that even the composition having low T_g exhibit higher degradation temperature as per TGA. Furthermore, storage modulus at elevated temperature sustains the material as per DMA. Generally, it is known that, regardless of the type of clay reinforcement, the thermal stabilities of the nanocomposites are improved as compared with the neat polymer blends. The improvement of thermal stability of the nanocomposites has been attributed to the inorganic nature of the clay particles and high thermal stability. After analyzing the experiment with clay as nanofiller in epoxy/CTBNPNC, furthermore study is planned to understand the behavior of other nanofiller. MWNT as a current decade modern filler material and its effect on the properties of epoxy/CTBN blends has been carried out.

7.7.7 TENSILE PROPERTIES OF HALF NEUTRALIZED ADIPIC ACID (HNAA) MODIFIED MWNT

MWNTs are found as a better filler material for PNC. It has challenges such as agglomeration and entanglement. To overcome these challenges, HNAA was used as a modifier which may help to reduce the agglomeration. As per the previous results reported, pure MWNTs does not exhibit enhancement in tensile properties, as a result of agglomeration as shown in Table 7.7. A very poor dispersion of pure MWNT in polymer matrix may deteriorate the mechanical properties of PNC. Hence, surface modification with functional group is performed for further study. HNAA has showed a little improvement than pure epoxy/CTBN samples in tensile properties. Results are tabulated as shown in Table 7.7.

TABLE 7.7 Tensile Test Results of MWNT Modified by HNAA.

Composition (MWNT 0.5 wt%)	Tensile strength (MPa)	Modulus (MPa)	Elongation at break (%)
0 phr MWNT:HNAA 1:1	28	3900	8
5 phr MWNT:HNAA 1:1	23	2890	8
10 phr MWNT:HNAA 1:1	21	2787	14
15 phr MWNT:HNAA 1:1	24	2972	26
20 phr MWNT:HNAA 1:1	17	2120	28

Previous reports[27,28,30,31] showed that MWNT does not have good interaction with epoxy/CTBN matrix. HNAA modified MWNT shows enhancement in tensile properties. Out of all compositions 15 phr combination proved to be the better one with HNAA modified MWNT. It may be due to the dispersion of MWNTs in 15 phr and may be due to the good interaction with MWNT. It is also observed that tensile strength is gradually decreasing with a decrease in the modulus as shown in Table 7.7 and sudden increase at 15 phr is observed along with an increase in modulus. Modulus is found to be decreased as rubber concentration is increased. Elongation may increase as an effect of rubber content is observed. The highest increment in 15 phr is observed and hence selected for further experimentation.

7.7.8 TENSILE PROPERTIES OF EPOXY/CTBN/SODIUM SALT OF HEXANOIC ACID (NA-AHA) MODIFIED MWNT

Further, Na-AHA is also tested as another modifier. It is found that the tensile strength is decreasing up to 10 phr, further in 15 phr composition enhancement of tensile strength may be due to improved adhesion of filler MWNT with the matrix epoxy/CTBN. Modulus was found to be decreasing till 10 phr and further sudden increase in modulus is observed at 15 phr. It is observed that elongation at break is increasing as a specific trend is shown in Table 7.8. It is also concluded that the elongation at break is comparatively less in Na-AHA than HNAA.

TABLE 7.8 Tensile Results of MWNT Modified by Na-AHA.

Composition (MWNT 0.5 wt%)	Tensile strength (MPa)	Modulus (MPa)	Elongation at break (%)
0 phr Na-AHA:MWNT (1:1)	39	3976	3
5 phr Na-AHA:MWNT (1:1)	32	3402	4
10 phr Na-AHA:MWNT (1:1)	24	3201	5
15 phr Na-AHA:MWNT (1:1)	33	3385	8
20 phr Na-AHA:MWNT (1:1)	20	2400	11

The interaction of modified MWNT with matrix and dispersion may be good in 15 phr and hence, shows enhanced tensile properties as compared to 15 and 15 phr MWNT:HNAA. Modulus of 0 phr is high but that is a brittle material. Hence, it is planned for tough material. Elongation trend was same as it was obtained in clay and HNAA modified MWNT. It is found to be increasing with increase in rubber concentration. Furthermore, it is found that there was good tensile property enhancement in 15 phr Na-AHA modified MWNT at 0.5 wt%. Hence, it is planned to optimize the wt% of Na-AHA modified MWNTs to get the maximum increment in tensile properties.

Na-AHA modified MWNTs of 0.1, 0.25, 0.5, 0.75, and 1 wt% in 15 phr system optimization study was carried out. The results obtained were the best one with 0.5 wt% Na-AHA modified MWNTs in 15 phr composition as shown in Table 7.9. It is observed that the tensile strength increases on increasing MWNT content addition up to 0.5 wt% and then starts to decrease may be due to agglomeration of MWNT. Same results are observed in the case of modulus. For elongation at break, it is decreasing as the wt% of reinforcement increased. It may be due to the chain breakage in the presence of Na-AHA content.

As shown in Table 7.9, tensile property improvement is clearly observed of Na-AHA samples over HNAA samples. It shows improvement in tensile properties to a great extent at 0.5 wt% MWNT composition.

TABLE 7.9 Tensile Results of MWNT Modified by Na-AHA Optimization.

Composition	Tensile strength (MPa)	Modulus (MPa)	Elongation at break (%)
15 phr	20	1506	58
15 phr Na-AHA:MWNT 0.1 wt% (1:1)	20	1902	42
15 phr Na-AHA:MWNT 0.25 wt% (1:1)	21	2014	33
15 phr Na-AHA:MWNT 0.5 wt% (1:1)	33	3385	8
15 phr Na-AHA:MWNT 0.75 wt% (1:1)	29	2971	6
20 phr Na-AHA:MWNT 1 wt% (1:1)	25	2809	5

MWNT:Na-AHA is added to epoxy/CTBN with different phr. This proved Na-AHA salt gives good improvement in tensile properties over HNAA salt. The ratio of MWNT to modifier (by weight) was maintained as 1:1. It was observed that all composites containing modified MWNT show higher tensile properties compared to neat epoxy and epoxy/CTBN, it is interesting to note that an improvement in tensile property is achieved in all aspects. This can be attributed to the increase in the level of dispersion of MWNT due to the addition of modifier and may be good interfacial adhesion with epoxy/CTBN as a result of the modification. The good dispersion of MWNT is believed to originate from cation–π interaction, cation of the modifier with π-electron clouds of MWNT. The unmodified MWNT composite does not show significant reinforcing effect due to poor dispersion as a result of agglomeration. As shown in Figure 7.14, Na-AHA modifier exhibits the best tensile property compared to other modifiers with HNAA. This can be explained by considering two factors namely polarity matching and the formation of the salts with the phr system. The formation of dicarboxylate is more for acids with higher number of carbon atoms. For achieving debundling and interfacial adhesion the formation of salt with one carboxylate and one carboxylic/sodium acid group is most desirable. The best performance of Na-AHA salts compared to HNAA salt is a result of combined effect of polarity matching and dicarboxylate formation. Further, a series of composites using various concentrations (wt%) of the (1:1) Na-AHA-modified MWNT is carried out to study the possible improvement in tensile property without any agglomeration. The tensile properties (tensile strength, tensile modulus, and elongations at break) increase with an increase in concentrations upto 0.5 wt% and thereafter decreases. The maximum value of tensile strength and tensile modulus is found to be 33 and 3385, respectively. This improvement in the tensile property can be attributed to uniform distribution of nanotubes and the strong interaction between epoxy/CTBN and modifier. The decrease in tensile property beyond the 0.5 wt% of MWNT is evident from microscopic studies which will be discussed in subsequent section.

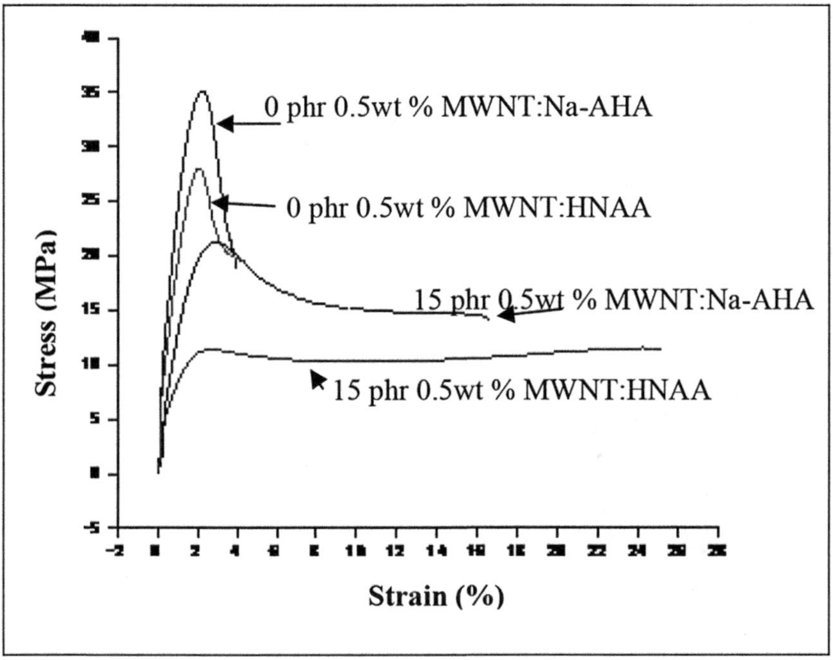

FIGURE 7.14 Stress–strain graph epoxy and epoxy/CTBN 0.5 wt% of MWNT modified by Na-AHA and HNAA.

7.7.9 DYNAMIC MECHANICAL THERMAL ANALYZER (DMA) OF EPOXY/CTBN/MODIFIED MWNT

To study the effect of modification of MWNT on epoxy/CTBN, a few selected nanocomposite samples were subjected to DMA analysis. Figure 7.15 exhibits storage modulus of epoxy and its nanocomposites containing 0.5 wt% of Na-AHA and HNAA modified MWNT. The storage modulus and T_g are evaluated by DMA. DMA was performed on the selected samples of epoxy, epoxy/CTBN and in presence of modified MWNT by Na-AHA and HNAA. It was observed that the T_g is increasing on addition of the modified MWNT. The storage modulus was found to be decreasing as compared with pure compositions with an increase in temperature. T_g obtained for HNAA was higher than Na-AHA modified compositions. It was also observed that the composite containing HNAA modified MWNT shows better storage modulus compared to Na-AHA modified MWNT due to more agglomeration. Significant improvement in storage modulus is achieved for Na-AHA modified MWNT composites compared to pure epoxy as evident in Figure 7.15. This dramatic increase in dynamic storage modulus is believed to originate from the interaction of Na^+ cation of Na-AHA with π-electron clouds of MWNT. This type of interaction reduces strong van der Waals interaction between nanotubes and promotes their debundling as evidenced by SEM analysis discussed later.

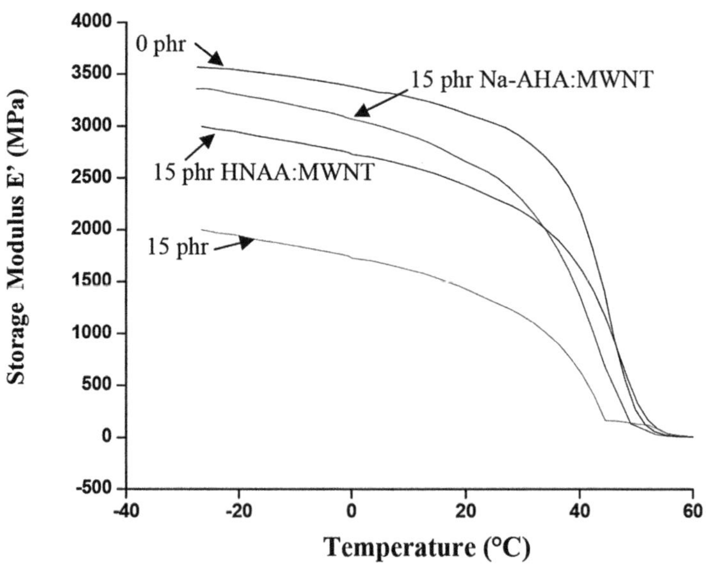

FIGURE 7.15 Storage modulus vs. temperature graph for epoxy, epoxy/CTBN, in presence of modified MWNT by Na-AHA and HNAA.

The strong interaction between the epoxy/CTBN and MWNT resulted in a higher loss because of the shearing action between the nanotubes and epoxy/CTBN chains leading to a higher loss modulus.

7.7.10 DIFFERENTIAL SCANNING CALORIMETRIC (DSC) ANALYSIS OF EPOXY/CTBN/MODIFIED MWNT

In case of epoxy/CTBN/MWNT nanocomposites, DSC analysis was carried out to calculate T_g of the compositions. Figure 7.16 exhibits the decrease in T_g as MWNT reinforcement and rubber CTBN concentration increases by wt%. It is observed that the range of the T_g is increased in Na-AHA modified MWNT samples as compared to HNAA modified MWNT samples. This is mainly because of the decrease in tensile strength and modulus.

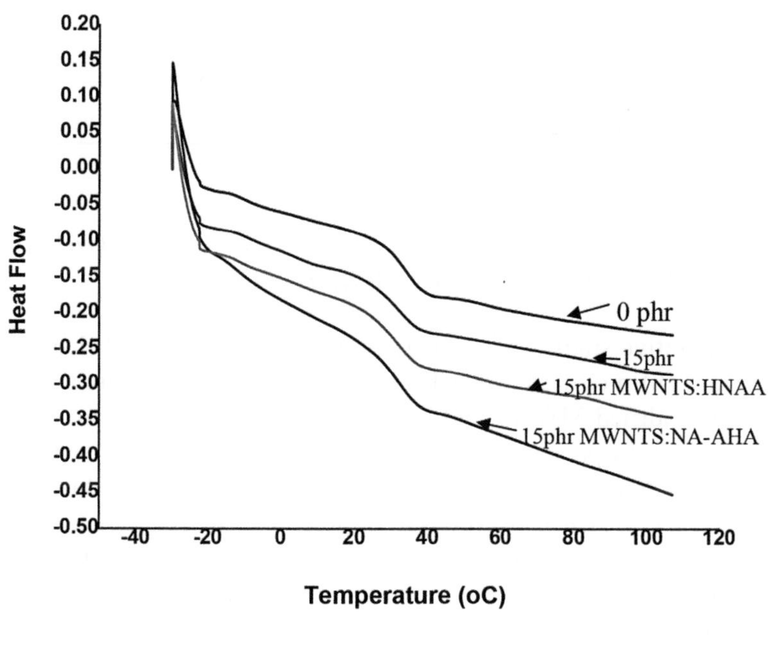

FIGURE 7.16 DSC plots for modified MWNT in presence of HNAA and Na-AHA.

DSC of selected MWNT samples was carried out and it was found that there was decrement in T_g. T_g was less in Na-AHA samples as compared to HNAA modified samples. DSC tests were carried out on "TA INSTRUMENTS"Q 100 DSC machine. The samples weighed about 6–10 g, −100 to 100°C was the temperature maintained in nitrogen atmosphere with single cycle and 10°C/min was the heating rate. DSC testing was also done with the temperature range of −40–120°C. And there was no change in the final results observed. T_g calculated by DSC and DMA were found to be almost having 10°C difference but this is considered under acceptance limit as shown in Table 7.10.

TABLE 7.10 T_g Compared from DSC and DMA.

Composition	T_g obtained by DSC (°C)	T_g obtained by DMA (°C)
0 phr	34.65	49.6
15 phr	31.93	48.8
0 phr HNAA:MWNT	34.54	55.7
15 phr HNAA:MWNT	32.95	52
0 phr Na-AHA:MWNT	36.45	51.9
15 phr Na-AHA:MWNT	33.35	49.8

7.7.11 SCANNING ELECTRON MICROSCOPIC (SEM) ANALYSIS OF EPOXY/CTBN/MODIFIED MWNT

A comparative study was carried out between HNAA modified MWNT and Na-AHA modified MWNT. SEM images of cryofracture surface for HNAA modified MWNT and Na-AHA modified MWNT are shown in Figure 7.17. In presence of HNAA, it shows a well-dispersed MWNT with agglomeration at some places are also observed. The white dots are showing the presence of MWNT in the epoxy matrix. The black holes are found less in number and size in 15 phr MWNT sample, it means filler has very good adhesion with the matrix. The extent of dispersion is good in Na-AHA modified samples. The bonding and interaction of the matrix with nanofiller is found to be good with Na-AHA modified MWNTs.

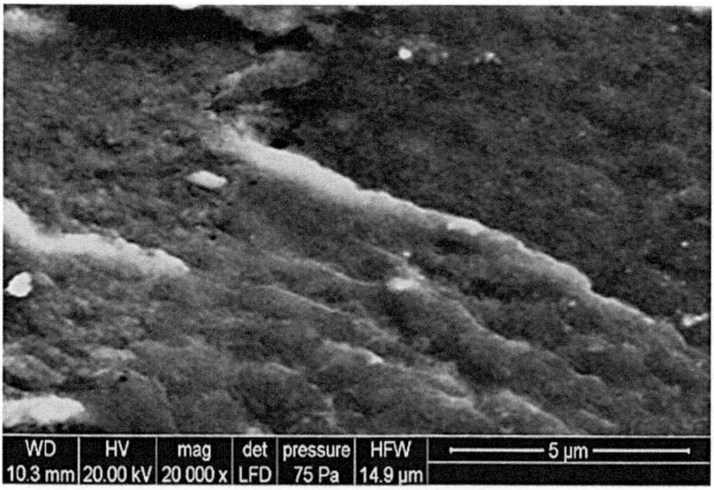

15 phr MWNT:HNAA

FIGURE 7.17 SEM photographs of 0.5 wt% modified MWNT composites (1:1).

7.7.12 THERMOGRAVIMETRIC ANALYSIS (TGA) OF EPOXY/CTBN/MODIFIED MWNT

TGA of the different composition samples was carried out to know the degradation temperature for PNC. TGA was carried out on "TA Instruments" machine. Then, 20°C ramp was given from 0 to 800°C and the conditions were monitored. It was found that the degradation temperature was more for Na-AHA modified MWNT samples as compared to HNAA modified MWNTs samples. While comparing phr systems no significant effect of the addition of rubber was seen on degradation temperature as shown in Figure 7.18.

Table 7.11 shows TGA analysis of PNC. The onset degradation temperature, offset degradation temperature and wt% at 500°C is evaluated. NA-AHA samples show improvement in TGA temperatures. Increase in rubber concentration shows increased onset and offset degradation temperature of the PNC. This is due to improved bonding within the matrix with MWNT. Material retained wt% at 500°C was increased for Na-AHA modified samples as compared to pure phr system and HNAA modified samples.

TABLE 7.11 Comparison of Degradation Temperatures.

Composition	Onset of degradation temperature (°C)	Maximum degradation temperature (°C)	Complete TGA weight loss at 500°C (%)
0 phr	328	450	8
15 phr	332	480	10
0 phr HNAA:MWNT	270	472	14
15 phr HNAA:MWNT	296	511	18
0 phr Na-AHA:MWNT	311	506	19
15 phr Na-AHA:MWNT	308	500	21

7.8 SUMMARY

Rubber toughened blends of epoxy LY556 and CTBN was selected as a system for this study. CTBN and epoxy are mixed and found no reaction with each other and CTBN remained as a separate phase in epoxy matrix giving toughness to the epoxy/CTBN composites. CTBN was used as toughening agent, and it was decided to prepare a mixture with different concentrations of CTBN with epoxy. 0, 5, 10, 15, and 20 phr were the combinations prepared by varying CTBN content in epoxy. Jeffamine 300 was used as curing agent for epoxy/CTBN blends. Through optimization, it was found that 15 phr shows improvement in mechanical properties (tensile and DMA).

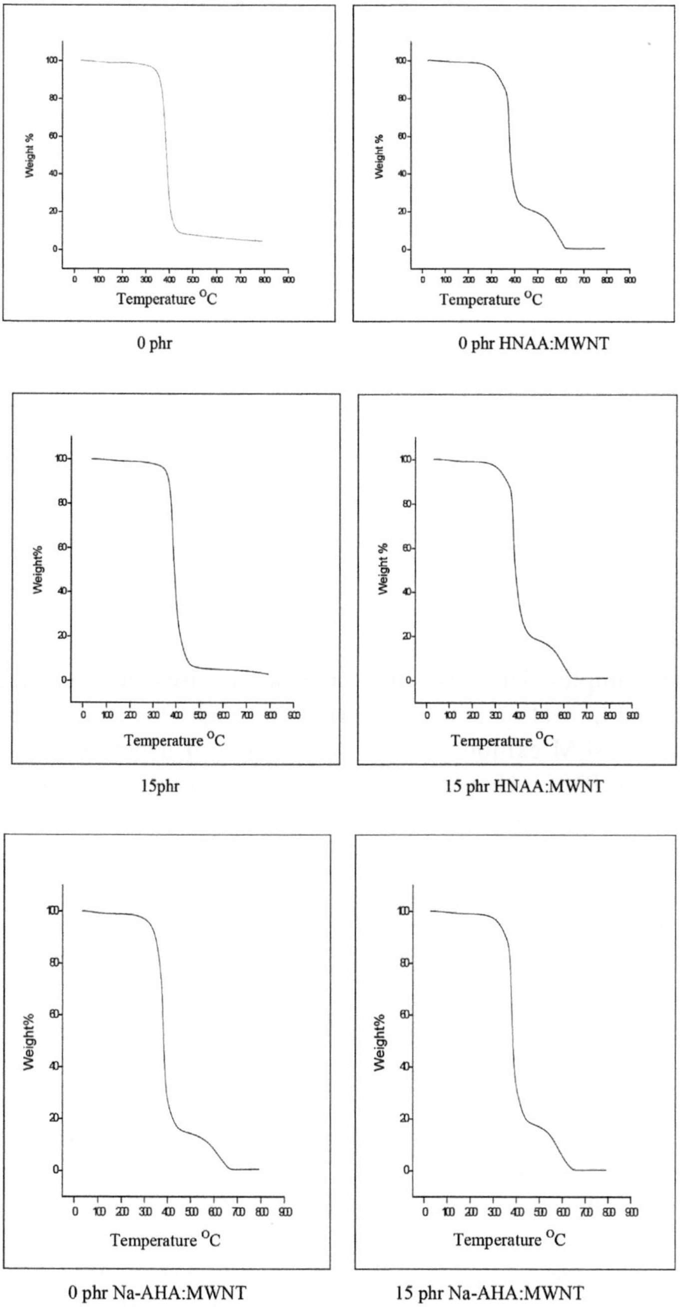

FIGURE 7.18 Comparison of TGA graphs of modified MWNT.

Furthermore, clay as nanofiller is optimized and found that 3 wt% of clay with 15 phr composition shows improvement in the mechanical properties. This supported the selection of 15 phr material, rubber content is high and also shows improved mechanical properties, may be due to good dispersion and high exfoliation of clay in PNC. The tensile testing of these samples was performed and 3 wt% was optimized. For further characterization 15 phr and 3 wt% showed good improvement in all respect except elongation at break which was decreasing after clay reinforcement. SEM was carried out for morphological interaction studies of few selective samples. In short, rubber toughened epoxy/CTBN polymer blend was prepared. The epoxy/CTBN/clay composites were successfully synthesized. The clay composites containing 3 wt% of clay reinforcement exhibited improved mechanical properties compared to other compositions. The addition of 3 wt% of clay improved tensile strength around 40%, tensile modulus around 86% and elongation at break decreased by 57% than pure epoxy/CTBN polymer blends. Storage modulus and loss modulus were found to be decreased as compared to pure samples on reinforcing clay as filler. For 15 phr (epoxy/CTBN) composition, 3 wt% of clay sample shows higher storage modulus and loss modulus for PNC as compared to other clay samples. The clay reinforcement increases the T_g of the system as compared to neat system. It is verified through DSC and DMTA characterization. SEM exhibit extent of dispersion of nanofiller into epoxy/CTBN based PNCs. Hence, it may be concluded that the epoxy/CTBN/clay polymer nanocomposite system is optimized.

Further, in the same system of epoxy/CTBN, MWNT as filler was studied. Pure MWNT does not give good mechanical properties due to poor dispersion and agglomeration. HNAA modified salt and MWNT of concentration 1:1 were prepared. Modified MWNT were optimized and used only 0.5 wt% to all compositions of epoxy/CTBN PNC. It was found that there was increment in the tensile properties of 15 phr composition. Modifier salt of Na-AHA was taken in 1:1 concentration and reinforcement of 0.5 wt% of modified MWNT was studied over all phr compositions. It exhibited more increment than in case of HNAA modified MWNT samples. Na-AHA is also optimized for blends. The modified MWNTs varied as 0.1, 0.25, 0.5, 0.75, and 1 wt% with PNC; and its tensile properties show the highest increment in 0.5 wt%, and 15 phr. The same composition is characterized and compared further by DMA, TGA, DSC, and SEM with pure phr system and also among modified MWNT samples.

It may be concluded from this study that the filler has a better effect on the mechanical properties of the matrix materials. However, nanofillers are found to be best suitable, if it is uniformly dispersed in the matrix material. According to the application point of view, the clay is a better filler for improvement of mechanical properties, as it is cheaper and more effective. However, MWNT is also suitable for improvement in mechanical properties for manufacturing multifunctional polymer nano composites. According to the desired applications, the suitable filler may be selected for PNC.

7.9 CONCLUSIONS

- Rubber toughened epoxy/CTBN polymer blend was prepared.
- The epoxy/CTBN clay composites were successfully synthesized. The clay composites containing 3 wt% clay reinforcement exhibited better tensile property compared to other wt% clay composites.
- The addition of 3 wt% of clay improved tensile strength around 40%, tensile modulus around 86%, respectively, and elongation at break decreased by 57% than pure epoxy/CTBN polymer blends.
- Storage modulus and loss modulus were found to be decreased as compared to pure samples on reinforcing clay as filler. For 15 phr, 3 wt% clay sample shows higher storage modulus and loss modulus for PNC as compared to other clay wt% composite samples.
- Addition of clay reinforcement increases T_g of the system as compared to neat system. It is verified through DSC and DMA characterization.
- Degradation temperature was found to be increasing on increment of rubber percentage. And it is found to be decreasing in presence of nanofiller verified by TGA analysis.
- The modified MWNT composites containing 0.5 wt% MWNT reinforcement exhibited better tensile property compared to other wt% MWNT composites. There were improvements in the tensile properties of epoxy/CTBN modified MWNT composites in all aspects. The addition of 0.5 wt% of NA-AHA modified MWNT reflect improved tensile strength around 65%, tensile modulus around 124% and elongation at break decreased by 57% than pure epoxy/CTBN composite.

- HNAA modified MWNT reinforcement exhibited an increase in the tensile properties but that improvement was not as high as compared to NA-AHA modified MWNT samples.
- Storage modulus and loss modulus were found to be decreased in MWNT reinforced composition as compared to neat samples.
- Addition of MWNT reinforcement increases T_g of the system as compared to neat system.
- Degradation temperature was found to be increasing on increment of rubber percentage and degradation temperature decreases on reinforcement of nanofiller.
- SEM exhibit extent of dispersion of nanofiller into epoxy/CTBN based PNCs.
- Multifunctional PNCs materials may be prepared by incorporating suitable filler as per the desired application.

7.10 FUTURE SCOPE

- The characterization of TEM, flame retardancy, and electrical conductivity, etc., may be evaluated.
- Clay modifier may be required to select and optimize for further work.
- New curing agent may be planned for further work.

ACKNOWLEDGMENTS

My special gratitude to Dr. Chakraborty (HOD of Polymer Department) and Dr. D. Ratna (Sc E') of Naval Material Research Laboratory (NMRL), Ambernath, Mumbai for their co-operation by providing the experimental facilities and invaluable suggestions and motivating guidance for carrying out this work. I am highly obliged to Dr. Pravin (Sc D'), Dr. Kanhai (Sc D'), Mr. Reddy, Mr. Vishal Dalvi, Mr. Siddheshwar Jagatap, and Mr. Jayendran Nayar for extending their co-operation during experimentation. I am highly thankful to Mr. P. Shivaraman (Sc E') for extending his co-operation during SEM investigations. I would also like to extend my heart-felt gratitude to all the staff members of NMRL, Ambernath for their help during the work. At last, I would like to thank all the researchers and publishers whose source of references are used in the literature part of this work.

KEYWORDS

- **clay**
- **CNT**
- **CTBN**
- **polymer nanocomposites**
- **rubber toughened epoxy**

REFERENCES

1. Pötschke, P.; Abdel-Goad, M.; Alig, I.; Dudkin, S.; Lellinger, D. Rheological and Dielectrical Characterization of Melt Mixed Polycarbonate-multiwalled Carbon Nanotube Composites. *Polymer* **2004,** *45,* 8863–8870.
2. Olaf, M.; Kaempfer, D.; Weickmann, H.; Friedrich, C.; Vathauer, M.; Warth, H. Mechanical Properties and Electrical Conductivity of Carbon-nanotube Filled Polyamide-6 and Its Blends with Acrylonitrile/Butadiene/Styrene. *Polym. Sci.* **2004,** *45,* 739–748.
3. Pötschke, P.; Frones, T. D.; Paul, D. R. Rheological Behavior of Multiwalled Carbon Nanotube/Polycarbonate Composites. *Polymer* **2002,** *43,* 3247–3255.
4. Pötschke, P.; Bhattacharya, A. R.; Janke, A.; Goering, H. Melt Mixing of Polycarbonate/Multi-wall Carbon Nanotube Composites. *Compos. Interface.* **2003,** *10,* 389–404.
5. Iijima, S. Helical Microtubules of Graphitic Carbon. *Nature* **1991,** *354,* 56–58.
6. Farahani, R. D. Three-dimensional Microstructures of Epoxy-carbon Nanotube Nanocomposites. Ph.D. Dissertation, ÉcolePolytechnique De Montréal, Canada, 2011.
7. Yu, M. F.; Lourie, O.; Dyer, M. J.; Moloni, K.; Kelly, T. F.; Ruoff, R. S. Strength and Breaking Mechanism of Multiwalled Carbon Nanotubes Under Tensile Load. *Science* **2000,** *287,* 637–640.
8. Yu, M. F.; Files, B. S.; Arepalli, S.; Ruoff, R. S. Tensile Loading of Ropes of Single Wall Carbon Nanotubes and Their Mechanical Properties. *Phys. Rev. Lett.* **2000,** *84,* 5552–5555.
9. Haggenmueller, R.; Gommans, H. H.; Rinzler, A. G.; Fischer, J. E.; Winey, K. I. Aligned Single-wall Carbon Nanotubes in Composites by Melt Processing Methods. *Chem. Phys. Lett.* **2000,** *330,* 219–225.
10. Lozano, K.; Barrera, E. V. Nanofiber-reinforced Thermoplastic Composites. I. Thermo Analytical and Mechanical Analyses. *J. Appl. Polym. Sci.* **2001,** *79,* 125–133.
11. Salvetat, J. P.; Briggs, G. A. D.; Bonard, J. M.; Bacsa, R. R.; Kulik, A. J.; Stöckli, T.; Burnham, N. A.; Forró, L. Elastic and Shear Moduli of Single-walled Carbon Nanotube Ropes. *Phys. Rev. Lett.* **1999,** *82,* 944–947.

12. Walters, D. A.; Ericson, L. M.; Casavant, M. J.; Liu, J.; Colbert, D. T.; Smith, K. A.; Smalley, R. E. Elastic Strain of Freely Suspended Single-wall Carbon Nanotube Ropes. *Appl. Phys. Lett.* **1999,** *74,* 3803–3805.
13. Li, Q.; Zaiser, M.; Koutsos, V. Carbon Nanotube/Epoxy Resin Composites Using a Block Copolymer as a Dispersing Agent. *Phys. Status Solidi.* **2004,** *201,* 89–91.
14. Jagtap, S.; Kushwah, R. K.; Ratna, D. Poly(Ethylene Oxide)-Multiwall Carbon Nanotubes Composites: Effect of Dicarboxylic Acid Salt-based Modifiers. *J. Appl. Polym. Sci.* **2013,** *127,* 5028–5036.
15. Gojny, F.; Nastalczyk, J.; Roslaniec, Z.; Schulte, K. Surface Modified Multi-wall Carbon Nanotubes in CNT/Epoxy-composites. *Chem. Phys. Lett.* **2003,** *370,* 820–824.
16. Nalwa, H. S. *Encyclopedia of Nanoscience and Nanotechnology;* American Scientific Publisher: Los Angeles, CA, 2004; Vol. 1.
17. Maria, D. M.; Salinas-Ruiz, A. A.; Skordos, I. K. P. Rubber-toughened Epoxy Loaded with Carbon Nanotubes: Structure–Property Relationships. *J. Mater. Sci.* **2010,** *45,* 2633–2639.
18. Kaganovskii, Y. S.; Paritskaya, L. N.; Nalwa, H. S. Diffusion in Nanomaterials. *Encycl. Nanosci. Nanotechnol.* **2004,** *2*(1), 399–427.
19. Ciprari, D. L. Mechanical Characterisation of Polymer Nanocomposites and the Role of Interphase. Ph.D. Dissertation, Georgia Institute of Technology, Atlanta, GA, 2004.
20. Osumi, Y. *Three Bond Technical News;* Three Bond Co. Ltd., 1456 Hazama-cho, Hachioji-shi, Tokyo 193-8533, Japan, Issued October 1, 1987.
21. Schaefer, W. D.; Justice, R. S. How Nano Are Nanocomposites? *Macromolecules* **2007,** *40,* 8501–8517.
22. Sanchez, C.; Julian, B.; Belleville, P.; Popall, M. Applications of Hybrid Organic–Inorganic Nanocomposites. *Mater. Chem.* **2005,** *15,* 3559–3592.
23. Ratna, D. *Handbook of Thermoset Resins;* Smithers Rapra Technology: London, UK, 2009.
24. Nayak, G. Mechanical Characterization of Epoxy Nanocomposites with Modified Clay. M. Tech. Dissertation, IIT, Kharagpur, 2008.
25. Jogi, B. F.; Sawant, M.; Kulkarni, M.; Brahmankar, P. K. Dispersion and Performance Properties of Carbon Nanotubes (CNTs) Based Polymer Composites: A Review. *J. Encapsulation Adsorpt. Sci.* **2012,** *2,* 69–78.
26. Bose, S.; Bhattacharyya, A. R.; Haußler, L.; Pötschke, P. Influence of Multiwall Carbon Nanotubes on the Mechanical Properties and Unusual Crystallization Behavior in Melt-mixed Co-continuous Blends of Polyamide6 and Acrylonitrile Butadiene Styrene. *Polym. Eng. Sci.* **2009,** *49,* 1533–1543.
27. Jogi, B. F.; Sawant, M.; Brahmankar, P. K.; Ratna, D.; Tarhekar M. C. Study of Mechanical and Crystalline Behavior of Polyamide6/Hytrel/Carbon Nanotubes (CNT) Based Polymer Composites. *Procedia Mater. Sci.* **2014,** *6,* 805–811.
28. Jogi, B. F.; Kulkarni, M.; Brahmankar, P. K.; Ratna, D. Some Studies on Mechanical Properties of Epoxy/CTBN/Clay Based Nanocompoaites (PNC). *Procedia Mater. Sci.* **2014,** *5,* 787–794.
29. Jogi, B. F.; Bhattacharyya, A. R.; Poyekar, A. V.; Pötschke, P.; Simon, G. P.; Kumar, S. The Simultaneous Addition of Styrene Maleic Anhydride Copolymer and Multi-wall Carbon Nanotubes During Melt-mixing on the Morphology of Binary Blends of

Polyamide6 and Acrylonitrile Butadiene Styrene Copolymer. *Polym. Eng. Sci.* **2015,** *55,* 457–465.

30. Subhani, T.; Latif, M.; Ahmad, I.; Rakha, S. A.; Ali, N.; Khurram, A. A. Mechanical Performance of Epoxy Matrix Hybrid Nanocomposites Containing Carbon Nanotubes and Nanodiamonds. *Mater. Sci. Des.* **2015,** *87,* 436–444.
31. Cha, J.; Jin, S.; Shim, J. H.; Park, C. S.; Ryu, H. J.; Hong, S. H. Functionalization of Carbon Nanotubes for Fabrication of CNT/Epoxy Nanocomposites. *Mater. Sci. Des.* **2016,** *95,* 1–8.
32. Li, C.; Chou, T. W. Elastic Moduli of Multi-walled Carbon Nanotubes and the Effect of Van Der Waals Forces. *Compos. Sci. Technol.* **2003,** *63,* 1517–1524.
33. Burton B..; Alexander, D.; Klein, H.; Garibay-Vasquez, A.; Pekarik, A.; Henkee, C. *Epoxy Formulations Using Jeffamine Polyetheramines;* Huntsman Corporation, 27 April 2005, EFB-0307, 10003 Woodloch Forest Drive, The Woodlands, Texas, 77380, 1–103.
34. Brostow, W.; Goodman, S. H.; Wahrmund, J. *Handbook of Thermoset Plastics;* William Andrew Applied Science Publishers: San Diego, CA, 2014.
35. Tavakoli, S. M. *An Assessment of Skin Sensitisation by the Use of Epoxy Resin in the Construction Industry;* HSE Books: Merseyside, UK, 2003.

CHAPTER 8

POSSIBLE LEAD-FREE NANOCOMPOSITE DIELECTRICS FOR HIGH-ENERGY STORAGE APPLICATIONS: POLYMER NANOCOMPOSITES

K. SRINIVAS*

Department of Physics, GITAM University, Bengaluru, Karnataka, India

E-mail: Srinkura@gmail.com

CONTENTS

ABSTRACT

There is an increasing demand to improve the energy density of dielectric capacitors for satisfying the next generation material systems. One effective approach is to embed high dielectric constant inclusions such as lead zirconia titanate in polymer matrix. However, with the increasing concerns on environmental safety and biocompatibility, the need to expel lead (Pb) from modern electronics has been receiving more attention. Using high aspect ratio, dielectric inclusions such as nanowires could lead to further enhancement of energy density. Therefore, this chapter focuses on the feasibility of development of a lead-free nanowire reinforced polymer matrix capacitor for energy storage application. It is expected that lead-free sodium niobate nanowires ($NaNbO_3$) and Boron nitride will be a future candidate to be synthesized using the simple hydrothermal method, followed by mixing them with polyvinylidene fluoride (PVDF)/divinyltetramethyldisiloxane bis (benzocyclobutene) matrix using a solution-casting method for nanocomposites fabrication. The energy density of $NaNbO_3$- and BN-based composites is also be compared with that of lead-containing ($PbTiO_3$/PVDF) nano composites to show the feasibility of replacing lead-containing materials from high-energy density dielectric capacitors. Further, this chapter explores the feasibility of these materials for space applications because of high-energy storage capacity, more flexibility, and high operating temperatures. This chapter is very much useful to researchers who would like to work on polymer nanocomposites for high-energy storage applications.

8.1 INTRODUCTION

The development of high energy density storage systems with reduced size is highly demanded in many applications, for example, consumer electronics, space-based and land-based pulsed power applications, commercial defibrillators, etc.[1] The electrostatic energy density that can be stored in a material is directly proportional to its dielectric permittivity at the local field and the square of the operational electric field. This necessitates that; modern materials for high energy density should not only possess high dielectric permittivity but also provide high operational electric fields with low dissipation factors. The apparent absence of one single-phase material exhibiting such a combination of properties emphasizes the need to integrate two or more materials with complimentary properties, thus, in turn, creating a composite with performance far better than that of its constituents. Ferroelectric oxides have high dielectric permittivity, but suffer from low dielectric strengths. Polymers, on the other hand, have high breakdown field tolerances, but are limited to low dielectric constants. A diphasic composite consisting of these two could provide a material with high dielectric constant[2] and high breakdown field, affording high storage density for a given thickness. The properties of such composites can be tailored through material selection and composition, as well as through percolation and connectivity of phases present within.[3,4] Recent efforts to obtain high energy density materials have primarily focused on randomly dispersed nano- or micron-sized ceramic particulates in a polymer matrix.

The dielectric properties of nanocomposites have been found to be better in comparison to microcomposites and neat polymers. Various proposed models with some experimental evidence have attributed this property enhancement to the interfacial effects related to filler-polymer interactions in nanocomposites. In this regard, this chapter has focused on finding the ways to enhance the energy storage capacity of polymer nanocomposites by concentrating on interfacial interactions between polymer and filler particle surfaces within an applied field. The main objectives of this research paper are: (1) To modify the surface of nanosized particles ceramic particles with bifunctional organophosphate coupling agents so as to achieve a covalent interface when used within an epoxy polymer matrix composites. (2) To study the influence of covalent interface on electrical properties, especially dielectric breakdown strength of polymer nanocomposite dielectrics compared to the physically adsorbed interface-based nanodielectrics and neat polymer. (3) To study the influence of electronic nature of surface functionalized filler particles on dielectric properties of their composites in epoxy. The electron donating and electron accepting functional groups were used as surface modifying reagents, attached via an organophosphate ligand on to the surface of filler particles. (4) To enhance the energy density of polymer nanocomposites by compromising the decrease in dielectric breakdown strength with increase in permittivity at higher filler particle volume concentration.[5]

It was found that interface layers in the nanocomposites might be more conductive than the polymer matrix, which mitigated the space charge accumulation and field concentration by fast charge dissipation. Zhang and coworkers demonstrated large enhancement in the electric energy density and electric displacement level in the nanocomposites of P(VDF-TrFE-CFE) terpolymer/ZrO_2 nanoparticles.[6] Through the interface effect, the presence of 1.6 vol% of ZrO_2 nanoparticles raised the maximum electric displacement D from 0.085 C/m^2 under 400 MV/m in the neat terpolymer to more than 0.11 C/m^2 under 300 MV/m in the nanocomposites. The dielectric nanocomposites composed of P(VDF-TrFECTFE)[7] and surface-functionalized TiO_2 nanoparticles with comparable dielectric permittivities and homogeneous nanoparticle dispersions were prepared.[8] It was found that the presence of the nanoscale filler favors the formation of smaller crystalline domains and a higher degree of crystallinity in the polymer. In drastic contrast to their weak-field dielectric behavior, substantial enhancements in electric displacement and energy density at high electric fields have been demonstrated in the nanocomposites.

Miniaturization and the current need for high-power density, high voltage capacitors, and power-storage devices have stimulated a new field of research interest in polymer nanocomposites as composite dielectrics (C).[9–15] By incorporating high permittivity inorganic nanoparticles into a polymer matrix with low dielectric loss and high breakdown strength, one may be able to develop new composite materials that have improved dielectric properties, dielectric strength, permittivity and dielectric losses, and retain unique attributes of polymers.

The most distinctive feature of polymer nanocomposites in comparison with conventional microcomposites is the participation of interfacial surface area between the nanoparticles and the polymer matrix. The smaller the size of the embedded nanoparticles, the larger is the surface area to volume ratio, which leads to larger interfacial regions.[16,17] For filler nanoparticles with modest loadings, the surface area associated with the internal interfaces becomes dominant in nanocomposites compared to microcomposites.[18] The properties of these interface areas may differ substantially from those of both the base polymer matrix and the nanoparticle material.[18,19] From literature, polymer nanocomposites with metal oxide nanoparticle fillers exhibited enhanced electrical breakdown strength and voltage endurance compared to their unfilled or micrometer-sized particle filled counterparts.

While the use of nanowire could lead to higher energy density capacitor, most of the high dielectric constant inclusions used have lead-(Pb-) containing materials, such as PZT and PLZT, which make the resulting nanocomposites toxic and not compatible with biological applications.

Expelling lead from commercial applications and materials such as solders, glass, and gasoline has been receiving extensive attention because of the concerns regarding its toxicity. Therefore, this work focuses on the development of a lead-free nanowire reinforced polymer capacitor with comparable dielectric properties to lead-containing capacitors.

The dielectric constant and breakdown strength of nanocomposites with volume fractions ranging from 5 to 30% were experimentally tested to determine the energy densities of both nanocomposites. Testing results have shown that the NNO/PVDF composites have higher dielectric constants, lower dielectric loss, and comparable energy density. Therefore, this may demonstrate the feasibility of developing lead-free high-energy polymer capacitors to ultimately replace lead-containing ones. Further, BN nanosheets are thus envisaged to be one of the best fillers in composites owing to the highly insulating and thermo conductive properties. The fabricated PMMA/BN composite plastics are, thus, envisaged to be valuable for diverse functional applications in many fields, especially for the new-generation thermo conductive insulating long-lifetime packaging materials.[21]

8.2 ROLE OF INTERFACE

Incorporation of nanoparticles into polymer matrix for developing nanocomposites, which has produced interesting results in last few years, has led researchers to investigate mechanisms for the improved dielectric properties. Researchers have emphasized the critical role of the interfacial region and present hypotheses for multiscale phenomena operating in polymer nanocomposites dielectrics.[18,19] Figure 8.1 shows the physical description of the interfacial region in polymer nanocomposites.[18] In thermoplastics, the interfacial polymer can exhibit changes in crystallinity, group mobility, chain conformation, molecular weight, and chain entanglement density. There is an additional complication of changes in cross-link density, in thermosets, due to small molecule migration to or from the interface.

The interfacial region has a direct impact on the dielectric properties of the composites. Therefore, it is important to study the interfacial region. A multi-core model was proposed, which tries to capture the charge behavior and structure of the interfacial region. The metal oxide nanoparticle has a surface charge, which creates a Stern layer at the 2D interface, which is screened by a charged layer in the polymer. The next layer is a diffuse double layer of charge with around 10 nm of radial depth in a resistive medium (polymer). Since, the diffusion double layer is a region of mobile charge; both the dispersion of nanoparticles and the resulting dielectric properties in the polymer nanocomposites have significant influence.

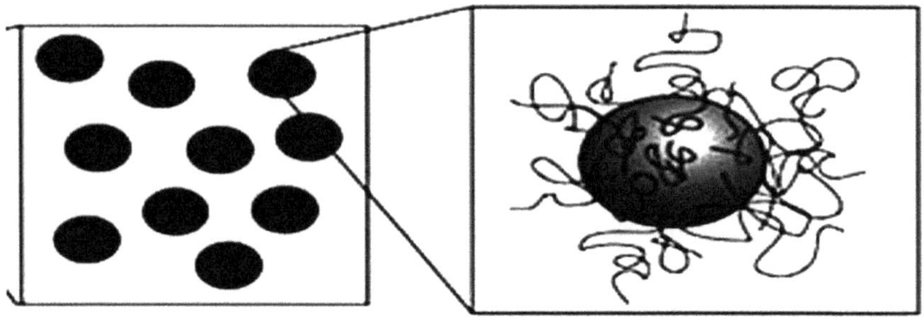

FIGURE 8.1 Physical depiction of the interface based on the interfacial structure and hypotheses to explain the impact of interfacial region on the dielectric properties.[18]

The interfacial region becomes more dominant, as the size of the filler is reduced. The density and perhaps the depth of trap sites are altered due to the change in local structure which affects the carrier mobility and energy. The carriers are accelerated over shorter distances and have reduced energy if they are trapped more often and same is the case for carriers that are scattered. As a result, the dielectric lifetime of the polymer is increased (Fig. 8.2).

The voltage required for charge injection is increased as the homocharge resulting from carrier trapping mitigates the electric field at the electrodes. Thus, the voltage required for the short-term breakdown is also increased. The breakdown strength becomes the function of the rate of measurement (A.C., D.C., or impulse) as the charge takes time to build up. As the nanocomposites have larger interfacial area, it increases the probability for scattering. Scattering may become the primary mechanism for the increase in the breakdown strength of nanocomposites during impulse test conditions, since significant shielding homocharge cannot be accumulated in such a short time. Because interfacial area is so large, while some of the above mechanisms may operate in micron-sized filler filled polymer composites, they are then overshadowed by the large defects the micron scale fillers introduce, and the field enhancements they create.

Microcomposites exhibit Maxwell–Wagner interfacial polarization, which is generally finite in nanocomposites and depends on filler concentration and filler material.[18,19] Electroluminescence, photoluminescence, thermally stimulated currents, X-ray secondary emission spectroscopy, and electron paramagnetic resonance provided experimental evidence to suggest the working of the hypothesis.[18,20]

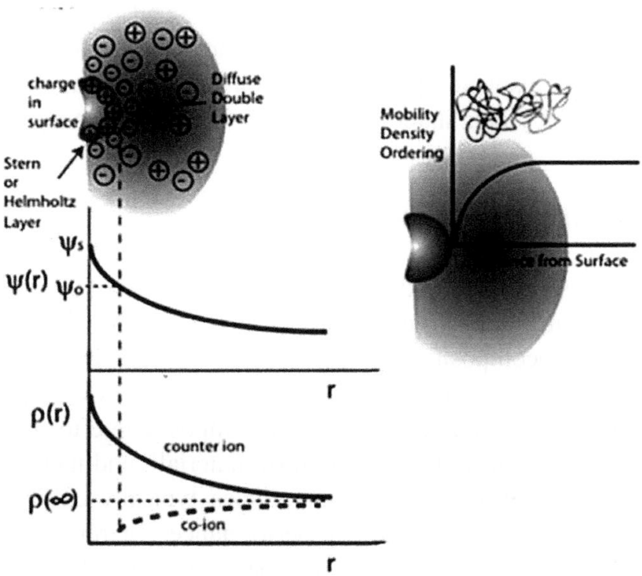

FIGURE 8.2 Physical depiction of the interface region in polymer nanocomposites.[19]

8.3 SURFACE MODIFICATION OF NANOPARTICLES

As the properties of polymer nanocomposites are often influenced by the interfacial region, control of interface becomes very important. Failure to control the interface results in aggregation or agglomeration of nanoparticles in polymer matrix, which leads undesirable properties due to poor film quality and inhomogeneities.[12] Thus, proper dispersion of nanoparticles in polymer nanocomposites plays an important role in polymer nanocomposites. Without proper dispersion and distribution of filler particles in polymer, the high surface area of nano-sized particles is compromised and the aggregates can act as defects, which limits properties.[22]

The most common method to achieve proper dispersion is to modify the surface of nanoparticles. The first aspect of modifying the surface of nanoparticles is to attain stabilization of particles against agglomeration to accomplish homogeneous nanocomposites. The second aspect is to render the guest material (nanoparticle filler) compatible with host material (polymer). The third interest in nanoparticles modification is to enable their self-organization.[23] Surface modification of nanoparticles can be obtained by using suitable surfactants that yield an adsorptive interface or by grafting organic groups on the surface of metal oxide nanoparticles, for example, using phosphates, phosphonates, or silanes as coupling agents or dispersants which yields stable and complex organic oxide interface.[24,25]

8.4 ROLE OF VOLUME FRACTION

Previous studies have shown that as the volume fraction of the high permittivity component, nano particle fillers, is increased, the effective permittivity of the nanocomposites also increases.[26–28] However, increasing the volume fraction of the nanoparticles typically decreases the apparent dielectric breakdown strength of the nanocomposite due to the enhancement of the local electric field in the host material[29] and nanocomposites with large volume fractions of nanoparticles typically exhibit porosity that is detrimental to their dielectric performance. Therefore, the role of volume fraction of high permittivity nanoparticles on the dielectric properties (permittivity, loss, and breakdown strength) is important and should be rationally chosen in order to maximize the stored electrical energy density.

Many mixing models like parallel model and series model, Lichteneckers rule exists which are able to predict electrical properties based on the dc conductivity/resistivity and work best for dilute composites at low volume fractions.[30] Another popular method of predicting the properties of composites is percolation theory, which is based on the assumption that the properties will change when the second phase is totally connected, that is, percolated, from one side of the composite to the other.[31] The volume fraction at which percolation occurs is called the "percolation threshold." Percolation threshold depends on many factors, including the connectivity of the phases, the size of each phase, the shape of each phase, and the wetting behavior of the phases. Percolation models allow for a large, orders of magnitude, and change of properties over a very small concentration range.[32]

As a mixing system, composites filled with inorganic fillers are ideal objects from the point of view of percolation theory. When the concentration of fillers is low, the composites will behave more like the insulating matrix. Once the volume fraction of fillers nears the percolation threshold, for example, 16 or 19% considering impurities, the electrical properties of the composites can be obviously changed by the channels formed in which charge carriers connect inorganic fillers.[33] The percolation threshold for a two-dimensional system is accurately predicted as 50% by effective medium theory and the predicted percolation threshold for three-dimensional system is at 33% by effective medium theory.[33,34]

8.5 DIELECTRIC BREAKDOWN

The dielectric material will suddenly begin to conduct current if the voltage across it becomes too high. This phenomenon is called "dielectric breakdown"[35] and the maximum voltage that can be applied without breaking is called "dielectric breakdown strength." In solid dielectrics, electrical breakdown usually results in permanent damage.

The breakdown in a dielectric material is controlled by several mechanisms shown in Figure 8.3.[36,37] Under a variety of field stresses, the breakdown suffered by dielectric materials presents a very strong time-dependent relationship and can be divided into five or more kinds by breakdown speed. Electrical, thermal, and electromechanical breakdown mechanisms are known as the short-term breakdown or degradation mechanisms[38] and the others are long term.

Electromechanical breakdown is controlled by mechanical properties of dielectric material under high electrical stress and structural parameters. Generally, large changes in dielectric breakdown strength at temperatures approaching glass transition temperature are attributed to be related to electromechanical breakdown mechanism (Fig. 8.3). Where gas is present inside any voids in the dielectric material, gas gets ionized leading to breakdown or discharge within the void under high electric fields leading to the phenomenon known as partial discharge breakdown. The discharge damages the structure of the materials and voids or cracks becomes larger, which can be considered as degradation, which erodes the dielectric resulting in breakdown. Discharge may also take place in the surface of the dielectric if the surface is contaminated by dirt, water, or any other impurities. Various factors influence the dielectric breakdown event, which include temperature; defects and inhomogeneity of material; thickness, area, and volume of the material; duration of time for which the dielectric is subjected to electric field; surface conditions and the method of placing the electrodes; area of the electrodes; composition of the electrodes; moisture (humidity) and other contaminations; and aging and mechanical stress.

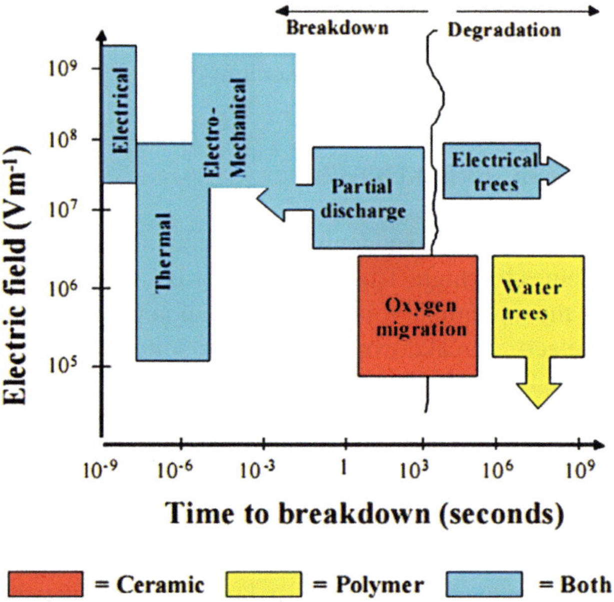

FIGURE 8.3 Schematic depicting times and electric fields at which various electrical breakdown in polymers and composite dielectrics is limited by high field carrier injection and charge trapping electrode–dielectric interface.[37]

8.6 FEASIBILITY OF NaNbO3/PVDF-BASED NANOCOMPOSITE

It was found that interface layers in the nanocomposites might be more conductive than the polymer matrix, which mitigated the space charge accumulation and field concentration by fast charge dissipation. Zhang and coworkers demonstrated large enhancement in the electric energy density and electric displacement level in the nanocomposites of (PVDF-TrFE-CFE) terpolymer/ZrO_2 nanoparticles. Through the interface effect, the presence of 1.6 vol% of ZrO_2 nanoparticles raised the maximum electric displacement D from 0.085 C/m^2 under 400 MV/m in the neat terpolymer to more than 0.11 C/m^2 under 300 MV/m in the nanocomposites. The dielectric nanocomposites composed of P(VDF-TrFECTFE) and surface-functionalized TiO_2 nanoparticles with comparable dielectric permittivities[38–40] and homogeneous nanoparticle dispersions were prepared. It was found that the presence of the nanoscale filler favors the formation of smaller crystalline domains and a higher degree of crystallinity in the polymer. In drastic contrast to their weak-field dielectric behavior, substantial enhancements in electric displacement and energy density at high electric fields have been demonstrated in the nanocomposites. However, with the increasing concerns on environmental safety and biocompatibility, the need to expel lead (Pb) from modern electronics has been receiving more attention. Using high aspect ratio dielectric inclusions such as nanowires could lead to further enhancement of energy density. Therefore, the present review focuses on the development of a lead-free nanowire reinforced polymer matrix capacitor for energy storage application. Lead-free sodium niobate nanowires ($NaNbO_3$) were synthesized using hydrothermal method, followed by mixing them with PVDF matrix using a solution-casting method for nanocomposites fabrication. Capacitance and breakdown strength of the samples were measured to determine the energy density. The energy density of $NaNbO_3$/PVDF composites was also compared with that of lead-containing ($PbTiO_3$/PVDF) nanocomposites and previously developed $Pb(Zr_{0.2}Ti_{0.8})O_3$/PVDF composites to show the feasibility of replacing lead-containing materials. The energy density of $NaNbO_3$/PVDF capacitor is comparable to those of lead-containing ones, indicating the possibility of expelling lead from high-energy density dielectric capacitors.

Ceramics are the other common material broadly used in dielectric capacitors.[41] Compared with polymers, ceramics such as $BaTiO_3$, $Pb(Zr_xTi_{1-x})O_3$ (PZT), and lanthanum doped PZT (PLZT) have significantly higher relative dielectric constants. However, the breakdown strength of ceramics is typically two orders of magnitude lower than dielectric polymers.[42] Moreover, due to its fragility and brittleness, it is difficult to manufacture ceramic capacitors with desired capacity for energy storage applications.[43]

Therefore, a good deal of research effort has been devoted to combining the high dielectric constant of ceramics and high breakdown strength of polymers through the composite approach. To date, many research groups increased the energy density of dielectric capacitors using a nano composite approach, including barium titanate ($BaTiO_3$) nano particles reinforced polycarbonate (PC)[44] and poly(vinylidene fluoride-co-hexafluoropropylene) (PVDF-HFP),[45] modified $BaTiO_3$ nanoparticle with PVDF,[46,47] tanium dioxide (TiO_2) nanoparticles with PVDF terpolymer,[48] calcium copper titanate (CCT) reinforced polyimide,[49] and silver nanoparticle/poly(vinyl pyrrolidone) core-shell structure for high dielectric constant and low loss epoxy matrix composite[50] and PZT or PLZT powders reinforced PVDF.[51,52]

Recently, Andrews et al. developed a micromechanics model to show that using higher aspect ratio nanowires instead of nanoparticle inclusions could lead to significant increase in the dielectric constant of the nanocomposites.[53] Further, Tang et al.[54] experimentally demonstrated that the use of nanowires instead of nanoparticles could significantly increase the dielectric energy density of the nanocomposites.

8.7 PREPARATION OF NNO/PVDF NANOCOMPOSITES

The fabrication process of the NNO/PVDF nanocomposites utilized a two-step procedure; NNO nanowires were first synthesized using the hydrothermal technique, and followed by dispersing nanowires in PVDF and dimethylformamide (DMF) solution to form nanocomposites by solution casting method. Samples with volume fractions ranging from 5 to 30% were prepared to study the nanowire volume fraction influence on energy density.

8.7.1 $NaNbO_3$ NANOWIRES SYNTHESIS

Sodium niobate nanowires were synthesized following a hydrothermal method developed by Nelson[20] In a typical process, a 12 M NaOH solution was prepared by dissolving 33.6 g of NaOH (Acros Organics, 98%) into 70 mL of deionized water (DI).

Subsequently, 3.5 g of Nb_2O_5 (Aldrich, 99.99%) were added into the NaOH solution. After stirring for a period of 30 min at room temperature, the mixture was transferred into a 160 mL Teflon lined stainless steel autoclave with a fill factor of 80%. The autoclave was placed inside an electric oven to undergo hydrothermal reaction at 180°C for 4 h. After cooling down to room temperature, white precipitate was filtered, washed with DI water for several times, and dried at 80°C for 12 h. Finally, NNO powders were annealed at 550°C for 4 h in order to obtain crystallized NNO nanowires.[55]

PVDF and DMF (99.8%) were mixed at a 1:10 weight ratio and heated up at 80°C for 30 min to fully dissolve the PVDF. Nanocomposites were prepared by dispersing NNO or PTO nanowires into PVDF/DMF solution by manual stirring and horn sonication until a homogeneous mixture was obtained. Subsequently, solution was cast onto a PTFE film and dried at 80°C for 6 h.[55]

In order to achieve a consistent thickness over the entire film, nanocomposites were hot pressed at 160°C for 15 min under a constant pressure of 1 ton. Finally, top and bottom surfaces of nanocomposites were coated with silver paint as electrodes for electrical testing. The fabrication process of the nanocomposites is schematically shown in Figure 8.4.[55]

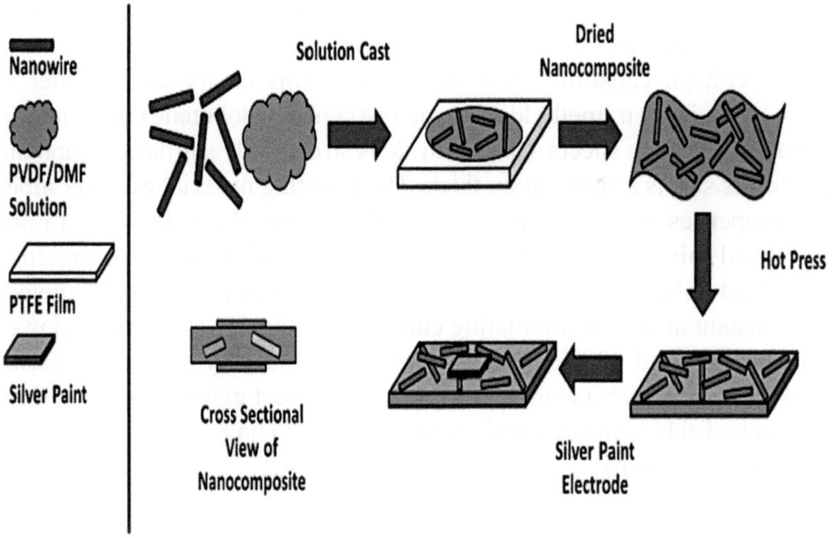

FIGURE 8.4 Preparation steps involved in nanocomposites.[55]

The breakdown voltages were measured according to standard (experimental setup utilized a 30 kV) high voltage power supply and a digital oscilloscope, both connected to a 20 MHz function/arbitrary waveform generator. In order to avoid arcing prior to breakdown, nanocomposites were held inside a 500 mL beaker filled with silicon oil. Breakdown voltage was recorded through a digital multimeter connected to the high voltage power supply.

The morphology of the synthesized NNO nanowires (NWs) shows that NNO NWs have an average diameter of 400 nm with length up to 20 μm. As indicated by the SEM images, homogeneous dispersion of the nanowires throughout the entire sample is achieved and no voids are shown in the samples. As expected, SEM images displayed random orientation of nanowires inside the polymer matrix. After the characterization of dielectric constant and loss of the nanocomposites, breakdown strength testing is carried out to fully understand the energy density.[55]

Recently, Qi et al.[56] developed the boron nitride polymer composite. It can handle high temperatures exceeding 480°F under high voltage application. This material is easy to manufacture by first combining the nanosheets and the polymer. Then, the polymer has to be cured using light or heat to generate crosslinks. As the nanosheets measure only around 2 nm in thickness and 400 nm in lateral size, they stay flexible. The combined material offers exceptional dielectric properties such as heat resistance, higher voltage capability, and bendability. This composite polymer is made of divinyltetramethyldisiloxane bis(benzocyclobutene) surrounding boron nitride nano sheets. Boron nitride is structurally similar to graphene and forms sheets a single atom thick. The resulting material exhibits superior properties compared to anything already out there. In order to see how good this new dielectric really is, the team compared its properties to the best polymeric dielectrics on the market. They examined the dielectric constant at 10^4 Hz alternating current, a frequency common in power conditioning, and 300°C. Under these conditions, they saw minor variations (less than 1.7%) compared to 8% for the best existing system. They also looked at how the dielectric constant changed under direct current and found similar results.[56]

8.8 CONCLUSIONS

In this chapter, we have made a thorough investigation of Pb free nanocomposites for high energy density applications. Expelling lead from commercial applications and materials such as solders, glass, and gasoline has been receiving extensive attention because of the concerns regarding its toxicity. Therefore, this work focuses on the development of a lead-free nanowire reinforced polymer capacitor with comparable dielectric properties to lead-containing capacitors. We have also emphasized the critical role of the interfacial region and present hypotheses for multiscale phenomena operating in polymer nanocomposites dielectrics. Further, increasing the volume fraction of the nanoparticles typically decreases the apparent dielectric breakdown strength of the nanocomposite due to the enhancement of the local electric field in the host material. The dielectric constant and breakdown strength of nanocomposites with volume fractions ranging from 5 to 30% were experimentally tested to determine the energy densities of both nanocomposites. Recent research suggests that the NNO/PVDF composites have higher dielectric constants, lower dielectric loss, and comparable energy density. Therefore, this may demonstrate the feasibility of developing lead-free high-energy polymer capacitors to ultimately replace lead-containing ones.

ACKNOWLEDGMENTS

I would like to acknowledge Prof. Vijaya Bhaskara Raju, Director of GITAM Institute of Technology, GITAM University Bengaluru Campus, India for his constant encouragement to finish this work.

KEYWORDS

- **nanocomposites**
- **interface**
- **nanowires**
- **nanoparticles**
- **thermoconductive properties**

REFERENCES

1. Slenes, K. M.; Winsor, P.; Scholz, T.; Hudis, M. Pulse Power Capability of High Energy Density Capacitors Based on a New Dielectric Material. *IEEE Trans. Magn.* **2001,** *37,* 324–327.

2. Randall, C. A.; Miyazaki, S.; More, K. L.; Bhalla, A. S.; Newnham, R. E. Structural-property Relations in Dielectrophoretically Assembled $BaTiO_3$ Nanocomposites. *Mater. Lett.* **1992,** *15,* 26–30.
3. Bowen, C. P.; Newnham, R. E.; Randall, C. A. Dielectric Properties of Dielectrophoretically Assembled Particulate–Polymer Composites. *J. Mater. Res.* **1998,** *13,* 205–210.
4. Randall, A.; Miller, D. V.; Adair, J. H.; Bhalla, A. S. Processing of Electroceramic-polymer Composites Using the Electrorheological Effect. *J. Mater. Res.* **1993,** *8,* 899–904.
5. Application Note 1217-1. *Basics of Measuring the Dielectric Properties of Materials*; Hewlett Packard Company, 1992.
6. Johnson, G. J. Lossy Capacitors. In *Solid State Tesla Coil;* December 10, 2001; Chapter 3, availabl http://www.ece.k-state.edu/people/faculty/gjohnson/files/tcchap3.pdf.
7. Gouda, O. E.; Thabet, A. M.; El-Tamaly, H. H. In *How to Get Low Dielectric Losses in Binary and Multi-mixtures Dielectrics at High Frequency,* 39th International Universities Engineering Conference, 2004, Vol. 3, pp 1237–1240.
8. Lanagan, M. *Glass Ceramic Materials for Pulsed Power Capacitor,* NSF Center for Dielectric Studies Meeting, Albuquerque, NM, May, 2004.
9. Xiaojun, Y.; Zhimin, Y.; Changhui, M.; Jun, D. Dependence of Dielectric Properties of BT Particle Size in EP/BT Composites. *Rare Metals* **2006,** *25,* 250–254.
10. Cao, Y.; Irwin, P. C.; Younsi, K. The Future of Nanodielectrics in the Electrical Power Industry. *IEEE. T. Dielect. El. In.* **2004,** *7,* 797–807.
11. Ciuprina, F.; Plesa, I.; Notingher, P. V.; Tudorache, T.; Panaitescu, D. Dielectric Properties of Nanodielectrics with Inorganic Fillers, *CEIDP Annual Report Conference on Electrical Insulation and Dielectric Phenomena*; 2008, pp 682–685.
12. Kim, P.; Jones, S. C.; Hotchkiss, P. J.; Haddock, J. N.; Kippelen, B.; Marder, S. R.; Perry, J. W. Phosphonic Acid-modified Barium Titanate Polymer Nanocomposites with High Permittivity and Dielectric Strength. *Adv. Mater.* **2007,** *19,* 1001–1005.
13. Ramesh, S.; Shutzberg, B. A.; Huang, C.; Gao, J.; Giannelis, E. P. Dielectric Nanocomposites for Integral Thin Film Capacitors: Materials Design, Fabrication and Integration Issues. *IEEE Trans. Adv. Packag.* **2003,** *26,* 17–24.
14. Bai, Y.; Cheng, Z. Y.; Bharti, V.; Xu, H. S.; Zhang, Q. M. High-dielectric-constant Ceramic-powder Polymer Composites. *Appl. Phys. Lett.* **2000,** *76,* 3804.
15. Liang, S.; Chong, S. R.; Giannelis, E. P. In *Barium Titanate/Epoxy Composite Dielectric Materials for Integrated Thin Film Capacitors,* Proceedings of 48th Electronic Components and Technology Conference, 1998, pp 171.
16. Lewis, T. J. Interfaces: Nanometric Dielectrics. *J. Phys D. Appl. Phys.* **2005,** *38,* 202–212.
17. Nelson, J. K.; Hu, Y. Nanocomposite Dielectrics–Properties and Implications. *J. Appl. Phys. D. Appl. Phys.* **2005,** *38,* 213–222.
18. Roy, M.; Nelson, J. K.; MacCrone, R. K.; Schadler, L. S.; Reed, C. W.; Keefe, R.; Zenger, W. Polymer Nanocomposite Dielectrics–The Role of the Interface. *IEEE T. Dielect. El. In.* **2005,** *12,* 629–643.
19. Smith, R. C.; Liang, C.; Landry, M.; Nelson, J. K.; Schadler, L. S. The Mechanisms Leading to the Useful Electrical Properties of Polymer Nanodielectrics. *IEEE T. Dielect. El. In.* **2008,** *15,* 187–196.

20. Nelson, J. K. Overview of Nanodielectrics: In *Insulating Materials of the Future*, IEEE Electrical Insulation Conference and Electrical Manufacturing Expo, Nashville, TN, Oct 22–24, 2007.
21. Wang, X.; Amir, P.; Zhang, J.; Qunhong, W.; Tianyou, Z.; Chunyi, Z.; Dmitri, G.; Yoshio, B. Large-surface-area BN Nanosheets and Their Utilization in Polymeric Composites with Improved Thermal and Dielectric Properties. *Nanoscale Res. Lett.* **2012,** *7,* 662.
22. Ajayan, P. M.; Schadler, L. S.; Brawn, P. V. *Nanocomposite Science and Technology*; Wiley-VCH: Weinheim, Germany, 2003.
23. Neouze, M. A.; Schubert, U. Surface Modification and Functionalization of Metal and Metal Oxide Nanoparticles by Organic Ligands. *Monatsh. Chem.* **2008,** *139* (3), 183–195.
24. Hosokawa, M., Nogi, K., Naito, M., Yokoyama, T. Ed.; *Nanoparticle Technology Handbook*; Elsevier: Oxford, UK, 2007.
25. Schuman, T. P.; Siddabattuni, S.; Cox, O.; Dogan, F. Improved Dielectric Breakdown Strength of Covalently-bonded Interface Polymer-particle Nanocomposites. *Compos. Interface.* **2010,** *17* (8), 719–731.
26. Jayasundere, N.; Smith, B. V. Dielectric Constant for Binary Piezoelectric 0–3 Composites. *J. Appl. Phys.* **1993,** *73* (5), 2462–2466.
27. Rao, Y.; Qu, J.; Marinis, T.; Wong, C. P. A Precise Numerical Prediction of Effective Dielectric Constant for Polymer–Ceramic Composite Based on Effective Medium Theory. *IEEE Trans. Adv. Packag. Technologies.* **2000,** *23* (4), 680–683.
28. Calame, J. P. Finite Difference Simulations of Permittivity and Electric Field Statistics in Ceramic–Polymer Composites for Capacitor Applications. *J. Appl. Phys.* **2006,** *99,* 084101.
29. Huang, C.; Zhang, Q. Enhanced Dielectric and Electromechanical Responses in High Dielectric Constant All-polymer Percolative Composites. *Adv. Funct. Mater.* **2004,** *14* (5), 501–506.
30. Head, J. G.; White, N. M.; Gale, P. S. In *Modification of the Dielectric Properties of Polymeric Materials,* 5th International Conference on Dielectric Materials, Measurement and Applications, 27–30, 1988, pp 61–64.
31. McLachlan, D. S.; Blaskiewicz, M.; Newnham, R. E. Electrical Resistivity of Composites. *J. Am. Ceram. Soc.* **1990,** *73* (8), 2187–2203.
32. Kokan, J. R.; Gerhardt, R. A.; Ruh, R.; McLachlan, D. S. In *Dielectric Spectroscopy of Insulator/Conductor Composites,* Materials Research Society Symposium Proceedings, Electrically Based Microstructural Characterization II, Gerhardt, R. A., Alim, M. A., Taylor, S. R., Eds.; Materials Research Society: Pittsburgh, PA 1998; Vol. 500, pp 341–346.
33. Zallen, R. *The Physics of Amorphous Solids*; John Wiley & Sons: New York, NY, 2004.
34. Kirkpatrick, S. Percolation and Conduction. *Rev. Mod. Phys.* **1973,** *45* (4), 574–588.
35. Mukherjee, C.; Bardhan, K.; Heaney, M. Predictable Electrical Breakdown in Composites. *Phys. Rev. Lett.* **1999,** *83* (6), 1215–1218.
36. Lanagan, M. High Power Capacitors and Energy Storage. Presented at *Materials Day,* University Park, Penn State University, State College, PA, April 14, 2008.
37. Dissado, L. A.; Fothergill, L. C. *Electrical Degradation and Breakdown in Polymers*; Peter Peregrins Ltd.: London, UK, 1992.

38. Li, S.; Yin, G.; Chen, G.; Li, J. Bai, S.; Zhong, L.; Zhang, Y.; Lei, Q. Short-term Breakdown and Long-term Failure in Nanodielectrics. *IEEE T. Dielect. El. In.* **2010,** *17* (5), 1523–1535.
39. Dwyer, J. J. O. *The Theory of Dielectric Breakdown of Solids;* Oxford University Press: London, UK, 1964.
40. Claude, J.; Lu, Y.; Wang, Q. Effect of Molecular Weight on the Dielectric Breakdown Strength of Ferroelectric Poly(Vinylidene Fluoride-Chlorotrifluoroethylene)s. *Appl. Phys. Lett.* **2007,** *91* (21) 212904.
41. Tian, Z.; Wang, X.; Shu, L. et al. Preparation of Nano $BaTiO_3$-based Ceramics for Multilayer Ceramic Capacitor Application by Chemical Coating Method. *J. Am. Ceram. Soc.* **2009,** *92* (4), 830–833.
42. Unruan, M.; Sareein, T.; Tangsritrakul, J. Prasertpalichatr, S.; Ngamjarurojana, A.; Ananta, S.; Yimniruna, R. Changes in Dielectric and Ferroelectric Properties of Fe^{3+}/Nb^{5+} Hybrid Doped Barium Titanate Ceramics Under Compressive Stress. *J. Appl. Phys.* **2008,** *104* (12), 124102.
43. Guillon, O.; Chang, J.; Schaab, S.; Kang, S. J. L. Capacitance Enhancement of Doped Barium Titanate Dielectrics and Multilayer Ceramic Capacitors by a Post-sintering Thermo Mechanical Treatment. *J. Am. Ceram. Soc.* **2012,** *95* (70), 2277–2281.
44. Ibrahim, S. S.; Al Jaafari, A. A.; Ayesh, A. S. Physical Characterizations of Three Phase Polycarbonate Nanocomposites. *J. Plast. Film Sheeting*. **2011,** *27*(4), 275–291.
45. Xie, L.; Huang, X.; Wu, C.; Jiang, P. Core-shell Structured Poly (Methyl Methacrylate)/$BaTiO_3$ Nanocomposites Prepared by In Situ Atom Transfer Radical Polymerization: A Route to High Dielectric Constant Materials with the Inherent Low Loss of the Base Polymer. *J. Mater. Chem.* **2011,** *21* (16), 5897–5906.
46. Kim, P.; Doss, N. M.; Tillotson, J. P. et al. High Energy Density Nanocomposites Based on Surface-modified $BaTiO_3$ and a Ferroelectric Polymer. *ACS Nano*. **2009,** *3* (9), 2581–2592.
47. Mao, Y. P.; Mao, S. Y.; Ye, Z. G.; Xie, Z. X.; Zheng, L. S. Sizedependences of the Dielectric and Ferroelectric Properties of $BaTiO_3$/Polyvinylidene Fluoride Nanocomposites. *J. Appl. Phys.* **2010,** *108* (1), 014102.
48. Dang, Z. M.; Yuan, J. M.; Zha, J. W.; Zhou, T.; Li, S. T.; Hu, G. H. Fundamentals, Processes and Applications of 8 ISRN Nanomaterials High-permittivity Polymer-matrix Composites. *Prog. Mater. Sci.* **2012,** *57* (4), 660–723.
49. Ni, L.; Chen, X. M. Dielectric Relaxations and Formation Mechanism of Giant Dielectric Constant Step in $CaCu_3Ti_4O_{12}$ Ceramics. *Appl. Phys. Lett.* **2007,** *91*(12), 122905.
50. Chen, A.; Kamata, K.; Nakagawa, M.; Iyoda, T.; Wang, H.; Li, X. Formation Process of Silver-polypyrrole Coaxial Nanocables Synthesized by Redox Reaction Between $AgNO_3$ and Pyrrole in the Presence of Poly (Vinylpyrrolidone). *J. Phys. Chem. B.* **2005,** *109* (39), 18283–18288.
51. Barber, P.; Balasubramanian, S.; Anguchamy, Y. et al. Polymer Composite and Nanocomposite Dielectric Materials for Pulse Power Energy Storage. *Materials*. **2009,** *2,* 1697–1733.
52. Das-Gupta.; Doughty, K. Polymer-ceramic Composite Materials with High Dielectric Constants. *Thin Solid Films* **1988,** *158* (1), 93–105.

53. Andrews, C.; Lin, Y.; Sodano, H. A. The Effect of Particle Aspect Ratio on the Electroelastic Properties of Piezoelectric Nanocomposites. *Smart Mater. Struct.* **2010,** *19*(2), 025018.
54. Tang, H.; Lin, Y.; Andrews, C.; Sodano, H. A. Nanocomposites with Increased Energy Density through High Aspect Ratio PZT Nanowires. *Nanotechnology* **2011,** *22*(1), 015702.
55. Mendoza, M.; Ashiqur Rahaman Khan, M. D.; Mohammad Arif, I. S.; Alberto, G.; Yirong, L. Development of Lead-free Nanowire Composites for Energy Storage Applications. *Int. Scholar. Res. Network ISRN Nanomater* **2012,** *e 2012,* Article ID 151748, 8.
56. Qi, L.; Lei, C.; Guangu, Z. et al. Flexible High-temperature Dielectric Materials from Polymer Nanocomposites. *Nature* **2015,** *523,* 576–579. DOI: 10.1038/nature14647

CHAPTER 9

ANTIBACTERIAL ACTIVITY OF SMART (AUXETIC) POLYURETHANE FOAMS

SUKHWINDER K. BHULLAR[1,3*], MEHMET ORHAN[2*], and M. B. G. JUN[3]

[1]*Department of Mechanical Engineering, Bursa Technical University, Bursa, Turkey*

[2]*Department of Textile Engineering, Uludag University, Bursa 16059, Turkey*

[3]*Department of Mechanical Engineering, University of Victoria, Victoria, BC, Canada*

[*]*Corresponding author. E-mail: sbhullar@uvic.ca; morhan@uludag.edu.tr*

CONTENTS

ABSTRACT

Polymeric biomaterials in their woven and non-woven form at macro to nano scale offer huge potentials for external and internal use in biomedical and other fields. Applications of polymers particularly in the form of foam such as PU foams are versatile. They are used in a variety of biomedical and other applications including wound care, tissue engineering, implants and prostheses, surgical masks, hospital bedding, packaging, sound insulation, air filtration, shock absorption, and as sponge materials. In this study, smart polyurethane (PU) foams called auxetic PU (APU) foams, are fabricated and characterized for antibacterial activities. APU foam samples are developed based on compression, heating, cooling, and relaxation method from the open literature. Antimicrobial agent's chitosan and silver are incorporated into the bulk in fabricated foam samples to test antibacterial activities against Gram-negative and Gram-positive bacteria qualitatively (agar diffusion test) and quantitatively (suspension test) according to EN ISO 20645 and ASTM 2149 standard methods, respectively. It was found APU foams incorporated with AgCl and Chitosan were more effective against *Staphylococcus aureus* than against *Escherichia coli* compared to conventional PU foams. Moreover, auxetic foam with silver exhibited the best antibacterial activity against both *S. aureus* and *E. coli*.

9.1 INTRODUCTION

Polyurethane (PU) foams like most of other materials have positive Poisson's ratio; however, smart PU foams called auxetic polyurethane (APU) foams having negative Poisson's ratio which were first developed in 1987.[1,2] The concept of negative Poisson's ratio in materials was known since 1927[3,4] but the structures with negative Poisson's ratio were first reported in 1982 in a two-dimensional silicon rubber and aluminum honeycomb.[5,6] Later, polymeric and metallic foams with negative Poisson's ratio (−0.7 and −0.8) were presented by researchers.[7–9] APU foams become wider on stretched, narrower on compressed, wraps around indenter due to indentation and shows synclastic (dome) behavior on bending.[9–12] Also, they have the ability to return their original shape as they have shape memory. In addition, they are more resilient,[10] vibration control,[13,14] energy absorption,[15–17] harder to indent,[18–21] better sound absorption capacity,[22] improved shear resistance,[23.24] fracture toughness,[25–27] and crack resistance[25,27–29] which make them superior than conventional PU foam for many biomedical and other applications. To name a few are—smart wound care devices,[30,31] smart implants, prostheses,[32–34] stents with minimum crimped diameter,[35–37] scaffolds with matching behavior of native tissues[39–41] and force sensors for rehabilitation[42,43] and smart filtration.[31] However, it has been reported by Tamborini et al.[44] (fungi and bacteria) develop in PU foam cells which can cause allergic reactions, infections, stains, surface degradation, and odors. Also, through studies and experiments, it has been demonstrated that antimicrobial agents to PU foam can be added to inhibit microorganism.[45] Chitosan, a natural and low-cost biopolymers due to their good compatibility with human tissues and antibacterial properties are well known for their vast application in biomedical and other industries.[46–48] Another antibacterial agent silver compounds have been used for a hundreds of years in their solid, solution, and cream form such as for wound care silver (silver metal and nanocrystalline silver) itself in solid form to control infection, solutions of silver salts to cleanse wounds and ointments containing antibiotic compound (silver sulfadiazine) for burns.[49,50] Silver chloride (AgCl) is one of the silver compounds which are most widely used in biological and biomedical applications as antibacterial agent.[51,52]

9.2 MATERIAL

9.2.1 FABRICATION OF APU FOAM

To fabricate APU foams, PU foam samples in cylindrical shape of length 150 m and diameter 18 mm were purchased from Sierra Foam Company, Turkey. APU foam samples were fabricated based on compression, heating, cooling, and relaxation method from the open literature[15] and microstructure of both conventional and auxetic foam are shown in Figure 9.1a,b.

In our study, chitosan and silver chloride are selected for antibacterial activities of APU. Silver chloride dispersion, iSys AG, with a silver concentration of 8.4 mg/g was used as silver source and this product was kindly donated by CHT/Bezema, Turkey and chitosan were purchased from Sigma-Aldrich. Chitosan solution was prepared by dissolving 1 g polymer in 100 mL 1% (v/v) acetic acid solution under magnetic stirrer for 24 h at room temperature. Samples were soaked in chitosan solution and silver separately.

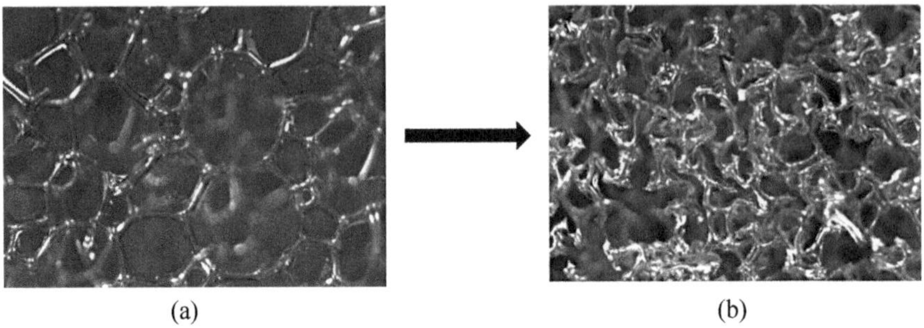

FIGURE 9.1 Microstructure of (a) conventional PU foam and (b) auxetic PU foam.

9.3 METHODS

The antibacterial activity of samples was qualitatively (agar diffusion test) and quantitatively (suspension test) determined following the procedure of two methods. The samples were tested against Gram-negative bacteria (*Escherichia coli* ATCC 35218) and Gram-positive bacteria (*Staphylococcus aureus* ATCC 6853). Stocks freeze-dried of these bacteria were suspended in triptychs soy broth (TSB) and incubated in 37°C for 24 h and then transferred to plates of nutrient agar to be reserved for examination.

9.3.1 QUALITATIVE TEST METHOD

The antibacterial efficiency of samples was evaluated by qualitative test method EN ISO 20645. This test is based on the assessment of bacterial growth around and under the samples that was laid into the particular bacterial strain inoculated agar. The samples were tested against both bacteria. Bacteria suspension was prepared in a nutrient broth and diluted with buffer solution at pH 7. The standardized concentration of about 1.53×10^5 cfu/mL was applied to agar plates and they were inoculated at 37°C for 24 h and then test samples were placed on it. After incubating, the evaluation was carried out by the assessment of bacterial growth under the sample and by the size of inhibition zone around the sample. Each cultivation test was made for two separate samples.

The test results in three possible outcomes for antimicrobial activity-insufficient effect (bacterial growth in agar under the sample), limit of efficacy (growth almost totally suppressed, restricted to small colonies); good effect (no bacterial growth underneath the sample, presence of inhibition zone around the sample). The purpose of the test is to identify the minimum concentration of antibacterial agent in order to achieve final assessment of good effect. According to this method's evaluation, extending or up to 1 mm inhibition zone and no growth under sample or no inhibition zone and no growth under sample, were accepted as effective.

The level of antibacterial activity was assessed by examining the extent of bacterial growth in the contact zone between the agar and the test sample. Inhibition zones were calculated using the following formula:

$$H = (D - d)/2 \quad (9.1)$$

where H is the inhibition zone in mm, D is the total diameter of sample and inhibition zone in mm, and d is the diameter of sample in mm.

It is given the detailed information on the evaluating of antibacterial effect in Table 9.1.

TABLE 9.1 Antibacterial Effect of the Antibacterial Treatment.

Inhibition zone (mm) mean value	Growth[a]	Description	Assessment
>1	None	Inhibition zone exceeding 1 mm, no growth[b]	Good effect
1–0	None	Inhibition zone up to 1 mm, no growth[b]	
0	None	No inhibition zone, no growth[c]	
0	Slight	No inhibition zone, only some restricted colonies, growth nearly totally suppressed[d]	Limit of efficacy
0	Moderate	No inhibition zone, compare to the control growth reduced to half[e]	Insufficient effect
0	Heavy	No inhibition zone, compare to the control growth reduction or only slightly reduced growth[e]	

[a]The growth of bacteria in the nutrient medium under the specimen.

[b]The extent of the inhibition shall only partly be taken into account. A large inhibition zone may indicate certain reserves of active substances or a weak fixation of a product on the substrate.

[c]The absence of growth, even without inhibition zone, may be regarded as a good effect, as the formation of such an inhibition zone may have been prevented by a low diffusibility of the active substance.

[d]"As good as no growth" indicates the limits of efficacy.

[e]Reduced density of bacterial growth means either the number of colonies or the colony diameter.

9.3.2 QUANTITATIVE TEST METHOD

The antibacterial efficiency of samples was quantitatively evaluated by the ASTM 2149 01 Standard Dynamic Contact Conditions. In this test, a homogenous suspension of both bacteria was prepared in a nutrient broth and diluted with buffer solution at pH 7. The standardized concentration of about 1.53×10^8 cfu/mL was applied to samples for the antibacterial testing. For each sample to be tested, 50 mL of the buffer solution and 1 mL of standardized microbial solution were placed into three sterile jars. One jar contained only bacterial suspension, another contained 1 g of treated test sample and the last contained 1 g of untreated test sample. All the jars were incubated at 37°C and shaken in a wrist-action shaker for 24 h. They were diluted with a series of dilutions using buffer solution. Bacteria cells were counted as time "0" by performing standard plate count techniques. The antimicrobial activity was expressed in % reduction of the organisms after contact with the test sample at time "0"compared to the number of bacterial cells surviving after contact with the sample after 24 h. The percentage reduction (R) of bacteria was calculated using following formula:

$$R = 100\ (B - A)/B \tag{9.2}$$

where A is the number of bacteria recovered from the inoculated treated test sample in the jar incubated for 24 h; and B is the number of bacteria recovered from the inoculated treated test sample at "0" contact time. During the antibacterial testing, all measurements were completed in microbiology laboratory environment of about 24°C and 55% relative humidity and repeated four times.

9.4 RESULTS AND DISCUSSION

9.4.1 QUALITATIVE TEST RESULTS

For the qualitative assessment, the antibacterial activities against both bacteria shown in Tables 9.2 and 9.3 were studied by EN ISO 20645. In this method, bactericidal activity was visually assessed to determine the zone of inhibition around the samples. The more detailed explanations about assessment were given in Table 9.1.

TABLE 9.2 The Antibacterial Assessment Against *S. aureus*.

Sample name	Inhibition zone (mm)	Growth	Description	Assessment
Conventional PU foam	0	Heavy	–	Insufficient effect
Auxetic PU foam	0	Heavy	–	Insufficient effect
Auxetic PU foam treated with chitosan	10	None	No inhibition zone, no growth	Good effect
Auxetic PU foam treated with silver	10	None	No inhibition zone, no growth	Good effect

TABLE 9.3 The Antibacterial Assessment Against *E. coli*.

Sample name	Inhibition zone (mm)	Growth	Description	Assessment
Conventional PU foam	0	Heavy	–	Insufficient effect
Auxetic PU foam	0	Heavy	–	Insufficient effect
Auxetic PU foam treated with chitosan	10	None	No inhibition zone, no growth[a]	Good effect
Auxetic PU foam treated with silver	10	None	No inhibition zone, no growth[a]	Good effect

[a]The absence of growth, even without inhibition zone, may be regarded as a good effect, as the formation of such an inhibition zone may have been prevented by a low diffusibility of the active substance.

As seen in Figures 9.2 and 9.3 for both bacteria species, there was heavy growth on contact surface of conventional and APU foam, as we expect. However, the results have shown that no growth of bacteria under auxetic foam with chitosan and silver and moreover, on the samples an inhibition zone of 10 mm. According to test method, antibacterial protection for $H \geq 10$ mm.

In test method, it is important to contact the agar surface to emerge and evaluate the antibacterial activity. Unfortunately, the foam samples were not heavy enough to contact completely the agar surface. So, it could not be clearly seen the formation of the zone through sample diameter while studying with treated foam samples. Honestly speaking, it is well known that there was antibacterial activity to both bacteria although the inhibition zones were not clearly seen in figures. The results were as good as we expect and were fairly comfortable to have a sufficient antibacterial effect according to test method.

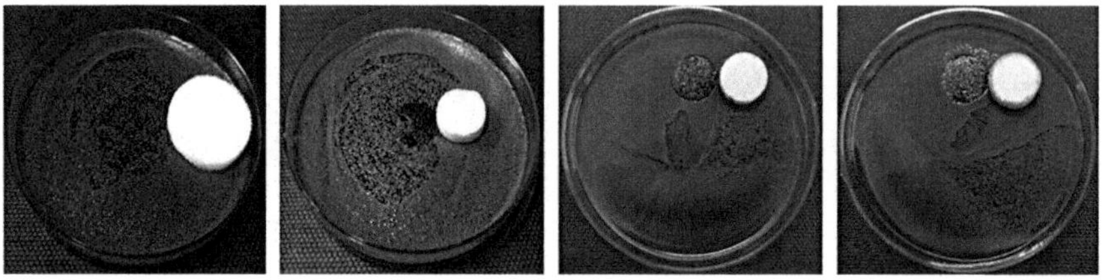

Conventional PU Foam	Auxetic PU foam	Auxetic PU foam treated with Chitosan	Auxetic PU foam treated with silver

FIGURE 9.2 The antibacterial assessment against *S. aureus.*

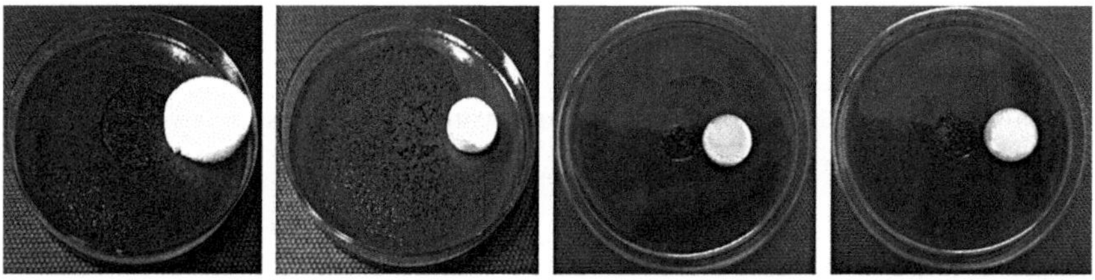

Conventional PU Foam	Auxetic PU foam	Auxetic PU foam treated with Chitosan	Auxetic PU foam treated with silver

FIGURE 9.3 The antibacterial assessment against *E. coli.*

These studies also showed that those antibacterial agents could not diffuse from the treated foam surfaces. Therefore, the results do not mean that both chitosan and silver have no antibacterial effect. The already published many studies on bactericidal activity have proved that chitosan and silver kill bacteria. Besides, the antibacterial activity on foam was specifically a result of chitosan and silver surface contact so that quantitative experimental data also confirm these results. It was also found that they had more effective against *S. aureus* than against *E. coli*.

Consequently, these results were enough to have a good antibacterial activity, which was significantly enhanced after treated with both chitosan and silver.

9.4.2 QUANTITATIVE TEST RESULTS

For the quantitative assessment, the antibacterial activities against both bacteria shown in Tables 9.4 and 9.5 were studied by ASTM E2149. This method is based on plate count method and consists of the determination of antibacterial activity of the treated material inoculated with bacteria, by numbering the bacteria colonies.

As seen in results for both bacteria species, the conventional and auxetic foams have sufficient antibacterial activities against *S. aureus* whereas they have no antibacterial activities against *E. coli*. However, it can be found that auxetic foam treated with chitosan exhibited good antibacterial properties, but auxetic foam treated with silver has the best effective against both *S. aureus* and *E. coli*. Gram-negative bacteria are generally more resistant to biocides than Gram-positive bacteria, because of an extra layer (outer membrane) surrounding the cell wall, composed of polysaccharides, proteins, and phospholipids.[53,54] It is clear from results auxetic PU foam with both chitosan and silver have the best antibacterial activity against both bacteria.

Furthermore, biocompatibility, high resistance against hydrolysis and body fluids, high tensile strength, and highly elastomeric properties make PUs good candidate for biomedical applications. They offer application in the form of foam, film, or bulk material. They have a variety of uses such as implants, prostheses, scaffolds, wound care, scleral sponge, decubitus pads, air purification filters, hospital and daily use mattresses, athletic pads, helmet liners, shoe insoles, infants and disabled patient's diapers, and cushioning furniture. However, literature and experiences demonstrate that smart foams called PU foams tailored with enhanced mechanical properties and unique deformation mechanism achieved under the effect of negative Poisson's ratio promote their candidature over conventional PU foams in biomedical industry and other fields. For example, their wrapping around the indenter behavior offers more cushioning and comfort which along with their synclastic behavior make them promising for knee and hip implants, implant liners, and support for the doubly curved human body which can improve quality of life of patients with disability. Also, for the treatment of an infected swollen wound on applying an APU foam dressing loaded with the drug will release drug as the push from wound toward dressing will open the pores to release the drug. Furthermore, APU foams have the ability to return to its original form as they have shape memory, therefore, as wound will heal and swelling come to normal, pores will be closed to seize drug delivery. In addition, antibacterial properties of auxetic foams are concerned it is noticed from Tables 9.4 and 9.5, bacteria reduction (%) though is almost same for both conventional and auxetic foams for the antibacterial assessment against *S. aureus* but % bacteria reduction (%) is higher in auxetic foams against *E. coli.* Furthermore, smart auxetic PU foams incorporated with antimicrobial agents chitosan and silver have the best antibacterial activity against both Gram-negative and Gram-positive bacteria.

TABLE 9.4 The Antibacterial Assessment Against *S. aureus.*

Sample name	The antibacterial activities after 24 h (bacteria reduction, %)
Conventional PU foam	−95.92
Auxetic PU foam	−96.60
Auxetic PU foam treated with chitosan	−00.00
Auxetic PU foam treated with silver	−100.00

Note: The concentration of bacteria was adjusted to 5.00×10^4 (log 4.70) cfu/mL for each sample.

TABLE 9.5 The Antibacterial Assessment Against *E. coli.*

Sample name	The antibacterial activities after 24 h (bacteria reduction, %)
Conventional PU foam	+252.94
Auxetic PU foam	+194.12
Auxetic PU foam treated with chitosan	−99.83
Auxetic PU foam treated with silver	−100.00

Notes: The concentration of bacteria was adjusted to 7.50×10^4 (log 4.88) cfu/mL for each sample. cfu: colony forming units. The positive values of bacterial reduction (%) demonstrate an increase in bacterial growth and also the negative values of bacterial reduction (%) demonstrate a decrease in bacterial growth. The value of (−) 100% indicates that all the bacteria on the surface were killed.

9.5 CONCLUSIONS

In this study, the fabrication and characterization of smart APU foam incorporated antibacterial agents; chitosan and silver were investigated for antimicrobial activities. The best antibacterial activity against Gram-negative bacteria (*E. coli* ATCC 35218) and Gram-positive bacteria results were observed in tested samples. They are the major microorganisms accounting for 25% of hospital infections and popular test organisms resistant to common antibacterial agents. Moreover, functional performance of auxetic foams is more beneficial over conventional foams in biomedical and many other applications due to their ability to return original shape and synclastic behavior.[32–42] Therefore, advantages of this research study include ease of fabrication of auxetic foams, good antibacterial properties and more effective against *S. aureus* than against *E. coli*. These APU foams with sufficient antimicrobial properties could be promising in implants, liners, and prostheses to provide more cushioning, comfort, and natural motion due to their ability to return to original shape. Also, they may offer more potential over the conventional PU foam as smart bandages impregnated with medicine for wound care, tissue engineering, smart filters, and as smart sponge materials in many other applications.

KEYWORDS

- **antibacterial activities**
- **auxetic behavior**
- **polyurethane foams**
- **biocompatibility**
- **bacterial suspension**

REFERENCES

1. Lakes, R. S. *Science* **1987,** *235,* 1038–1040.
2. Lakes, R. S. International Patent WO 88/00523, 1988.
3. Love, A. E. H. *A Treatise on the Mathematical Theory of Elasticity;* Cambridge University Press: Dover, NY, 1927.
4. Fung, Y. C. *Foundations of Solid Mechanics;* Prentice-Hall: Englewood Cliffs, NJ, 1968.
5. Gibson, L. J.; Ashby, M. F.; Schajer, G. S.; Robertson, C. I. *Proc. Royal Soc. London.* **1982,** *382,* 25–42.

6. Gibson, L. J.; Ashby, M. F. *Cellular Solids: Structure and Properties;* Pergamon Press: London, 1988.
7. Friis, E. A.; Lakes, R. S.; Park, J. B. *J. Mater. Sci.* **1988**, *23,* 4406–4414.
8. Choi, J. B. Lakes, R. S. *J. Compos. Mater.* **1995,** *29,* 113–128.
9. Chan, N.; Evans, K. E. *J. Mater. Sci.* **1997,** *32,* 5725–5736.
10. Chan, N.; Evans, K. E. *J. Cell Plast.* **1999,** *35,* 130–165.
11. Smith, C. W.; Lehman, F.; Wootton, R. J.; Evans, K. E. *Cell Polym.* **1999,** *18,* 79–101.
12. Chan, N.; Evans, K. E. *J. Cell Plast.* **1998,** *34,* 231–260.
13. Evans, K. E.; Nkansah MA, *J. Appl. Phys.* **1990,** *67,* 6654–6660.
14. Scarpa, F.; Giacomin, J.; Zhang, Y.; Pastorino, P. *Cell Polym.* **2005,** *24,* 253–268.
15. Bianchi, M.; Scarpa, F. L.; Smith, C. W. *J. Mater. Sci.* **2008,** *43,* 5851–5860.
16. Roach, A. M.; Jones, N.; Evans, K. E. *Compos. Struct.* **1998,** *42,* 135–152.
17. Roach, A. M., Jones, N.; Evans, K. E. *Compos. Struct.* **1998,** *42,* 119–134.
18. Burke, M. *New Sci.* **1997,** *154,* 36–39.
19. Smith, C. W.; Grima, J. N.; Evans, K. E. *Acta Mater.* **2000,** *48,* 4349–4356.
20. Alderson, K. L.; Fitzgerald, A.; Evans, K. E. *J. Mater. Sci.* **2000,** *35,* 4039–4047.
21. Bhullar, S. K.; Wegner, J. L.; Mioduchowski, A. *J. Mater. Sci. Eng.* **2012,** *2,* 436–441.
22. Howell, B.; Prendergast, P.; Hansen, L.; *Appl. Acoustics* **1994,** *43,* 141–148. doi: 10.1016/0003-682X(94)90057-4.
23. Evans K. E.; Alderson, A. *Adv. Mater.* **2000,** *12,* 617–628.
24. Evans K. E.; Alderson, A. *Eng. Sci. Educ. J.* **2000,** *9,* 148–154.
25. Yang, W.; Li, Z. M.; Shi, W.; Xie, B. H. Yang, M. B. *J. Mater. Sci.* **2004,** *39,* 3269–3279.
26. Scarpa, F.; Ciffo, L. G.; Yates, J. R. *Smart Mater. Struct.* **2000,** *13,* 49–56.
27. Chan, N.; Evans, K. E. *J. Mater. Sci.* **1997,** *32,* 5945–5953. doi: 10.1023/A:1018606926094
28. Scarpa, F.; Bullough, W. A.; Lumley, P. *Proc. IME CJ Mech. Eng. Sci.* **2004**, *218,* 241–244.
29. Bhullar, S. K.; Wegner, J. L.; Mioduchowski, A. *J. Eng. Tech. Res.* **2010,** *2,* 118–126.
30. White, L. *Urethanes Technol. Int.* **2009,** *26,* 34–36.
31. Abdelrahman, T.; Newton, H. *Surgery* **2011,** *29,* 491–495.
32. Baker, C. E.; Auxetic Spinal Implants: Consideration of Negative Poisson's Ratio in the Design of an Artificial Intervertebral Disc. Ph.D. Thesis, The University of Toledo, May 2011.
33. Scarpa, F. *IEEE Signal Proc. Mag.* **2008,** *25,* 125–128.
34. Bhullar, S. K. *e-Polymers* **2014,** *14,* 383–390. doi: 10.1515/epoly-2014-_0137
35. Ali, M. N.; Rehman, I. U. *J. Mater. Sci. Mater. Med.* **2011,** *22,* 2573–2581. doi: 10.1007/s10856-011-4436-y
36. Bhullar, S. K.; Ko, J.; Cho, Y; Jun, M. B. G. *J. Polym-Plast. Techno. Eng.* **2015,** *54,* 1553–1559. doi: 10.1080/03602559.2014.986812
37. Ramachandran, J. *Comp. Methods Biomech. Biomed. Eng.* **2007,** *10,* 245–255.
38. Hengelmolen, R., Patent WO GB2003/003393, 2004.
39. Fozdar, D. Y.; Soman, P.; Lee, J. W.; Han, L. H.; Chen, S. *Adv. Funct. Mater.* **2011,** *21,* 2712–2720.
40. Jun, Y.; Jeong, K. *Adv. Mater. Sci. Eng.* **2013,** *2013,* 853289. doi: org/10.1155/2013/853289.

41. Soman, P.; Lee, J. W.; Phadke, A.; Varghese S.; Chen, S. *Acta Biomater*. **2012,** *8,* 2587–2594.
42. Ko, J.; Bhullar, S. K.; Cho, Y.; Jun M. B. G. *Smart Mater. Struct.* **2015,** *24,* 075027. doi: 10.1088/0964-1726/24/7/075027
43. Evans, K. E.; Alderson, A. *Adv. Mater.* **2000,** *12,* 617–628.
44. Tamborini, S. M.; Mahoney, J.; McEntee, T. C. In *Proceedings of the SPI 27th Annual Technical/Markting Conference*, 1982.
45. Patarcity, R.; Stern, E.; Murthy, U. *Proceedings of the SPI 6th Annual Technical/ Marketing Conference*, 1983.
46. Rui, Z.; Xiang, L.; Bolun, S.; Ying, Z.; Dawei, Z.; Zhaohui, T. X.; Chenb, C. W. *Int. J. Biol. Macromol.* **2014,** *68,* 92–97.
47. Archana, D.; Singh, B. K.; Dutta, J.; Dutta, P. K. *Int. J. Biol. Macromol.* **2015,** *73,* 49–57.
48. Mehta, A. S.; Singh, B. K.; Dutta, J. *Asian Chitin. J.* **2014,** *10,* 25–28.
49. Lansdown, A. B. G. *Br. J. Nurs.* **2004,** *13,* 6–19.
50. White, R. J.; Cooper, R. *Wounds* **2005,** *1,* 51–61.
51. Yun, O. K.; Taek, S. L.; Won, H. P. *J. Mater. Sci. Mater. Med.* **2014,** *25,* 2629–2638.
52. Min, S. H.; Yang J. H.; Kim, J. Y.; Kwon, Y. U. *Micropor. Mesopor. Mat.* **2010,** *128,* 19–25.
53. Orhan, M.; Kut, D.; Gunesoglu, C. *J. Appl. Polym. Sci.* **2009,** *111,* 1344–1352.
54. Orhan, M.; Kut, D.; Gunesoglu, C. *Plasma Chem. Plasma Process.* **2012,** *32,* 293–304.

CHAPTER 10

OPTICALLY TUNED MDMO-PPV/ PCBM BLEND FOR PLASTIC SOLAR CELLS

ISHWAR NAIK[1*], RAJASHEKHAR BHAJANTRI[2], and JAGADISH NAIK[3]

[1]*Government Arts and Science College, Karwar, Karnataka, India*

[2]*Department of Physics, Karnatak University, Dharwad, Karnataka, India*

[3]*Department of Physics, Mangalore University, Mangalore, Karnataka, India*

[*]*Corresponding author. E-mail: iknaik@rediffmail.com*

CONTENTS

ABSTRACT

This work is focused to optimize the photoactive blend of Poly[2-methoxy-5-(3',7'-dimethyloctyloxy)-1,4-phenylenevinylene] (MDMO-PPV) and [6,6]-Phenyl C61 butyric acid methyl ester (PCBM) for maximum absorption of the solar energy. MDMO-PPV:PCBM blends of weight ratio 3:1, 1:1, and 1:3 are prepared in chloro-benzene (CB) as the common solvent and glass-coated samples are prepared by solution cast method. Samples are scanned for absorption by JASCO UV–Vis NIR V-670 spectrometer. The spectral analysis revealed that 1:3 blend of MDMO-PPV with PCBM has the broad spectral sensitivity for absorption and can be used as the best photoactive material. Band gap of the samples is determined using Tauc's plot and the optical band gap of the optimized blend is 2.225 eV. The morphological study of the optimized blend has been carried out using scanning electron microscopy (SEM) and atomic force microscopy (AFM). SEM images reveal the formation of numerous hetero P–N junctions due to incorporation of N-type PCBM in the P-type MDMO-PPV polymer matrix. Elemental analysis of the SEM images is also carried out using energy dispersion spectroscopy (EDS), indicating the presence of only carbon and oxygen as expected AFM studies remarked the presence of considerable smoothness accounting for the intense inter-chain effects needed for good transport properties. Crystal structural investigations are made through X-ray diffraction studies where the semi crystalline nature of the photoactive blend is depicted in the spectrum. Diffused bands indicate the amorphous state dominance in the blend, suggesting the need for improvement of crystallinity that can be achieved through thermal annealing. We end up with consideration of 1:3 MDMO-PPV:PCBM blend to be the best active blend for practical construction of the solar cell and the work is in progress.

10.1 INTRODUCTION

The solar cells are classified as first-, second-, third-, and fourth-generation solar cells. The silicon-based first-generation solar cells have enough efficiency, enabling them to be put into practical use for the society. Wide absorption spectrum and high charge motilities have endowed them with good efficiency. However, these cells suffer from fabrication complications, material costs, and installation cost. The second-generation cells are the thin film cells based on Si, CdTe, etc. The advantage of thin-film solar cells is that they can be fabricated on all substrates, including flexible substrates, and material cost is drastically reduced relative to the first-generation cells. The efficiency of these cells is lower than the first-generation solar cells and processing cost is more. The third-generation solar cells make use of organic semiconductors like conducting polymers or small molecules. The branch has been developed since the invention of conjugated polymers. The use of nanoparticles and dyes has improved the performance of the third-generation organic photovoltaic devices. A lot of reports are available related to organic solar cells, and the research on polymeric cells has taken a global importance in seeking a solution for energy crisis. Low cost, flexibility, tunability, and lightweight are the merits of these cells over the conventional solar cells but efficiency and stability must be highly improved to make it available for public.

Fourth-generation cells are the hybrid solar cells, fabricated from organic polymers along with inorganic semiconductors. The active material is prepared by incorporating inorganic semiconducting crystals into the conducting polymer matrix. In this work, we focused on the preparation and characterization of an organic solar cell active material so that a glance on organic conducting polymers is meaningful at these steps. Generally, conventional polymers such as plastic and rubber offer resistance to electrical conduction and are dielectrics. However, these beliefs have changed with the discovery of conducting polymers by Allan J Heeger, Alan G MacDiarmid, and Hideki Shikawa in 1976 who showed that the poly acetylene is conductive almost like a metal with conductivity increased by 10^9 times the original value on doping with halogen and they are awarded with Nobel prize for this discovery. Some of the examples of conducting polymers are poly acetylene, poly aniline, poly thiophene, PEDOT-PSS, etc. These conducting polymers can conduct electricity like semiconductors and metals. They are inexpensive and more advantageous due to lightweight, flexibility, and the ease of processing ability.[1–3] One early explanation of conducting polymers used band theory as a method of conduction according to which half-filled valence band would be formed from a continuous delocalized pi system which is the ideal condition for conduction of electricity. Also, the conductivity can be increased by doping the polymer with an electron acceptor or electron donor. This is reminiscent of doping of a silicon-based semiconductor with arsenic or boron. However, while the doping of silicon produces a donor energy level close to the conduction band or acceptor level close to the valence band, this is not the case with conducting polymer. The high conductivity upon doping finds wide application in organic electronic devices like LED, solar cells, sensors, etc.

10.1.1 ORGANIC SOLAR AND THEIR CLASSIFICATION

Organic solar cell (plastic solar cell or polymer solar cell) is a specialized P–N junction in which P and N-type semiconductors used are the organic conducting polymers, being able to convert light in to direct current. These conducting polymers have energy gap (separation between HOMO and LUMO) in the range 1–4 eV. Donor (P-type) polymer absorbs visible light to form exciton which then splits into electron and hole by the effective field across hetero junction (PN junction) of active layer. Separated electrons and holes are collected by electrodes to generate emf.

a) **Single layer cell.** It is the simplest form of organic photovoltaics (OPVs), consisting of a layer of conducting polymer sandwiched between two metallic electrodes. When the P-type layer absorbs light, electrons will be excited to the LUMO and leave holes in the HOMO, thereby forming excitons. Exciton will dissociate into electron and hole, pulling electrons to the positive electrode, and holes to the negative electrode. Single layer organic solar cells do not work well. They have low quantum efficiencies (< 1%) and low power conversion efficiencies (< 0.1).[4,5]
b) **Bilayer cells.** It contains hetero junction of donor and acceptor conductive polymer between two electrodes. Donor and acceptor layers are properly chosen to have internal field across the junction sufficient enough to create electron and hole separation. Efficiency is more than that of single layer cells. In order to have exciton dissociation, the layer thickness must be in the range of about 10 nm or otherwise electron–hole recombination will dominate the charge separation.
c) **Bulk hetero junction cell.** Electron donor and acceptor are blended and interposed between electrodes. Numerous P–N junctions will be formed at the donor–acceptor interphases of the blend. The domain sizes of this blend are on the order of nanometers, allowing excitons to easily reach an interface and dissociate due to the large donor–acceptor interfacial area.[6] Bulk hetero junctions

have an advantage over layered photoactive structures because they can be made thick enough for effective photon absorption. Most bulk hetero junction cells use two components, although three-component cells have been explored.

d) **Multilayer or tandem solar cells.** These cells consist of multiple P–N junctions each one tuned to specific frequency of the spectrum. High band gap solar cells are at the top followed by lower band gap cells. Short wavelength is absorbed by top cell and longer wavelengths by successive lower cells. Efficiency can be crossed over Shockley–Queisser limit of a single cell (two-layer cell).

e) **Plasmonic enhanced tandem OPV.** This is based on the collective oscillation of electrons of the nanoparticles under the influence of light called Plasmon resonance. Metal nanoparticles like gold and silver show Plasmon resonance in the visible region allowing large scattering and absorption of light. Yang et al. reported the first Plasmon-enhanced cell with 20% efficiency.

10.1.2 WORKING PRINCIPLE

Conversion of light energy in to electrical energy is a multi-step process.

i) Photon absorption
ii) Exciton formation
iii) Exciton migration
iv) Exciton dissociation
v) Charge collection

i) **Photon absorption.** It is the primary and most important aspect of the solar cell. The photoactive blend must have a broad absorption spectrum for visible light. This can be achieved by proper selection of P and N-type pair and also their weight ratio must be tuned for maximum absorption of the incident energy. Active material should have low band gap matching well with the visible spectrum.

ii) **Exciton formation.** Incident photon will be absorbed by the P-type material due to which an electron in the highest occupied molecular orbit (HOMO) will be excited to the lowest unoccupied molecular orbit (LUMO). Unlike the case of an inorganic system, here the hole and electrons are not free from each other but it is the bound state of electron and hole called exciton.

iii) **Exciton migration.** After the formation of exciton, it has to migrate toward the P–N junction for dissociation as free electron and hole. The diffusion length of excitons are of the order 5–10 nm and they have a finite life time so that care must be taken to make them migrate to the junction before they decay or recombine.

iv) **Exciton dissociation (charge separation).** As the exciton migrate and reach the P–N junction, it is influenced by the electric field of the depletion layer and gets dissociated as electron and hole. Electron will be in the LUMO of acceptor, holes in the HOMO of the donor.

v) **Charge collection.** Free charges created will migrate toward the respective electrodes creating a potential difference to drive the current in the circuit. Charge collection at the electrodes will be efficient only when the mobility of electron and holes is high. Therefore the choice of P- and N-type polymers having considerable mobility of electron and holes is an important aspect in the construction of solar cells.[7] As described above, dispersed hetero junctions of donor–acceptor organic materials have high quantum efficiencies compared to that of the planar hetero junction, because in dispersed hetero junctions it is more likely for an exciton to find an interface within its diffusion length.

10.2 MOTIVATION BEHIND THE WORK

The first-generation silicon-based cells and even the second-generation thin film solar cells suffer from material cost, installation cost, and fabrication complications.[8] Organic solar cells are promising because of their low cost, simple processing, flexibility, and tunability of lightweight. Low efficiency being the main drawback of these solar cells, the search for an efficient, stable organic solar cell is the most challenged and demanded research problem.[9] In the constructional hierarchy, single layer, bi-layer, bulk hetero junction, tandem cells and Plasmon-enhanced cells follow in steps in an attempt to achieve more efficiency. The main difficulty with plastic solar cells is the electron–hole recombination before the exciton migration to the P–N junction interface. The problem is overcome by introducing the concept of bulk hetero junction blend of donor–acceptor pair. Due to the formation of numerous P–N junctions in the blend, the excitons can easily encounter the junction interfaces to get separated as charge carriers. This bulk hetero junction concept of the donor–acceptor system has proved remarkable progress in the efficiency enhancement. In this work, we tried to optimize the bulk hetero junction (mixed junction) photoactive blend of P-type conducting polymer MDMO-PPV and the N-type acceptor PCBM (modified fullerene). Optical energy gap of the blends is also calculated from the absorption data. This work is focused to optimize the photoactive blend of poly[2-methoxy-5-(3',7'-dimethyloctyloxy)-1,4-phenylenevinylene] (MDMO-PPV) and [6,6]-phenyl C61 butyric acid methyl ester (PCBM) for maximum absorption of the solar energy. MDMO-PPV:PCBM blends of weight ratio 3:1, 1:1, and 1:3 are prepared in CB as the common solvent, and glass-coated samples are prepared by solution cast method. Samples scanned for absorption by JASCO UV–Vis NIR V-670 spectrometer. The spectral analysis revealed that 1:3 blend of MDMO-PPV with PCBM has the broad spectral sensitivity for absorption and can be used as the best photoactive material. Band gap of the samples is determined using Tauc's plot and the optical band gap of the optimized blend is 2.225 eV. Further enhancement of absorption can be done by doping the optimized blend by dye or metal nanoparticles like gold or silver.

10.3 EXPERIMENTAL

The P-type donor polymer poly[2-methoxy-5-(3',7'-dimethyloctyloxy)-1,4-phenylenevinylene] (MDMO-PPV) and the N-type material [6,6]-phenyl C61 butyric acid methyl ester (PCBM) are purchased from Sigma-Aldrich Corporation. The solvent CB is procured from Rankem Chemicals. All chemicals are used as received without further purification. The specifications of the chemicals are mentioned here.

- MDMO-PPV: Molecular weight – 120,000 : HOMO – 5.4 eV: LUMO – 3.2 eV.
- PCBM: Functionalized fullerene : HOMO – 6.1 eV: LUMO – 3.7 eV.

Then 10 mg of MDMO-PPV and 8 mg of PCBM are dissolved in 50 and 100 cc of CB, respectively, in separate beakers, magnetically stirred for 48 h at room temperature until clear solutions are formed. The resulting solutions are of concentrations 0.2 and 0.08 mg/cc, respectively. The solutions are blended with MDMO-PPV:PCBM weight ratios of 3:1, 1:1, and 1:3. Detailed chemical compositions were tabulated in Table 10.1. The mixtures are magnetically stirred for three days at room temperature to ensure optimum blending, then transferred to 3 cm diameter petri-plates, dried at room temperature, and then at about 50°C in an oven.

TABLE 10.1 Detailed Chemical Composition.

P-type solution	N-type solution	P:N
2 mg (10 cc)	–	Pure P
1.5 mg (7.5 cc)	0.5 mg (6.25 cc)	3:1
1 mg (5 cc)	1 mg (12.5 cc)	1:1
0.5 mg (2.5 cc)	1.5 mg (18.75 cc)	1:3
–	2 mg (25 cc)	Pure N

10.4 CHARACTERIZATION, RESULT, AND DISCUSSION

10.4.1 UV–VISIBLE SPECTROSCOPY

The UV–Visible optical absorption spectroscopy is used for investigating optically induced transitions and helps to find out band structure, molecular constitution, and energy gap. It is based on the principle of electronic transition in atoms or molecules on absorbing a proper radiation in the ultraviolet and visible regions. The radiation that can be absorbed is specific for each element and varies with the chemical structure. A beam of UV–Visible radiation is split into component wavelengths by grating or prism and then the monochromatic beam is split into two beams; one is passed through the reference sample and the other through the sample. Some of the radiation will be absorbed resulting in as the electron transition to a higher energy orbital. The spectrometer records the wavelengths at which absorption occurs, together with the degree of absorption at each wavelength. The resulting spectrum is presented as a graph of absorbance (A) versus wavelength λ UV spectroscopy obeys the Beer–Lambert law $A = \log (I_0/I)$.

A spectrophotometer can be either single beam or double beam. In a double-beam instrument, the light is split into two beams before it reaches the sample. One beam is used as the reference; the other beam passes through the sample. The reference beam intensity is taken as 100% transmission and the measurement displayed is the ratio of the two beam intensities. The absorption of UV or visible radiation corresponds to the excitation of outer electrons. Possible excitations are s→s* transitions, n→s* transitions, n→p* or p→p* transitions. Most absorption spectroscopy of organic compounds is based on transitions of n or p electrons to the p^* excited state. This is because the absorption peaks for these transitions fall in an experimentally convenient region of the spectrum (200–700 nm).

In this work, the samples are analyzed by JASCO UV–Vis NIR V-670 spectrometer. The V-670 double-beam spectrophotometer utilizes a unique, single monochromatic design covering a wavelength range from 190 to 2500 nm. A PMT detector is provided for the UV–Vis region and a Peltier-cooled PbS detector is employed for the NIR region. Both gratings and detector are automatically exchanged within the user selectable 750–900 nm range. The absorption spectra are presented in Figures 10.1–10.6.

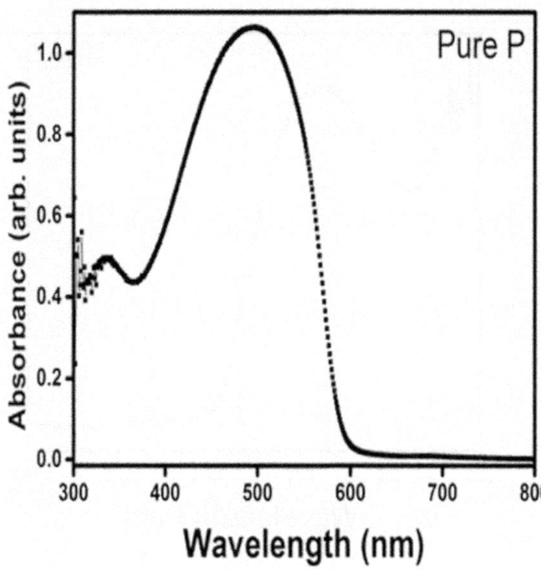

FIGURE 10.1 PURE P.

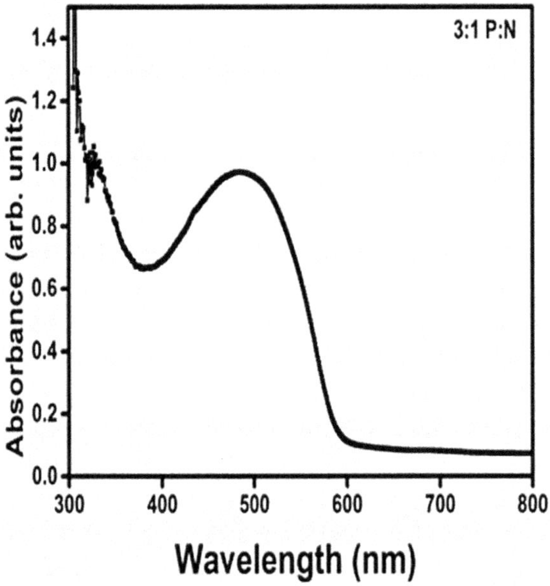

FIGURE 10.2 3:1 P:N.

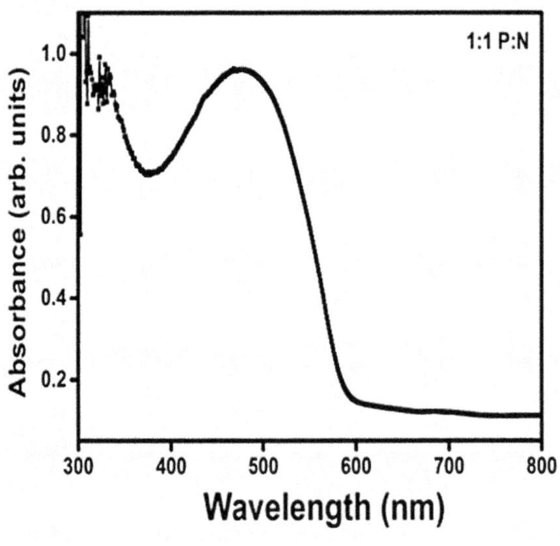

FIGURE 10.3 1:1 P:N.

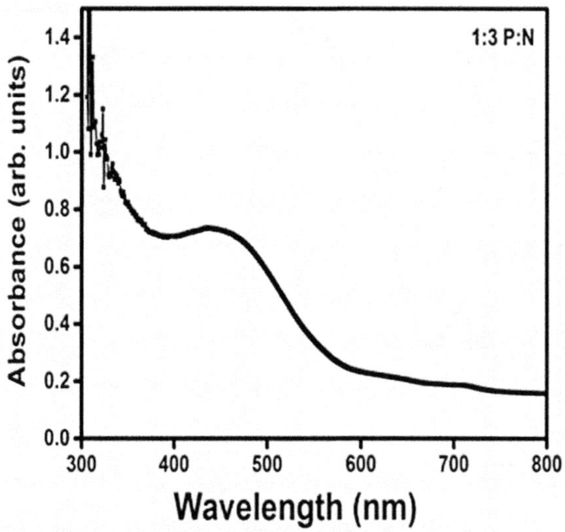

FIGURE 10.4 1:3 P:N.

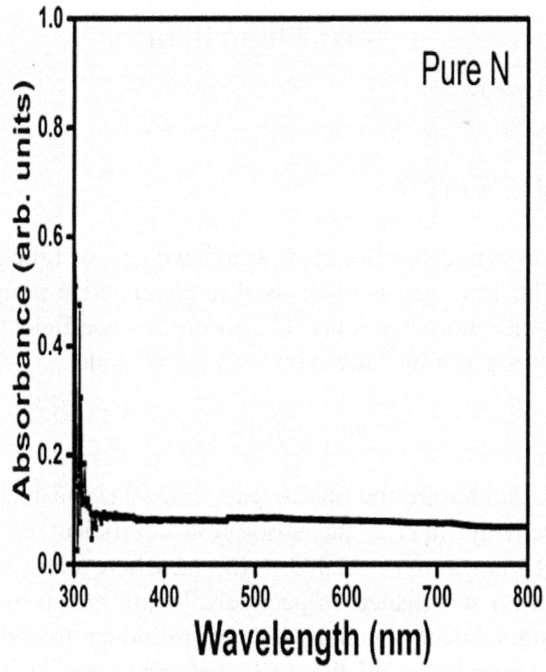

FIGURE 10.5 Pure N.

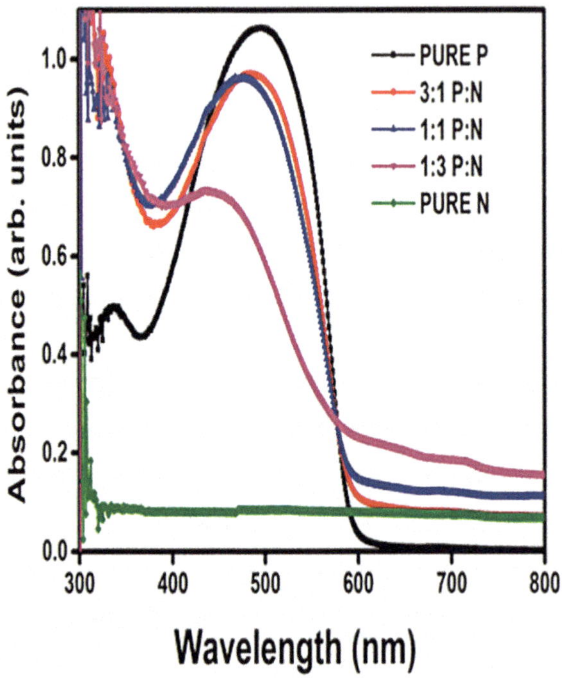

FIGURE 10.6 Overlay.

10.4.2 TAUC'S PLOT

A Tauc plot is used to determine the optical band gap, or Tauc gap, in semiconductors. The Tauc gap is often used to characterize practical optical properties of amorphous materials. The absorption coefficient of an amorphous material is given by Tauc and Davis–Mott model:

$$\alpha(v)hv = B\,(hv - E_{gap})^{m}$$

where E_{gap}, B, and hv are the optical gap, constant, and incident photon energy, respectively; $\alpha(v)$ is the absorption coefficient defined by the Beer–Lambert's law as $\alpha(v) = 2.303 \times A(\lambda)\,/d$, where d and A are the film thickness and film absorbance, respectively. In amorphous materials, four types of transitions occur: namely, indirect forbidden transition, indirect allowed, direct forbidden, and direct allowed transitions. These transitions are characterized by the m values 1/3, 1/2, 2/3, and 2, respectively.[10,11]

For indirect allowed transition, $m = 1/2$

$$\alpha(v)hv = B\,(hv - E_{gap})^{1/2}$$

on squaring,

$$[\alpha(v)hv]^2 = B\,(hv - E_{gap})$$

$$[\alpha(v)\,hC/\lambda\,]^2 = B\,(hC/\lambda - E_{gap})$$

$$[2.303\times \text{Abs}\,(\lambda)/\text{d}\ \text{hC}/\lambda]^2 = B\,(hC/\lambda - E_{gap})$$

or

$$(A/\lambda)^2 = B\,(hC/\lambda) - B\,(hC/\lambda_{gap})$$

Graph of $(A/\lambda)^2$ with $1/\lambda$ plotted,

If $(A/\lambda)^2 = 0$, then $B\,(hC/\lambda) - B\,(hC/\lambda_{gap}) = 0$; therefore, $1/\lambda = 1/\lambda_g$, that is, $\lambda = \lambda_{gap}$.

Thus extrapolating the straight-line curve to zero absorption gives the onset wavelength λ_{gap}. Then, energy gap is determined using the relation

$$E_{gap} = hC/\lambda_{gap}.$$

Analysis of the absorption spectra (Figs.10.1–10.6) reveals that donor polymer is the main light-harvesting component of the polymer/fullerene active blends. The contribution from the acceptor PCBM in harvesting the light is relatively small compared to the polymer MDMO-PPV. Although PCBM lacks good light absorption, its good transport property and stability has opened a new path of research to use it as a good donor in the active material preparation. In the case of the active materials made up of polymer/polymer blends, light harvesting is relatively more because of the high absorption coefficient of both donor and acceptor polymers. But these active materials suffer from poor carrier transport properties that hinder the rate of charge collection and hence the reduced efficiency. Thus the study of polymer /fullerene blend system has significant research importance. The poly[2-methoxy-5(3’,7’-dimethyloctyloxy)-1,4-phenylenevinylene] (MDMO-PPV) is a conjugated and light emitting polymer easily soluble in organic solvents like chloroform, toluene, CB, 1,2, dichlorobenzene. It is a conducting polymer with a high level of crystallinity.[12] Pure MDMMO-PPV film has absorption extending from 300 to 600 nm with $\lambda_{max} = 500$ nm. The onset of absorption is 600 nm indicates that there is no absorption above 600 nm. Pure PCBM has the strongest absorption in the UV region with a broad tail of absorption extending beyond the visible region. The strong absorption in the UV region arises from HOMO to LUMO (highest occupied molecular orbital to lowest unoccupied molecular orbital) transitions[12] or, in other words, the strongest absorption is attributed to the formation of higher excited singlet states.[13] Spectra for all other blends are the superposition of component spectra. Effect of PCBM in the blend is to reduce the absorption of MDMO-PPV in the visible region followed by an increase of absorption in UV region and in the region beyond 600 nm. Although the 3:1and 1:1 blends have enough absorption in the visible region, 1:3 blend can be considered to be the optimized blend due to enhanced absorption beyond the visible region of the spectrum. The onset wavelength and energy gap are calculated using Tauc's plot (Figs. 10.7–10.10) and the results are depicted in Table 10.2.

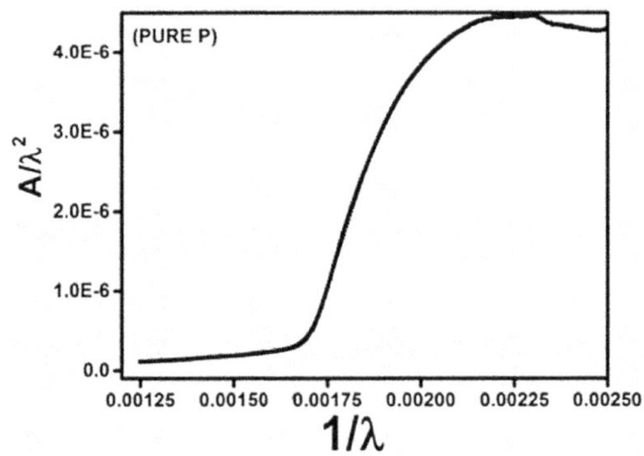

FIGURE 10.7 Pure P.

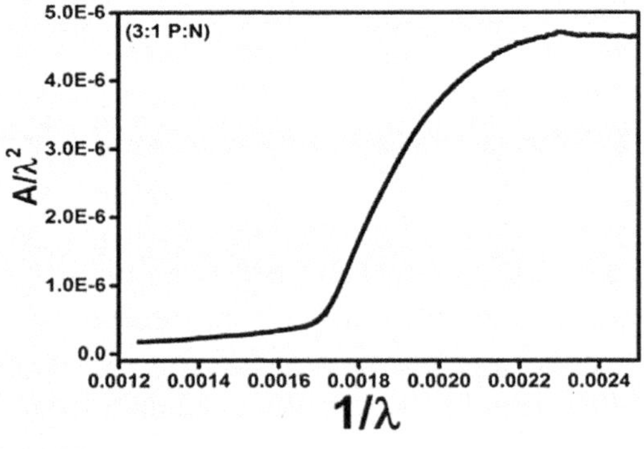

FIGURE 10.8 3:1 P:N.

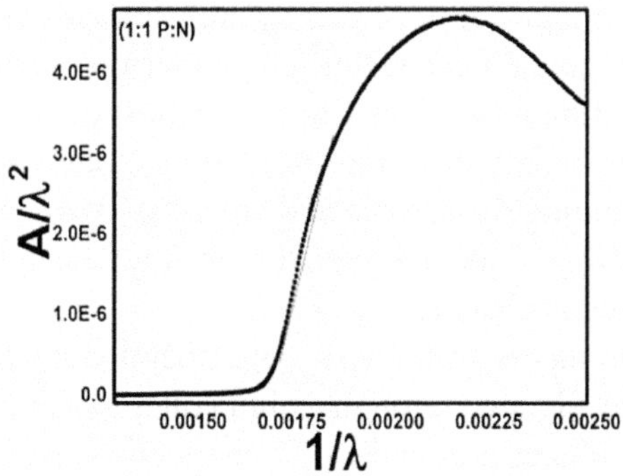

FIGURE 10.9 1:1 P:N.

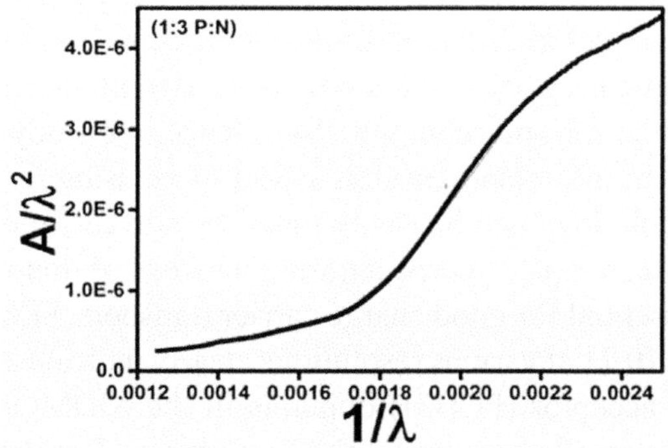

FIGURE 10.10 1:3 P:N.

TABLE 10.2 Optical Parameters of the Prepared Samples.

Sample	λ_{max} (nm)	λ_{onset} (nm)	E_g^{opt} (eV)
Pure P	501.3	578.70	2.143
3:1	486	578.37	2.144
1:1	473	575.04	2.156
1:3	444	557.41	2.225

10.4.3 SCANNING ELECTRON MICROSCOPY (SEM)

SEM is a type of electron microscope that produces images of a sample by scanning it with a focused beam of electrons. The electrons interact with atoms in the sample, producing various signals that contain information about the sample's surface topography and composition. The electron beam is generally scanned in a raster scan pattern produce an image. SEM can achieve resolution more than 1 nm. The most common SEM mode is the detection of secondary electrons emitted by atoms excited by the electron beam. By scanning the sample and collecting the secondary electrons that are emitted using a special detector, an image displaying the topography of the surface is created.

A normal scanning electron microscope operates at a high vacuum. The electron beam is accelerated through a high voltage (e.g., 20 kV) and passes through a system of apertures and electromagnetic lenses to produce a thin beam of electrons, then the beam scans the surface of the specimen by means of scan coils. Electrons are emitted from the specimen by the action of the scanning beam and collected by a suitably positioned detector. SEM images of the samples at different resolutions are given in Figure 10.11.

Optoelectronic properties of the polymer-based photoactive blends are determined by the film morphology also. Hence, the study of active film morphology is an important research aspect in devising the organic solar cells. Film morphology can be easily tuned by solvent selection and also by preparation technique. Interpenetrating network of donor and acceptor molecules is essential for good charge carrier transport. SEM images indicated in Figure 10.11 at various resolutions clearly indicate almost uniform distribution of acceptor (PCBM) domains in the MDMO-PPV polymer matrix. The very small nanoscale grain sizes observed are because of good solubility of both MDMO-PPV and PCBM in CB. The SEM images also indicate almost uniform dispersion and intimate mixing accounting for strong inter-chain interaction which will increase the efficiency.[14]

10.4.4 EDS

Energy-dispersive spectroscopy (EDS) is a supplementary tool of SEM to determine the elemental analysis along with their composition in the SEM imaged sample. It is based on the principle that the electron beam hitting the sample will generate X-rays within the sample. The emitted X-rays have the varying energies determined by the elements from which it is emitted. A careful investigation of the energies of the emitted X-ray will reveal the elemental analysis along with the percentage composition. Figure 10.12 indicates the EDS images representing the elemental composition of the samples and the results signify the purity of the samples prepared.

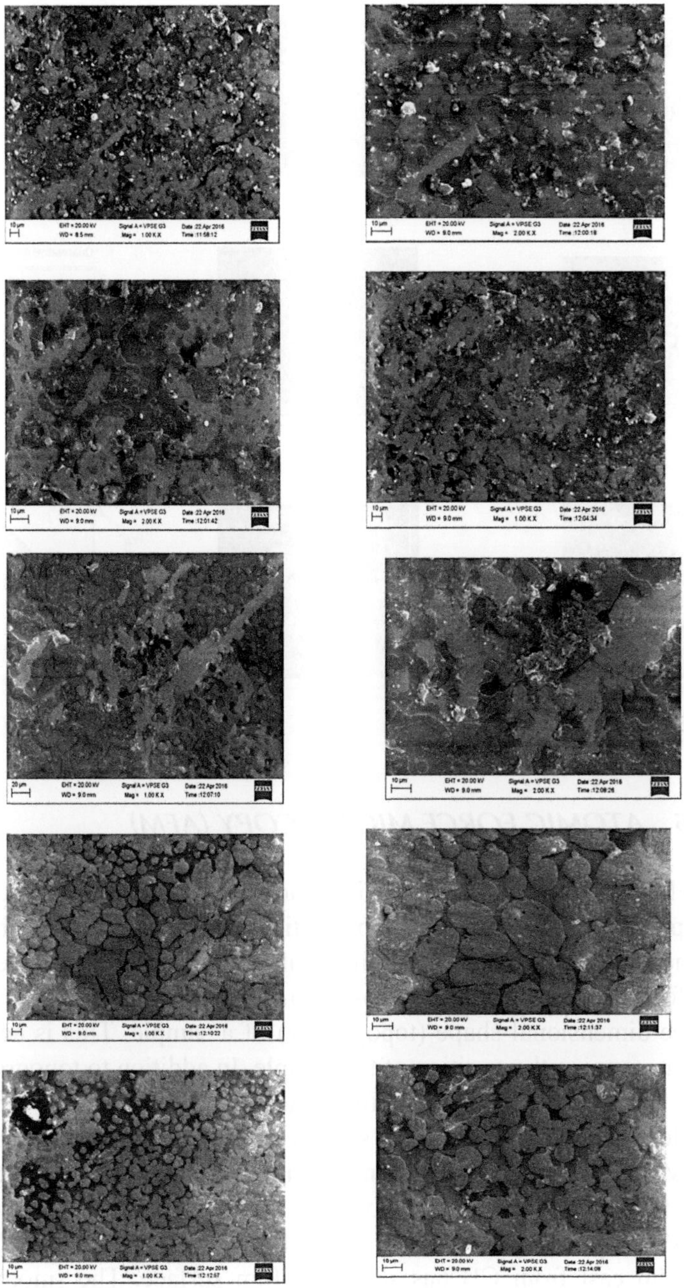

FIGURE 10.11 SEM images.

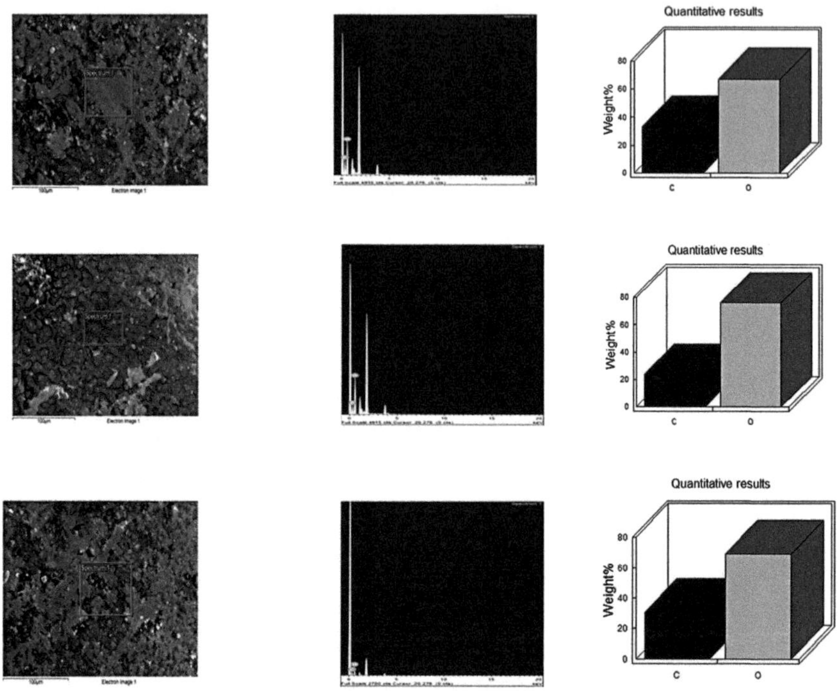

FIGURE 10.12 EDS results.

10.4.5 ATOMIC FORCE MICROSCOPY (AFM)

The AFM enables to get the three-dimensional surface profile, unlike SEM which provides two- dimensional projections. It has three major abilities: force measurement, imaging, and manipulation. The reaction of the probe to the forces that the sample imposes on it, is used to form an image of the three-dimensional shape (topography) of a sample. This is achieved by raster scanning the position of the sample. In addition to topographical images, other properties like stiffness, adhesion strength, and electrical properties such as conductivity or surface potential can also be measured. An AFM can operate in different modes like contact mode, intermittent or tapping mode, and non-contact mode.

AFM images (Figs. 10.13–10.18) indicate that MDMO-PPV/PCBM active film cast from CB has smoother film surface. The smooth film surface is the result of intimate intermixing and interchain interaction needed for efficiency enhancement. The AFM study justifies the selection of CB as the solvent in preparing MDMO-PPV/PCBM bulk junction active blend in the present experiment.[14]

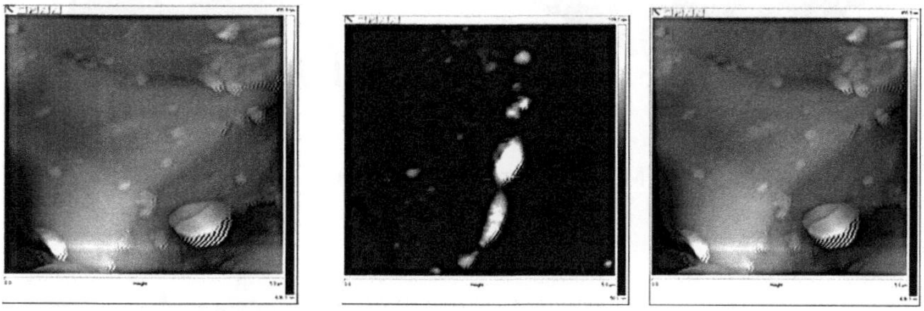

FIGURE 10.13 5 μm AFM images (two dimensional).

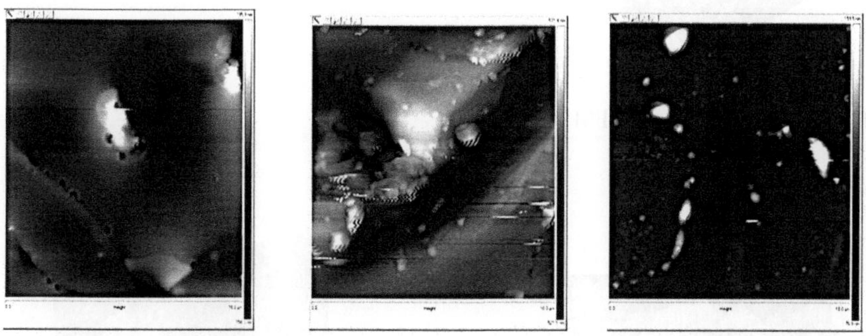

FIGURE 10.14 10 μm AFM images.

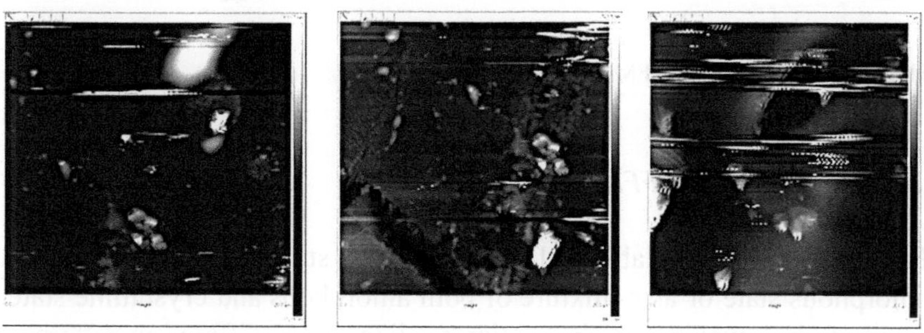

FIGURE 10.15 20 μm AFM images.

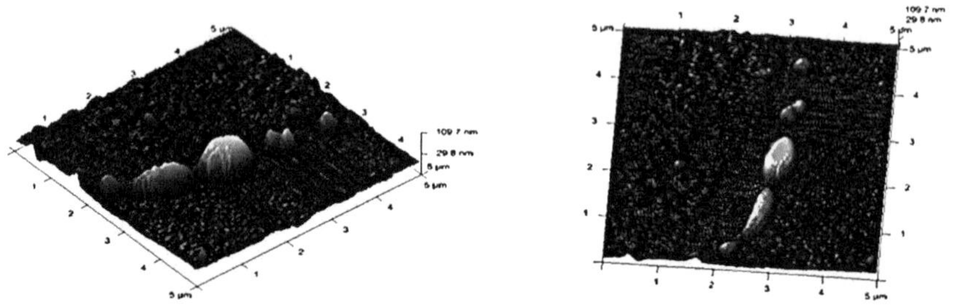

FIGURE 10.16 5 µm AFM images (three dimensional).

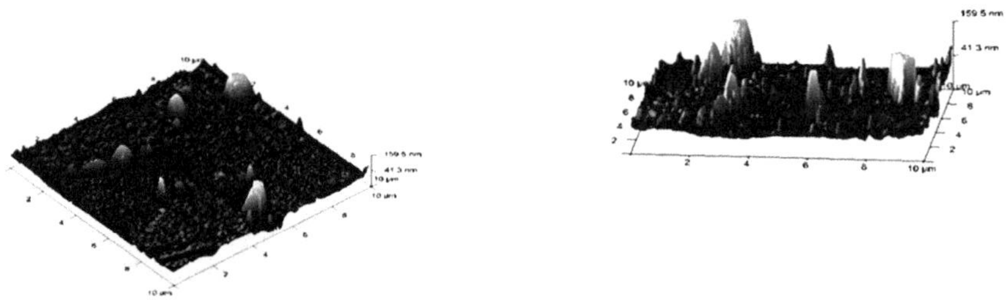

FIGURE 10.17 10 µm AFM images.

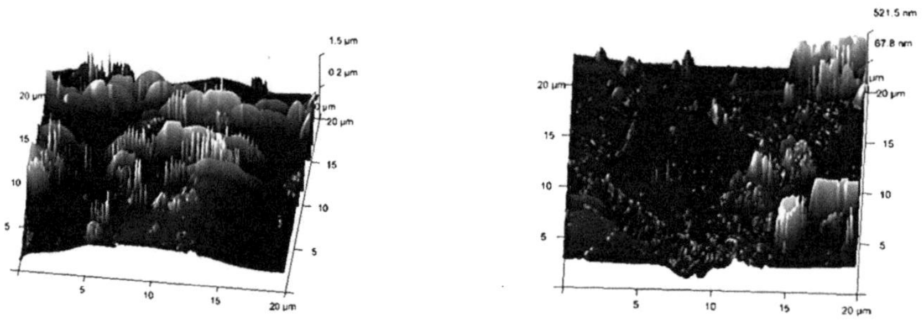

FIGURE 10.18 20 µm AFM images.

10.4.6 X-RAY DIFFRACTION (XRD)

Polymers are not available in 100% crystalline state, they are available in amorphous state or as a mixture of both amorphous and crystalline state. They can be considered to be semicrystalline. X-ray diffraction pattern of polymers contains both sharp and defused bands. Sharp bands correspond to crystalline orderly regions and defused bands correspond to amorphous regions.[15] The fraction of the material that is crystalline, the crystallinity or crystalline index, is an important parameter in the two-phase model. Crystallinity can be determined from a wide-angle X-ray diffraction (WAXD) scan by comparing the area under the crystalline peaks to the total scattered intensity. the crystalline parameters of prepared composites were presented in Table 10.3.[16]

Percentage of crystallinity, X_c % is measured as the ratio of crystalline area to total area.

$$X_c\% = \{(A_c/A_a) + A_c\} \times 100(\%)$$

TABLE 10.3 Crystalline Parameters of the Prepared Samples.

Sample	2θ	d-spacing	Size <P> nm	R (A)
PURE MDMO-PPV	14.3714	6.15815	0.938245855	7.697727635
	22.2891	3.98528	3.662464421	4.981635745
3:1 MDMO-PPV:PCBN	15.3498	5.76777	1.205319644	7.2097361
	22.4525	3.95666	4.353504477	4.945841542
1:1 MDMO-PPV:PCBN	14.768	5.99365	1.029139789	7.492102428
	22.3804	3.96923	4.534802329	4.961570825
1:3 MDMO-PPV:PCBN	15.3182	5.7796	1.068745	7.224520134
	22.1794	4.00475	3.409473268	5.005964331
Pure PCBN	15.0415	5.88527	1.003760717	7.356635411
	22.4284	3.96085	4.630866164	4.951087854

Analysis of XRD (Fig. 10.19) reveals that the active samples are neither purely crystalline nor amorphous. It is partially crystalline and partially amorphous with dominance of amorphous nature as indicated by the diffused spectra. Crystallinity gives rise to near infrared (NIR) absorption, hole mobility. Reports indicate that crystalline structure tends to reduce the charge recombination leading to increased charge collection rate. XRD study stresses upon the need for improvement in the crystallinity of the blend. The thermal annealing can be carried out on the active blend for better device performance.

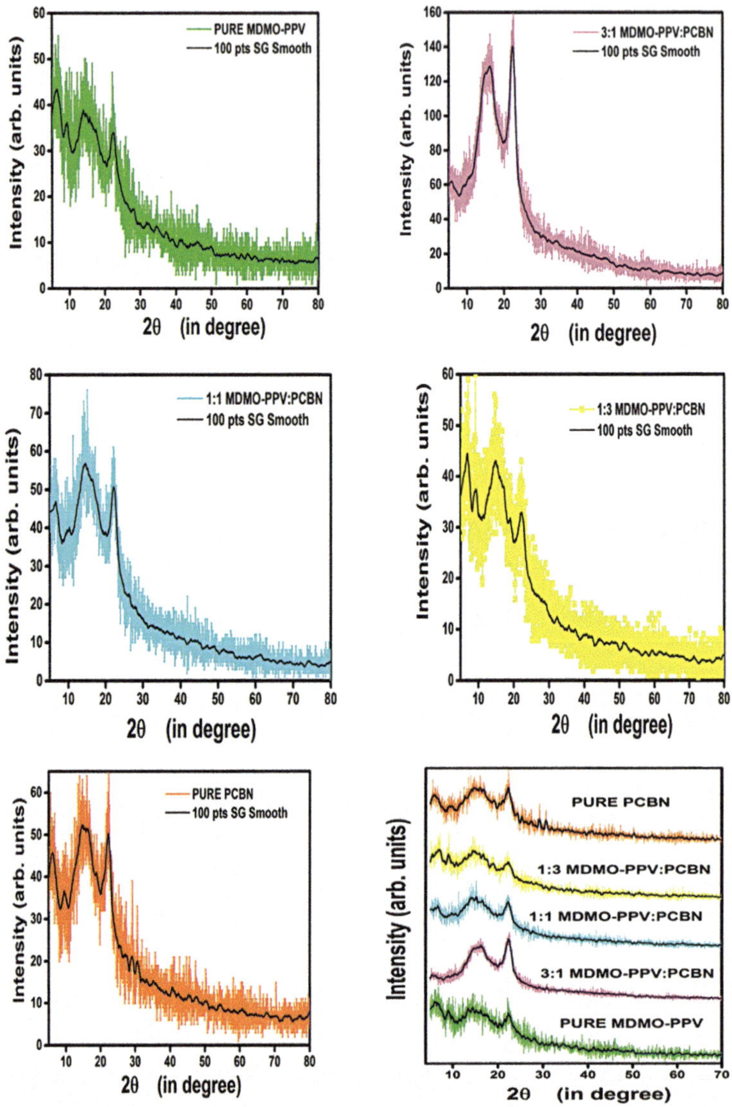

FIGURE 10.19 XRD spectra of polymer composites.

10.5 CONCLUSIONS

We have investigated UV–Visible absorption spectra for 3:1, 1:1, and 1:3 blends of MDMO-PPV:PCBM mixed P–N junction photoactive material along with their prestrine glass-coated films. Spectral analysis indicated that increased weight percentage of PCBM in the blend has enhanced the spectral sensitivity of the material 1:3 blend of MDMO-PPV:PCBM shows a broad spectral absorption covering UV, visible region, and extending beyond the visible region. The optical band gap of 1:3 blend is determined to be 2.225 eV. By doping with dye or metal nanoparticles like gold or silver, absorption can be enhanced to harvest maximum solar energy. The SEM images confirmed the formation of multi P–N junctions in the blend and almost uniform distribution of the PCBM domains in the polymer matrix. Small grain sizes of the dispersed phase justify the use of CB as the proper solvent. AFM studies confirm the smoothness of the surface accounting for intensive inter chain effects leading to good transport property. Diffused spectra of X-ray diffraction study confirm the dominance of amorphous structure in the blend and suggests the need for improvement in the crystallinity. Thermal annealing can be carried out on the blend during device fabrication for improved device performance. Finally, we conclude that 1:3 blend of MDMO-PPV:PCBM can be used as the photoactive material for constructing a plastic solar cell and the construction of the solar cell is under progress.

ACKNOWLEDGMENTS

The authors are grateful to UGC for financial assistance by sanctioning the minor research project entitled "Construction and Characterization of an Organic Solar Cell (OPV) devised from a Self-made, Low cost Spin coating machine." No. 1419-MRP/14-15/KAKA088/UGC- SWRO, dated 4-2-2015.

KEYWORDS

- **MDMO-PPV**
- **PCBM**
- **CB**
- **solar cell**

REFERENCES

1. Alexander, P.; Nikolay, O.; Alexander, K.; Galina, S. *Prog. Polym. Sci.* **2003,** *28,* 1701–1753.
2. Annamali, P. K.; Dilip, D.; Namrata, S. T.; Raj, P. S. *Prog. Polym. Sci.* **2009,** *34,* 479–515.
3. Khan, M. F. S.; Alexander, A. B. *Nano Lett.* **2012,** *12,* 861–867.
4. *McGehee, D. G.; Topinka, M. A. Nat. Mater.* **2006,** *5 (9), 675–676.*
5. *Nelson, J. Curr. Opin. Solid State Mater. Sci.* **2002,** *6, 87–95.*
6. Cao, W.; Xue, J. Recent Progress in Organic Photovoltaics: Device Architecture and Optical Design. *Energy Environ. Sci.* **2014,** *7* (7), 2123.
7. Li, G.; Shrotriya, V.; Yao, Y.; Haung, J.; Yang, Y. *J. Mater. Chem.* **2007,** *17,* 3126–3140.
8. Tipnis, R.; Laird, D.; Mathai, M. Polymer Based Materials for Printed Electronics: Enabling High Efficiency Solar Power and Lighting. *Mater Matter.* **2008,** *3* (4), 92.
9. Organic Solar Cell–Wikipedia. Bilayer-Organic-Photovoltaic Cells. http://en.wikipedia.org/wiki/Organic solar cell#
10. Ghobadi, N. *Int. Nano Lett.* **2013,** *3,* 2.
11. Alias, A. N.; Kudin, T. I.; Zabidi, Z. M.; Harun, M. K.; Ali, A. M. M.; Yahya, M. Z. *Adv. Mater. Res.* **2013,** *652–654,* 527–513.
12. Iwan, A.; Tazbir, I.; Sibinski, M.; Boharewicz, B.; Pasciak, G.; Balcerzak, E. S. *Mater. Semicond. Process.* **2014,** *24,* 110–116.
13. Brabec, J. C.; Gowrisanker, S.; Halls, J. J. M.; Laird, D.; Jia, S.; Williams, S. P. *Adv. Mater.* **2010,** *22,* 3839–3856.
14. Chen, L.; Hong, Z.; Li, G.; Yang, Y. *Adv. Mater.* **2009,** *21,* 1434–1449.
15. Gowariker, V. R.; Viswanathan, N. V.; Sreedhar, J. *Polymer Science;* New Age International (P) Ltd.: New Delhi, 1986; p 173.
16. Murthy, N. S. *Rigaku J.* **2004,** *21* (1), 15–24.

PART III

Nanostructured Materials for Biomedical Applications

CHAPTER 11

SYNTHESIS OF WELL-DISPERSED NANOPARTICLES FOR NANOFINISHING TEXTILES AND BIOMEDICAL APPLICATIONS

SIDHHARTH SIROHI[1*], RATYAKSHI NAIN[1], KRISHNA DUTT[1], BALARAM PANI[1], NISHANT JAIN[1], RAVINDER SINGH[1], PIYUSH WADHWA[1], GAJENDRA SAINI[2], and KAMLESH PANWAR[3]

[1]*Bhaskaracharya College of Applied Sciences, University of Delhi, Dwarka 110075, Delhi, India*

[2]*Advanced Instrumentation Research Facility, Jawaharlal Nehru University, New Delhi 110067, Delhi, India*

[3]*Department of Textile Technology, Indian Institute of Technology, Hauz Khas, New Delhi 110016, India*

[*]*Corresponding author. E-mail: siddharth.sirohi@)bcas.du.ac.in*

CONTENTS

ABSTRACT

ZnO nanoparticles (NPs) find tremendous applications in the area of electronics, optoelectronics, bioengineering, catalysis, biosensors, and nanofinishing textiles. Various methods of synthesis are described in the literature like hydrothermal, chemical, solvothermal, sol–gel, and electrochemical synthesis. The aggregation tendency of NPs restricts the complete use of their potential. Earlier, the use of surfactants and capping agents has been reported for improving the dispersion behavior of NPs. However, the results are not promising and need further improvement. In this work, the synthesis of ZnO NPs is carried out and the effect of surfactant has been investigated. The resulting NPs were characterized using particle size analyzer, ultraviolet/visible (UV/VIS) spectrophotometer, and field emission-scanning electron microscopy (FE-SEM). It was observed that sodium dodecyl sulfate (SDS) acted as an effective stabilizer for the dispersion of NPs and resulted in the improvement of particle size and enhanced the UV properties. These well-dispersed NPs may be used as potential antimicrobial agent and nanofinishing textiles.

11.1 INTRODUCTION

The requirement for the development of a material with enhanced properties is to control its structure at smaller dimension in order to create a range consisting of novel characteristics, functions, and applications. The larger surface area (per unit mass) compared to the same mass of material produced in bulk form is the primary factor that differentiates the properties of nano-materials over bulk materials. The larger surface area of nanomaterials results in increased chemical reactivity and electrical properties. Since most of the functionalization procedures and chemical reactions take place on the surface of nanomaterials, they possess greater reactivity than the similar mass of larger-scale materials. ZnO NPs have drawn much attention in recent times due to their potential technological application namely solar energy conversion, catalysis, nonlinear optics, gas sensors, UV-blockings, etc. The properties of ZnO NPs may be tuned by controlling conditions of synthesis including surfactants, solvents, and concentration of precursor. This study aims at investigating the effect of sodium dodecyl sulfate (SDS) on the size, dispersion behavior, and properties of ZnO NPs.

Zinc oxide, being a wide band gap semiconductor (3.37 eV), has attracted tremendous interest in last few decades. ZnO could be in the form of transparent film materials and is used as transparent electrodes for solar cells, ZnO NPs applications in gas detection sensors, electrode and piezoelectric devices, etc., which have been exploited largely. More recently, ZnO nanostructures offer applications in optoelectronic devices. Great efforts have been devoted to synthesize ZnO nanostructures, such as the chemical vapor deposition (CVD) process, the thermal evaporating and pyrolysis process, wet chemical methods, etc. Among these, the chemical solution method was found to be best-exploited method to prepare nanostructured materials because of its low cost, versatility, and potential large-scale production. The wet chemical methods include hydrothermal approaches, reflux, microemulsion, etc. Polymer nanocomposites are a unique class of materials where nanomaterials are uniformly dispersed in the polymer matrix. The hybrid properties are offered by both the components, that is, the polymer as well as NPs. Interaction of both the components, that is, NPs and polymer result in not only reinforcement of the polymer matrix, but also the additional functionalities like enhancement in the other properties such as mechanical, electrical, thermal or optical properties which can be obtained while retaining the polymer property of easy processability. These polymer nanocomposites can further be converted into various structures and offer applications in aerospace, biomedical, automotive, and protective textiles.

Although a significant amount of work has already been done using micron-sized particles for reinforcement, these composites are not easy to process. NPs, on the other hand, when dispersed effectively in polymer melt or form solution, resulting in good interfacial interaction with polymer because of their very high specific surface area. NPs may influence the bulk properties efficiently even when present in a very small fraction. The enhanced properties of these nanostructures can, however, be exploited only if they are homogeneously dispersed and embedded inside the polymer matrix.

Nanomaterials have very large surface area per unit mass. Hence, most of the atoms are at the surface resulting in enhanced functionality. Further, the quantum effect dominates at nanoscale affecting the optical, electrical, and magnetic behavior of the materials. Nanocomposites are multiphase solid materials with at least one of its phase having one, two, or three dimensions in the range of 100 nm.

The NPs are available in various shapes such as spherical, hexagonal cone, hexagonal plate, and rod shape. Being 3D confinement at nanoscale, the NPs offer a very large surface area for interaction with the matrix and hence, lead to significant improvement in the properties.

Nanoparticle shape and size can be controlled by controlling the synthesis conditions, that is, the rate of nucleation and growth. Different shapes of nanostructures including spherical, rod/tube, flower, rugby, needle, etc., can be synthesized depending upon whether nucleation or growth is the rate-determining step. If the growth rate dominates, nanostructures having larger aspect ratio can be synthesized. This can be achieved by changing the synthesis conditions such as precursor concentration, temperature, polarity of the solvent, etc. The shape, size, and aspect ratio of NPs affect the interaction and adhesion of polymer chains with incorporated NPs. The smaller particle size offers a higher surface area for matrix–NP interaction and provides better properties.

11.2 SYNTHESIS OF ZnO

Synthesis of ZnO particles can be classified into two main categories on the basis of production scale/yield, namely, large-scale metallurgical processes and small-scale chemical processes. These processes differ in synthesis conditions including production temperature and precursors.

11.2.1 METALLURGICAL METHODS

11.2.1.1 INDIRECT PROCESS

Indirect process or French process is most widely used industrial method for ZnO particle synthesis. In this method, zinc metal is heated in crucible at about 1273 K. The vapors of zinc metal are allowed to pass through an orifice into the atmosphere with controlled rate where rapid oxidation of zinc vapors leads to ZnO formation. Zinc oxide obtained from indirect method must be purified by means of physical processes in order to remove ash content. ZnO particles from indirect process have particle size from 20 to 3000 nm and the shape of particle was found to be nodular.[1]

11.2.1.2 DIRECT PROCESS

In direct or American process instead of oxidized zinc, zinc metals are used. Raw materials are fed into a heated chamber where these oxides are reduced back to zinc metal to improve their purity. Zinc metal obtained from reduction is then processed as in indirect method. The main difference between ZnO particles formed by the direct and indirect methods is in the degree of purity. Zinc oxide produced from the direct method is inferior to zinc oxide obtained from the indirect process due to impurities present in raw materials. Generally, direct processed zinc oxide is utilized in paints and ceramics.

11.2.2 CHEMICAL PROCESSES

The control of size, increased purity, and hence, improved properties need highly controlled synthesis conditions. Although various methods are available for synthesis of ZnO, the chemical route is the widely used method due to ease of production, cost-effectiveness, and tailoring the size and shape of obtained ZnO NPs.

11.2.2.1 CONTROLLED PRECIPITATION

In controlled precipitation zinc, precursor is reduced rapidly with the help of a reducing agent. In the next step, the precursors are calcined and impurities are removed by milling. Chemical precipitation can facilitate the synthesis of tailor-made ZnO NPs by varying the synthesis conditions like temperature, pH, and time of reaction. The main advantage of controlled precipitation method is the ability to synthesize ZnO NPs with reproducible properties.

The precursors for ZnO NPs synthesis could be zinc chloride, zinc acetate, and zinc sulfate. Zinc oxide precipitated from aqueous solutions of zinc chloride and zinc acetate yield high surface area NPs. However, lower surface area (<30 m^2/g) ZnO is obtained using zinc sulfate as precursor.[1,2]

11.2.2.2 SOL–GEL METHOD

Sol–gel method provides straightforward and economical way for ZnO NPs synthesis. Reproducible properties can be obtained from sol–gel synthesis. The properties of ZnO can easily be controlled by playing with process parameters such as temperature and nucleation time, etc. Scheme 11.1 shows the process of ZnO NP synthesis by sol–gel method. The ZnO NPs obtained from sol–gel process were reported to have better optical properties.[3]

Benhebal et al.[4] synthesized zinc oxide powder using sol–gel process. The precursors used for the synthesis were zinc acetate dehydrate and oxalic acid and ethanol was used as solvent. The obtained ZnO NPs were found to be spherical in shape with hexagonal wurtzite structure and high surface area of 10 m^2/g. This range of surface area is a character of a material with low degree of porosity.

The nanocrystalline ZnO has also been prepared using sol–gel process.[5] Tetramethylammonium hydroxide (TMAH) solution was added into zinc 2-ethylhexanoate (ZEH) solution in propan-2-ol. After 30 min thus, obtained colloidal suspension was washed with water and ethanol. The ZnO NPs obtained when characterized by transmission electron microscope (TEM) were found to be of 20–50 nm size.

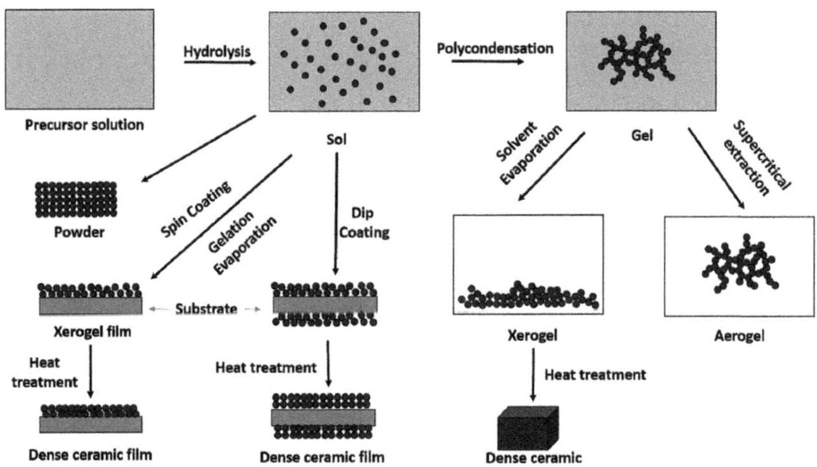

SCHEME 11.1 Synthesis of ZnO NPs by sol–gel method.

11.2.2.3 SOLVOTHERMAL METHOD

As the name suggests, solvothermal process involves thermal treatment of precursor–solvent mixture (when solvent is water then method is called hydrothermal) at 100–300°C in an autoclave for several hours (12–24 h). The process-involved nucleation followed by growth of ZnO NPs. The particles prepared by solvothermal process do not need post processing such as calcination or grinding. Apart from this there are other advantages such as reaction can be carried out at low temperature, shape, and size of particles can be varied by changing concentration of precursors, pressure, and temperature and highly crystalline particles can be obtained.[6,7]

Chen et al.[8] proposed a scheme for the ZnO NPs synthesis using hydrothermal process. Zinc chloride ($ZnCl_2$) and sodium hydroxide (NaOH) (molar ratio of 1:2) were used as precursors.

$$\mathrm{Cl{-}Zn{-}Cl} + 2\mathrm{NaOH} \longrightarrow \mathrm{OH{-}Zn{-}OH} + 2\mathrm{Na}^{\oplus} + 2\mathrm{Cl}^{\ominus}$$

The white precipitate of $Zn(OH)_2$ was filtered and washed. Thus, obtained ZnO particles were then redispersed in deionized (DI) water, and the pH was maintained at 5–8 using hydrochloric acid (HCl). In the autoclave, heating for 3 h was carried out followed by cooling at room temperature. The end product ZnO was obtained by the following step:

$$\mathrm{OH{-}Zn{-}OH} \xrightarrow{\text{Heating}} \mathrm{ZnO} + \mathrm{H{-}O{-}H}$$

X-ray diffraction (XRD) and TEM were used to determine the average size and the structure of the obtained ZnO particles. The process temperature and reaction time significantly influenced the size and structure of the particles. An increase in the pH of solution leads to increase in crystallinity and size.

11.2.2.4 EMULSION OR MICRO-EMULSION METHOD

Vorobyova et al.[9] reported the preparation of ZnO by emulsion method. Sodium hydroxide solution in water or ethanol was added to zinc oleate solution in decane and ZnO was obtained as precipitate. Solvents were removed and particles were dried at room temperature. The obtained particles were characterized by XRD and SEM. It was found that reaction may be carried out both in aqueous and in organic phases. The process conditions such as temperature, molar ratio of components, etc., affect particles size and their phase location.

Li et al.[10] prepared nanometric ZnO using microemulsion. The microemulsion can be prepared by adding an alcohol into an emulsion of water and oil with an emulsifier. This mixture becomes transparent on microemulsion formation. In this work, the microemulsion contains heptane and hexanol solution and a non-ionic surfactant was used (Triton X-100). The reaction and growth take place in droplets of microemulsion involving precursor exchange between droplets and polyethylene glycol (PEG) (medium). The concentration of PEG varied from 0–50%, which affected the structure of obtained ZnO NPs.

11.2.3 GREEN SYNTHESIS

The synthesis of NPs from the physical and chemical methods is a time-consuming task. To prevail over this problem, green synthesis, in which biological systems are used for synthesis, has also been used.[11] Furthermore, green synthesis is an ecofriendly and economical process.[12]

Awwad et al. synthesized ZnO nanosheets using *Olea europaea* leaf extract.[13] The size of NPs was found to be 20 nm. The various biomolecules present in the synthesis system are responsible for reduction of NPs.[11]

Nano catalyst obtained from *Averrhoa bilimbi* fruit extract broth were utilized in gold and silver NPs where salt chloroauric acid and $AgNO_3$ were precursors for Au and Ag NPs, respectively.[14] Formation of yellow and violet color indicated Ag and Au NPs formation. Lower UV absorbance was observed at lower concentration due to the slower rate of NPs formation.

Vidya C et al. used leaf extract of *Calotropis gigantea* for ZnO NPs synthesis (Scheme 11.2).[15] The prepared particles were found to be spherical in shape with average particle size 10–15 nm confirmed by SEM and XRD, respectively. It was found that rate of synthesis is dependent on temperature, and higher temperature favors rapid growth rate.

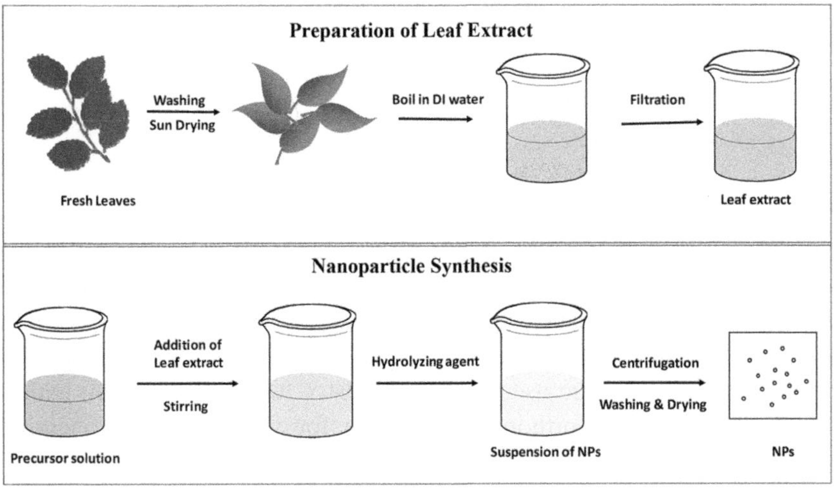

SCHEME 11.2 Synthesis of ZnO NPs using green synthesis.

11.3 APPLICATIONS

11.3.1 AGRICULTURE

NPs help in efficient distribution of pesticides, fertilizers, and herbicides. They not only minimize the fertilizer consumption but also decrease environmental pollution. Nanotechnology plays a vital role in controlled delivery of pesticides and fertilizers, disease detection, and disease control.[16] This method not only minimizes nutrient loses in fertilization but also increases yield through optimal nutrient management.

Zinc oxide treated peanuts helped in increasing its stem and root. NPs also promote seed germination, plant growth, and seeding vigor.[17] The zinc oxide colloidal solution was used as fertilizer. These nanofertilizers are used in very small amount acting as plant nutrient and soil revitalizer. The requirement of nanofertilizer for mature tree is about 40–50 kg while an amount of 150 kg of ordinary fertilizers would be required.[18,19] The yield of wheat plants grown from seeds which were treated with metal NPs on average increased by 20–25%.[20] ZnO NPs were studied for the control of Grasserie disease in silk worm caused by the *Bombyx mori* nuclear polyhedrosis virus (NPV) and Rice weevil in rice.[21] Application of these NPs improves and increases crop growth.

Therefore, the commercialization of such NPs is imperative for sustainable agriculture.[22]

11.3.2 MEDICINE

Nanotechnology is the most sophisticated technology which assures accuracy and sensitivity in vitro and in vivo diagnostics. This technology is widely used in early detection, prevention, and treatment of diseases such as cancer, atherosclerosis, etc. It follows targeted drug delivery system in which time of drug release and specific targeting of diseased cell plays an important role. It is also used in nanodentistry for treating Dentin hypersensitivity (DH) and orthopedic implants for fixing internal fractured bones.[23–25]

ZnO and other metal oxide nanomaterials are used as biomarkers for cancer diagnosis, screening, and imaging. Studies depict that ZnO NPs possess a high degree of cancer cell selectivity and can also be capped with poly(methyl methacrylate) for detecting low abundant biomarker.[26–28]

Atherosclerosis is an inflammatory disease associated with cardiovascular complications such as stroke, ischemia, and myocardial infarctions. Inflammation in human aortic endothelial cells was observed after exposure to engineered iron oxide and zinc oxide NPs.[29]

11.3.3 ANTIMICROBIAL ACTIVITY

Due to the small size and high surface to volume ratio, metal NPs interacts with microbial membrane and facilitates penetration into the cell thus, excluding internal components of the cell and making the microorganism inactive by inhibiting the cell growth, in turn, causing cell death.

Structure of cell wall, rate of bacterial growth, and type of NPs are some of the major factors influencing the susceptibility or the tolerance of bacteria to NPs. NPs are used as antibacterial agent in textile industry, water disinfection, medicine, and food packaging.[30,31]

In Kirby Bauer method, the solution whose antimicrobial activity has to be checked is made to diffuse through a solid medium on which bacterial cells are cultured (Scheme 11.3). The area in which the solution gets diffused appears to be clear upon incubation, that is, no cells are seen due to antimicrobial activity there. In this method, subculture of the required bacterial species is prepared using a mother culture by spreading and before incubation a muslin cloth, dipped in the NP solution is placed at the center of the sub-cultured plate. This plate is then incubated at 37°C for 24 h. During incubation, the NP solution diffuses away from the muslin cloth. If the solution is intended to show the antimicrobial activity, it will lyse or kill the cells present in the area of its diffusion and form a "Zone of Inhibition." Larger the diameter of the zone of inhibition, stronger is the antimicrobial of the NP solution.

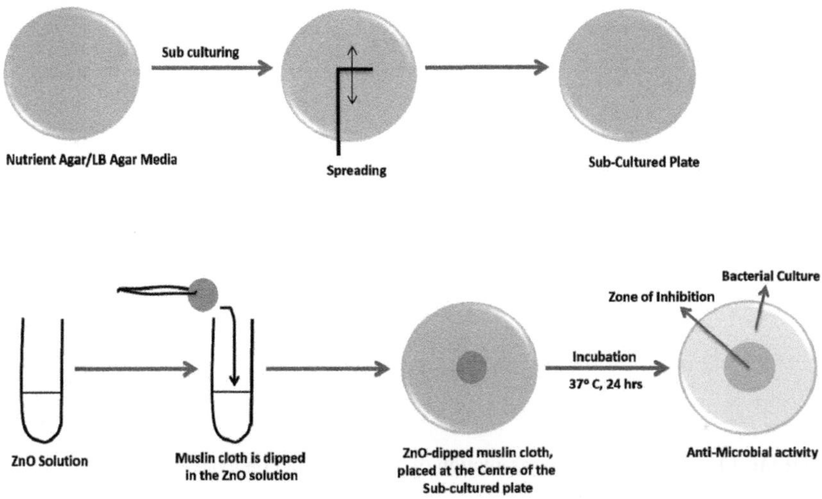

SCHEME 11.3 Antimicrobial susceptibility testing by Kirby Bauer disc diffusion.

ZnO NPs possess strong bactericidal activity as it tends to disrupt the cell membrane and cause cytoplasmic leakage (Scheme 11.4). It inhibits and kills major food borne and human pathogens like *Escherichia coli DH5a, Staphylococcus aureus, Pseudomonas aeruginosa,* and *Fecal coliform.*

ZnO NPs are more powerful antidiabetic agent than silver NPs. ZnO NPs lead to the reduction of blood glucose, increased insulin level, and expression. Therefore, they are widely used in controlling diabetes and hyperglycemia.[32]

ZnO NPs synthesized via wet chemical process exhibits high optical transparency and UV absorption therefore, it is used in cosmetic industry mainly in facial creams and sunscreens. It reduced the pathogen contamination toward *P. aeruginosa* and *E. coli* which were isolated from mint leaf and frozen ice cream.[33] One more feature of ZnO NPs, as stated earlier, is their ability to induce reactive oxygen species (ROS) generation, which can lead to cell death when the antioxidative capacity of the cell is exceeded.[34–37]

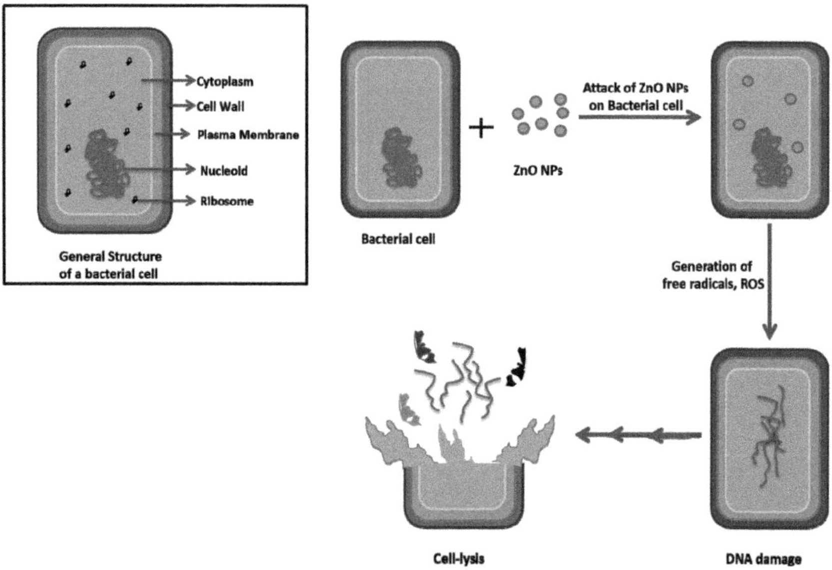

SCHEME 11.4 Schematic representation of action of ZnO on a bacterial cell.

ZnO NPs are widely used in food processing and food preservation industry as sterilizing and disinfectant agents for food containers against pathogens. The NPs of ZnO showed toxicity on both pathogenic bacteria (e.g., *E. coli* and *S. aureus*) and beneficial effects on microbes, such as *Pseudomonas putida*, which has bioremediation potential and is a strong root colonizer.[38] ZnO being non-toxic in nature, it can be used as photocatalytic degradation materials of environmental pollutants. Bulk and thin films of ZnO have demonstrated high sensitivity for many toxic gases.[39]

11.3.4 FOOD PACKAGING

Nanotechnology serves potential application in food packaging and food processing. The antibacterial property of NPs is majorly exploited in active, smart, and intelligent packaging to increase the shelf-life and also protect the food from spoilage. They are also used as nanosensors for the detection of food decay and tracking of fresh packed food items. Food items wrapped with smart safety packaging also detect the microbial spoilage. The use of polymer nanotechnology in packaging was introduced by Silvestre et al.[40] to achieve a novel way of packaging that mainly meets the requirements of protection against bacteria. NPs also increase the barrier property, mechanical property, and constancy of the material. ZnO NPs are used in food lining in packaged food for its protection.

Active packaging system is the replacement of conventional packaging which used passive barriers approach to protect food from surroundings whereas active packaging generates effective antimicrobial atmosphere and saves the product from foodborne pathogens.[41]

Active packaging is also called as antimicrobial packaging due to its bacteriostatic or bactericidal property of NPs. Gradual diffusion of these NPs halts the growth of bacteria.[42] The major advantage of this type of packaging is that it enhances safety and increases shelf-life.[43]

Intelligent packaging offers smart functions, such as sensing, tracing, recording, detection, and communication.[44] This system exploits a number of indicators for monitoring the food quality in terms of microbial growth, temperature, and packing integrity.[45,46]

11.3.5 WATER PURIFICATION

High concentration of toxic contamination and harmful microorganism in natural resources of drinking water is constantly causing serious environmental pollution. Advances in nanoscale science and engineering suggest that many of the current problems involving water quality could be resolved or greatly diminished by using nanoabsorbent, nanocatalysts, bioactive NPs, and nanostructured catalytic membranes.[47] NPs are expected to play a crucial role in water purification.[48]

Heterogeneous photocatalysis is an efficient approach for water purification in comparison to other conventional methods,[17,18] as it can break up complex long chained toxic organic molecules into simpler pieces as well as rupture cell walls of microbes thereby inhibiting them.

Photocatalysis, using nanostructures of metal oxide semiconductors like zinc oxide (ZnO), titania (TiO_2), etc., is an effective way of water purification as it is capable of removing not only chemical but also biological contaminants.[49,50] This method is highly beneficial due to its low cost, simplicity, ease of controlling parameters, and its ability to degrade organic and inorganic substances in aqueous medium. It has been found that cetylpyridinium chloride (CPC) is readily and rapidly degraded in aqueous solution by UV/ZnO NPs in a relatively short time of about 60 min after selection of desired operational parameters (pH = 8.0, ZnO NPs = 40 mg/100 mL, [AB] = 9.0×10^{-5}, [H_2O_2] = $8 \times 10^{-}5$ mol/dm^{-3}).[51]

ZnO is the most suitable material for water purification as it can be altered to absorb visible light. It is much superior to other metal oxides like TiO_2 in water purification as it is soluble in water and ends up as metallic zinc in the ecosystem,[52] whereas TiO_2 is insoluble in water and the NPs remains in the environment which can pose threat to the environment.[21]

11.3.6 TEXTILE

The use of NPs in textile is quite extensive due to its distinctive and significant properties.[53] Traditional textile finishes are short lived as their properties deteriorate after repeated washes. Therefore, a long lasting finish is required for making the fabric durable. NPs adhere to the fiber surface by van der Waals forces and increases the fabrics wash fastness. The wash fastness can also be increased by dipping the fabric in a specific binder containing NPs.[54,55]

ZnO NPs exhibits UV blocking, antibacterial, and self-cleaning activity toward fabric. Nano-sized ZnO on polyester/cotton blended fabric increased UV light absorbance in the region between 300 and 400 nm due its high UV absorbance capacity on the fabric.[56] Water repellence is imparted to the cotton material simply by coating of a nano plasma over it.[57]

11.3.7 RUBBER INDUSTRY

A major portion of globally produced ZnO is used in manufacturing cross-linked products.[58] The use of ZnO in silicon rubber not only increases thermal conductivity but also retains its electrical resistance. Similar thermal conductivity of compounds can be obtained at relatively low concentration using nanofillers. But ZnO NPs have a tendency to form agglomerate due to high surface energy.

Surface modification of ZnO NPs may lead to better interaction with rubber as carried out by Yuan et al.[59] The modification of sol–gel synthesized NPs (particle size <10 nm) by hydrosilylation. The structure, properties, and morphology of the Si rubber–ZnO (SR/ZnO) were studied. In this work, vinyltriethoxysilane (VTES) was used as coupling agent.

ZnO is the most common crosslinking agent for carboxylated nitrile rubber (XNBR).[60,61] Products with high mechanical properties such as tensile, tear, hardness, etc., can be obtained. There are several disadvantages also like poor scorch resistance, low flex properties, and very high degree of compression set. Scorch resistance can be improved using ZnO/zinc peroxide system and the formed crosslinks are ionic in nature. This combined system, however, requires longer vulcanization time to produce a product with higher crosslinking density in comparison to those cured with ZnO alone. Surface area, morphology, and particle size significantly affects reactivity of ZnO. The size of the interphase (rubber crosslinking agent) is determined by these factors.[62]

Przybyszewska et al.[63] studied the effect of surface area, morphology, and particle size on zinc oxide activity. They found that nano-sized ZnO crosslinking agent gives a product with high degree of crosslinks and hence better mechanical properties than products cured with microsized ZnO. The order of tensile strength with NPs was four times than with microparticles. Thus, only 60% ZnO NPs are required to obtain comparable properties. ZnO is found to be toxic to aquatic species and ZnO cured XNBR shrinks on heating. Therefore, ZnO NPs cured elastomers not only give a product with better properties but are also eco-friendly. Przybyszewska et al.[63] found that only morphology affects the activity of ZnO in curing reactions, whereas surface area and particle size do not play a significant role. 3D Snowflake ZnO having a surface area 24 m^2/g was found to have maximum activity.

Snowflake shaped particles provide high elastomer-ZnO interfacial area due to specific and complex shape. As a result, stress relaxation is observed due to the high concentration of ionic domains producing multifunctional crosslinks, which changes the order when subjected to an external stress leading to better mechanical properties. ZnO NPs used by them had a lesser tendency to agglomerate which only concentrates stresses to a smaller extent while other ZnO forms a larger mass.

Sabura et al.[64] synthesized ZnO via solid-state pyrolytic route and the average particle size was in the range of 15–30 nm and the surface area was found to be 12–30 m^2/g. This ZnO NPs were used for neoprene curing. The researchers found that this nano ZnO consumption for comparable properties is lower than industrial grade ZnO.

As discussed earlier, ZnO NPs can easily be synthesized using various techniques and offer tremendous applications in different areas. NPs however, when synthesized exist as elementary elements that can form aggregates due to particle–particle interaction. These aggregates, if present in polymer matrix, lead to reversible secondary structures called agglomerates. The dispersion of NPs in polymer matrix should be homogeneous to have good polymer–particle interaction. Homogeneous dispersion and distribution of NPs increases the specific surface area available for polymer–particle interaction and hence, better properties can be achieved at comparatively lower loadings. Various interactions existing are van der Waals forces, dipole–dipole interaction, and chemisorptions at the surface of NPs.

Dispersion of nanostructures is an uphill task because of their high surface energy and aggregation tendency. Various methods are reported in the literature for breaking the aggregates which include mechanical stirring, sonication, and the surface modification of NPs. The aggregation tendency of NPs limits their potential and hence the improvement in dispersion is the major challenge in this area. In this work, the synthesis of ZnO NPs is carried out and the effect of surfactant and solvent system on dispersion behavior of ZnO NPs has been investigated.

11.4 EXPERIMENTAL METHODS

11.4.1 MATERIALS

Zinc acetate dihydrate, NaOH, SDS, ethanol, methanol, polyvinyl alcohol (PVA), and dimethylformamide (DMF) were procured from Merck (Bangalore, India). All the reagents were used as provided. Double DI water was used in all the synthesis processes. All glassware were first rinsed with aqua regia and thoroughly with water followed by DI water before use.

11.4.2 SYNTHESIS OF ZnO NANOSTRUCTURES

The ZnO nanostructures were synthesized using the chemical route. Zinc acetate dehydrates ($Zn(Ac)_2.2H_2O$) was used as precursor for the synthesis and NaOH as an oxidizing agent. The Zn^{2+}/OH^{-} ratio was kept 10 and effect of solvent (water, ethanol, methanol, and DMF:water (10:1)) and surfactant (SDS, PVA using water as solvent) on the shape and dispersion behavior of ZnO nanostructure was investigated. In a typical procedure, the precursor, that is, Zinc acetate dehydrate ($Zn(Ac)_2.2H_2O$) was added in 5 mL of solvent water, ethanol, methanol, and DMF: water). In the above precursor solution, the oxidizing agent, that is NaOH dissolved in 5 mL solvent, was added dropwise with a rate of 5 mL/min at 60°C and 500 rpm. This mixture was then stirred further for 2 h at 60°C. The nucleation and hence growth of nanostructure was tailored using different solvents. In order to investigate the effect of surfactant, 1 wt% of the surfactant, namely SDS, low molecular weight PVA (Mw 14,000) and high molecular weight PVA (Mw 125,000) was dissolved in water. The precursor was dissolved in this solution followed by addition of NaOH and growth for 2 h at 60°C. The nanostructures, thus obtained as dispersion in the solvent, were then analyzed for stability and particle size. The morphology of the obtained nanostructures was characterized using particle size analyzer, FE-SEM, and UV spectrophotometer.

11.5 RESULTS AND DISCUSSION

11.5.1 EFFECT OF SURFACTANT ON DISPERSION

The visual assessment of the obtained ZnO particles was done just after synthesis and after one day to observe the stability of dispersion as shown in Figure 11.1. The surfactant used for the study were SDS and low and high molecular weight (HMW) PVA, and it has been observed that in the absence of surfactant and with 1 wt% high molecular weight PVA (Mw 125,000) the dispersion was nonhomogeneous and particle tend to settle down within 30 min. Addition of 1 wt% of surfactant (SDS and low molecular weight (LMW) PVA, Mw 14,000) resulted in the homogeneous dispersion of NPs. Further, when the dispersion was allowed to settle for one day, the dispersion with LMW, PVA also tend to settle down.

ZnO particles obtained using SDS, on the other hand, were stable for a week. This clearly indicated that ZnO particles obtained without SDS are either of large size or in the form of aggregates. The morphology of ZnO NPs obtained with and without SDS was analyzed using particle size as shown in Figure 11.2.

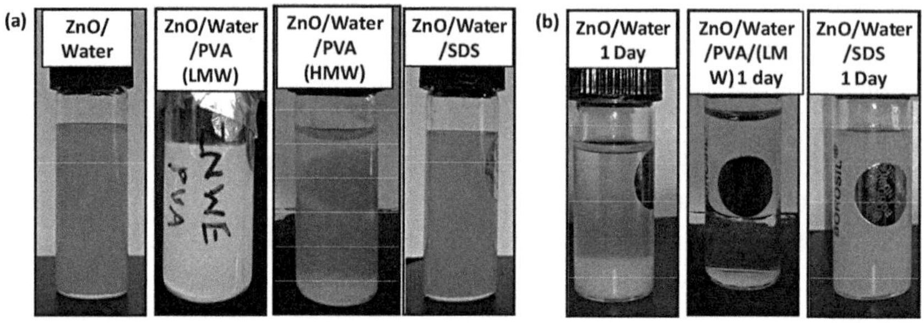

FIGURE 11.1 Visual assessment of ZnO NPs (a) just synthesized and (b) after 1 day of synthesis.

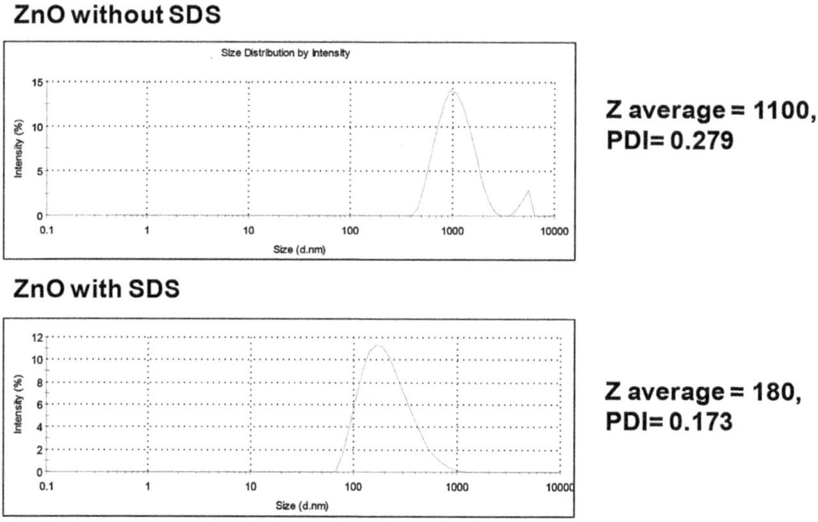

FIGURE 11.2 Particle size of ZnO NPs obtained without and with SDS.

It has been observed that without surfactant the average particle size of ZnO is 1.1 μm. However, ZnO obtained using SDS as surfactant was observed to have reduced average particle size of 180 nm. This clearly indicates that addition of SDS resulted in ZnO particles with nano-confinement. However, the exact particle size was determined using FE-SEM (Fig. 11.3).

FIGURE 11.3 FE-SEM micrographs of ZnO NPs obtained without and with SDS.

The FE-SEM study revealed that without SDS the ZnO NPs have an average diameter of 205 ± 25 nm. Addition of SDS resulted in smaller NPs with an average diameter of 80 ± 10 nm. The surfactant not only helped in confinement of ZnO NPs to smaller diameter but also helped in the formation of stable dispersion.

11.5.2 EFFECT OF SOLVENT ON DISPERSION

The solvents used for the study were water, ethanol, methanol, and DMF. It has been observed that dispersion was homogeneous for ethanol, methanol, and DMF:water, but not stable and tend to settle down after one day. However, the dispersion with DMF:water was easily redispersible. The FE-SEM analysis (Fig. 11.4) revealed the particle size was 100 ± 25 nm, 30 ± 5 nm, and 28 ± 5 nm in ethanol, methanol, and DMF, respectively.

Further, the ZnO nanostructures were found to have UV absorbance, and the maxima of wavelength depends upon the size (Fig. 11.5).

The ZnO NPs with water as solvent without surfactant were found to have λ_{max} at 373 nm. The blue shift was observed at 355 nm as the size was decreased to 80 nm in ZnO NPs with the addition of SDS, which may be attributed to quantum confinement of the ZnO NPs. Further, the use of DMF as solvent resulted in a shift at 352 nm as the particle size was further reduced to 28 nm.

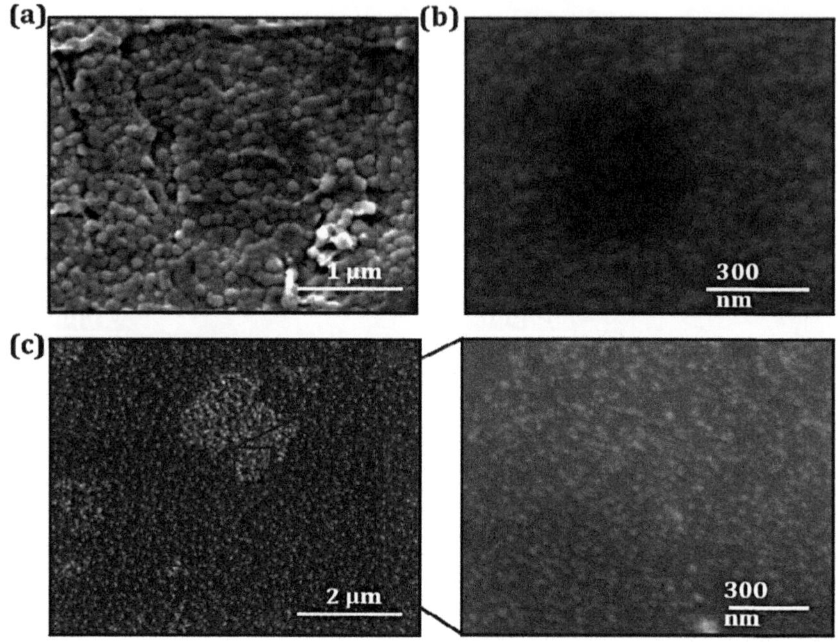

FIGURE 11.4 FE-SEM micrographs of ZnO NPs obtained with different solvents (a) ethanol, (b) methanol, and (c) DMF:water (10:1).

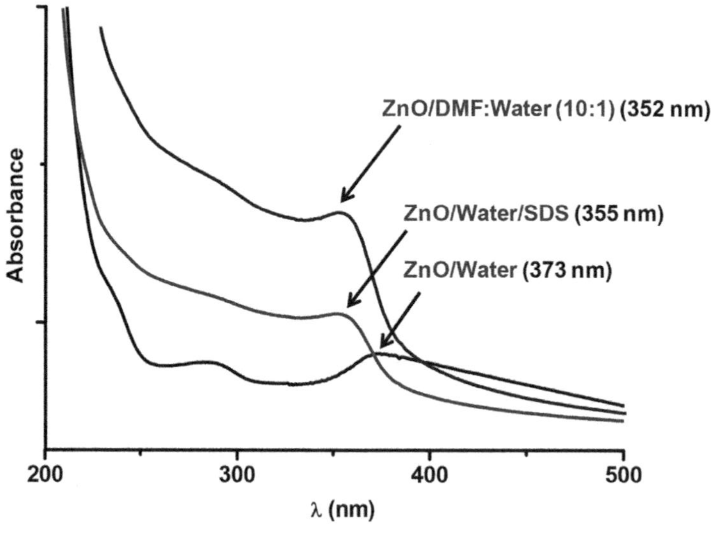

FIGURE 11.5 UV/VIS spectra of ZnO NPs.

11.6 CONCLUSIONS

The ZnO nanostructures were tailored for their size and dispersion behavior by using SDS as surfactant during synthesis. The morphology and dispersion behavior of the ZnO NPs were analyzed. The particle size and SEM analysis revealed that ZnO NPs obtained without addition of SDS, had a higher average diameter of 205 nm. However, being in aggregated form showed higher average diameter in particle size analyzer and are unstable on storage. The addition of surfactant not only reduced the particle size to 80 nm, but also improved the dispersion in ZnO NPs. The UV/VIS spectrum further confirmed the nano confinement of ZnO NPs obtained with SDS. Further, the use of DMF:water as solvent resulted in homogeneous, easily dispersible, and smaller size ZnO NPs (28 nm).

ACKNOWLEDGMENTS

The authors would like to thank University of Delhi for providing financial support under Innovation projects.

KEYWORDS

- **ZnO**
- **nanoparticles**
- **dispersion**
- **surfactant**
- **UV/VIS spectrophotometer**

REFERENCES

1. Auer, G.; Woditsch, P.; Westerhaus, A.; Kischkewitz, J.; Griebler, W. D.; De Liedekerke, M. Pigments, Inorganic, 2. White Pigments. In *Ullmann's Encyclopedia of Industrial Chemistry;* Wiley-VCH Verlag GmbH & Co. KGaA: Weinheim, Germany, 2000.

2. Kołodziejczak-Radzimska, A.; Jesionowski, T.; Krysztafkiewicz, A. Obtaining Zinc Oxide from Aqueous Solutions of KOH and Zn (CH3COO) 2. *Physicochem. Probl. Mi. Process.* **2010,** *44,* 93–102.
3. Mahato, T. H.; Prasad, G. K.; Singh, B.; Acharya, J.; Srivastava, A. R.; Vijayaraghavan, R. Nanocrystalline Zinc Oxide for the Decontamination of Sarin. *J. Hazard. Mater.* **2009,** *165,* 928–932.
4. Benhebal, H.; Chaib, M.; Salmon, T.; Geens, J.; Leonard, A.; Lambert, S. D., et al. Photocatalytic Degradation of Phenol and Benzoic Acid Using Zinc Oxide Powders Prepared by the Sol–Gel Process. *Alexandria Eng. J.* **2013,** *52,* 517–523.
5. Ristić, M.; Musić, S.; Ivanda, M.; Popović, S. Sol–Gel Synthesis and Characterization of Nanocrystalline ZnO Powders. *J. Alloys Compd.* **2005,** *397,* L1–L4.
6. Djurisic, A. B.; Chen, X. Y.; Leung, Y. H. Recent Progress in Hydrothermal Synthesis of Zinc Oxide Nanomaterials. *Recent Pat. Nanotechnol.* **2012,** *6,* 124–134.
7. Innes, B.; Tsuzuki, T.; Dawkins, H.; Dunlop, J.; Trotter, G.; Nearn, M., et al. Nanotechnology and the Cosmetic Chemist. *Aust. Asian J. Cosmet Sci.* **2012,** *15,* 10–24.
8. Chen, D.; Jiao, X.; Cheng, G. Hydrothermal Synthesis of Zinc Oxide Powders with Different Morphologies. *Solid State Commun.* **1999,** *113,* 363–366.
9. Vorobyova, S.; Lesnikovich, A.; Mushinskii, V. Interphase Synthesis and Characterization of Zinc Oxide. *Mater. Lett.* **2004,** *58,* 863–866.
10. Li, X.; He, G.; Xiao, G.; Liu, H.; Wang, M. Synthesis and Morphology Control of ZnO Nanostructures in Microemulsions. *J. Colloid Interface Sci.* **2009,** *333,* 465–473.
11. Iravani, S. Green Synthesis of Metal Nanoparticles Using Plants. *Green Chem.* **2011,** *13,* 2638–2650.
12. Cao, G.; Wang, Y. Nanostructures and Nanomaterials: Synthesis, Properties, and Applications.; World Scientific: Singapore, 2004; Vol. 2, pp 143–228.
13. Awwad, A. M.; Albiss, B.; Ahmad, A. L. Green Synthesis, Characterization and Optical Properties of Zinc Oxide Nanosheets Using *Olea europea* Leaf Extract. *Adv. Mater. Lett.* **2014,** *5,* 520–524.
14. Isaac, R. S. R.; Sakthivel, G.; Murthy, C. Green Synthesis of Gold and Silver Nanoparticles Using *Averrhoa bilimbi* Fruit Extract. *J. Nanotechnol.* **2013,** *2013,* 6.
15. Vidya, C.; Hiremath, S.; Chandraprabha, M.; Antonyraj, I.; Gopal, V.; Jai, A., et al. Green Synthesis of ZnO Nanoparticles by *Calotropis gigantea. Int. J. Curr. Eng. Technol.* **2013,** *1,* 118–120.
16. Rai, M.; Ingle, A. Role of Nanotechnology in Agriculture with Special Reference to Management of Insect Pests. *Appl. Microbiol. Biotechnol.* **2012,** *94,* 287–293.
17. Prasad, T. N. V. K. V.; Sudhakar, P.; Sreenivasulu, Y.; Latha, P.; Munaswamy, V.; Reddy, K. R., et al. Effect of Nanoscale Zinc Oxide Particles on the Germination, Growth and Yield of Peanut. *J. Plant Nutr.* **2012,** *35,* 905–927.
18. Selivanov, V.; Zorin, E. Sustained Action of Ultrafine Metal Powders on Seeds of Grain Crops. *Persp. Mater.* **2001,** *4,* 66–69.
19. Raikova, O.; Panichkin, L.; Raikova, N. Studies on the Effect of Ultrafine Metal Powders Produced by Different Methods on Plant Growth and Development. Nanotechnologies and Information Technologies in the 21st Century. In *Proceedings of the International Scientific and Practical Conference*, **2006**, pp 108–111.
20. Batsmanova, L.; Gonchar, L.; Taran, N. Y.; Okanenko, A. In *Using a Colloidal Solution of Metal Nanoparticles as Micronutrient Fertiliser for Cereals,* Nanomaterials:

Applications & Properties (NAP-2013): 2nd International Conference, September 17–22, 2013; Alushta, Crimea, 2013.
21. Goswami, A.; Roy, I.; Sengupta, S.; Debnath, N. Novel Applications of Solid and Liquid Formulations of Nanoparticles against Insect Pests and Pathogens. *Thin Solid Films.* **2010,** *519,* 1252–1257.
22. Sabir, S.; Arshad, M.; Chaudhari, S. K. Zinc Oxide Nanoparticles for Revolutionizing Agriculture: Synthesis and Applications. *Sci. World J.* **2014,** *2014,* 8.
23. Wickline, S. A.; Neubauer, A. M.; Winter, P.; Caruthers, S.; Lanza, G. Applications of Nanotechnology to Atherosclerosis, Thrombosis, and Vascular Biology. *Arterioscler. Thromb. Vasc. Biol.* **2006,** *26,* 435–441.
24. Freitas, R. A., Jr. What is Nanomedicine? *Nanomedicine.* **2005,** *1,* 2–9.
25. Sahoo, S. K.; Parveen, S.; Panda, J. J. The Present and Future of Nanotechnology in Human Health Care. *Nanomedicine.* **2007,** *3,* 20–31.
26. Cory, H.; Janet, L.; Alex, P.; Reddy, K. M.; Isaac, C.; Andrew, C., et al. Preferential Killing of Cancer Cells and Activated Human T Cells Using ZnO Nanoparticles. *Nanotechnology.* **2008,** *19,* 295103.
27. Wang, H.; Wingett, D.; Engelhard, M. H.; Feris, K.; Reddy, K. M.; Turner, P., et al. Fluorescent Dye Encapsulated ZnO Particles with Cell-specific Toxicity for Potential Use in Biomedical Applications. *J. Mater. Sci. Mater. Med.* **2008,** *20,* 11–22.
28. Shen, W.; Xiong, H.; Xu, Y.; Cai, S.; Lu, H.; Yang, P. ZnO–Poly(methyl Methacrylate) Nanobeads for Enriching and Desalting Low-abundant Proteins Followed by Directly MALDI-TOF MS Analysis. *Anal. Chem.* **2008,** *80,* 6758–6763.
29. Andrea, G.; Guo, B.; Rama, S. K.; John, C. R.; Kennedy, I. M.; Abdul, I. B. Induction of Inflammation in Vascular Endothelial Cells by Metal Oxide Nanoparticles: Effect of Particle Composition. *Environ. Health Perspect.* **2007,** *115,* 403–409.
30. Kaur, P.; Thakur, R.; Kumar, S.; Dilbaghi, N. Interaction Of ZnO Nanoparticles with Food Borne Pathogens *Escherichia coli* DH5α and *Staphylococcus aureus* 5021 & Their Bactericidal Efficacy. In *International Conference on Advances in Condensed and Nano Materials,* AIP Proceedings, Chandigarh, India, Feb 23–26, 2011, p 153.
31. Narayanan, P.; Wilson, W. S.; Abraham, A. T.; Sevanan, M. Synthesis, Characterization, and Antimicrobial Activity of Zinc Oxide Nanoparticles Against Human Pathogens. *BioNanoScience* **2012,** *2,* 329–335.
32. Alkaladi, A.; Abdelazim, A. M.; Afifi, M. Antidiabetic Activity of Zinc Oxide and Silver Nanoparticles on Streptozotocin-induced Diabetic Rats. *Int. J. Mol. Sci.* **2014,** *15,* 2015–2023.
33. Nohynek, G. J.; Lademann, J.; Ribaud, C.; Roberts, M. S. Grey Goo on the Skin? Nanotechnology, Cosmetic and Sunscreen Safety. *Crit. Rev. Toxicol.* **2007,** *37,* 251–277.
34. Xia, T.; Kovochich, M.; Brant, J.; Hotze, M.; Sempf, J.; Oberley, T., et al. Comparison of the Abilities of Ambient and Manufactured Nanoparticles to Induce Cellular Toxicity According to an Oxidative Stress Paradigm. *Nano Lett.* **2006,** *6,* 1794–1807.
35. Ryter, S. W.; Kim, H. P.; Hoetzel, A.; Park, J. W.; Nakahira, K.; Wang, X., et al. Mechanisms of Cell Death in Oxidative Stress. *Antioxid. Redox Signal.* **2007,** *9,* 49–89.
36. Long, T. C.; Saleh, N.; Tilton, R. D.; Lowry, G. V.; Veronesi, B. Titanium Dioxide (P25) Produces Reactive Oxygen Species in Immortalized Brain Microglia (BV2):

Implications for Nanoparticle Neurotoxicity. *Environ. Sci. Technol.* **2006,** *40,* 4346–4352.

37. Lewinski, N.; Colvin, V.; Drezek, R. Cytotoxicity of Nanoparticles. *Small.* **2008,** *4,* 26–49.
38. Molina, M. A.; Ramos, J. L.; Espinosa-Urgel, M. A Two-partner Secretion System is Involved in Seed and Root Colonization and Iron Uptake by *Pseudomonas putida* KT2440. *Environ. Microbiol.* **2006,** *8,* 639–647.
39. Ryu, H. W.; Park, B. S.; Akbar, S. A.; Lee, W. S.; Hong, K. J.; Seo, Y. J., et al. ZnO Sol–Gel Derived Porous Film for CO Gas Sensing. *Sens. Actuator B Chem.* **2003,** *96,* 717–722.
40. Silvestre, C.; Duraccio, D.; Cimmino, S. Food Packaging Based on Polymer Nanomaterials. *Prog. Polym. Sci.* **2011,** *36,* 1766–1782.
41. de Azeredo, H. M. Antimicrobial Nanostructures in Food Packaging. *Trends Food Sci. Technol.* **2013,** *30,* 56–69.
42. Nildade FátimaFerreira, S.; CleuberAntônio de Sá, S.; Paula, S. S.; PaulaJudith-Pérez, E.; G. MariaPaulaJunqueiraConceiç, O.; MariaJoséGalotto, L., et al. Active and Intelligent Packaging for Milk and Milk Products. In *Engineering Aspects of Milk and Dairy Products;* Jane Selia dos Reis, C., Jose Teixeira, A., Eds.; CRC Press: Boca Raton, FL, 2009; pp 175–199.
43. Ahvenainen, R. *Novel Food Packaging Techniques;* Elsevier: Amsterdam, the Netherlands, 2003.
44. Otles, S.; Yalcin, B. Intelligent Food Packaging. *LogForum.* **2008,** *4* (4), 3.
45. Kruijf, N. D.; Beest, M. V.; Rijk, R.; Sipiläinen-Malm, T.; Losada, P. P.; Meulenaer, B. D. Active and Intelligent Packaging: Applications and Regulatory Aspects. *Food Addit. Contam.* **2002,** *19,* 144–162.
46. Yam, K. L.; Takhistov, P. T.; Miltz, J. Intelligent Packaging: Concepts and Applications. *J. Food Sci.* **2005,** *70,* R1–R10.
47. Diallo, M. S.; Savage, N. Nanoparticles and Water Quality. *J. Nanopart. Res.* **2005,** *7,* 325–330.
48. Stoimenov, P. K.; Klinger, R. L.; Marchin, G. L.; Klabunde, K. J. Metal Oxide Nanoparticles as Bactericidal Agents. *Langmuir* **2002,** *18,* 6679–6686.
49. Baruah, S.; Dutta, J. Nanotechnology Applications in Pollution Sensing and Degradation in Agriculture: A Review. *Environ. Chem. Lett.* **2009,** *7,* 191–204.
50. Kim, J.; Yong, K. Facile, A. Coverage Controlled Deposition of Au Nanoparticles on ZnO Nanorods by Sonochemical Reaction for Enhancement of Photocatalytic Activity. *J. Nanopart. Res.* **2012,** *14,* 1–10.
51. Asthana, S.; Pal, S. K.; Monika, S. Solar Light Assisted Nano ZnO Photo Catalytic Mineralization–The Green Technique for the Degradation of Detergents. *Int. J. Chem. Pharm. Anal.* **2014,** *1,* 141–147.
52. Han, J.; Qiu, W.; Gao, W. Potential Dissolution and Photo-dissolution of ZnO Thin Films. *J. Hazard. Mater.* **2010,** *178,* 115–122.
53. Wong, Y.; Yuen, C.; Leung, M.; Ku, S.; Lam, H. Selected Applications of Nanotechnology in Textiles. *AUTEX Res. J.* **2006,** *6,* 1–8.
54. Kathirvelu, S.; D'souza, L.; Dhurai, B. UV Protection Finishing of Textiles Using ZnO Nanoparticles. *Indian J. Fibre Text. Res.* **2009,** *34,* 267–273.

55. Vigneshwaran, N.; Kumar, S.; Kathe, A.; Varadarajan, P.; Prasad, V. Functional Finishing of Cotton Fabrics Using Zinc Oxide–Soluble Starch Nanocomposites. *Nanotechnology.* **2006,** *17,* 5087.
56. Yadav, A.; Prasad, V.; Kathe, A.; Raj, S.; Yadav, D.; Sundaramoorthy, C., et al. Functional Finishing in Cotton Fabrics Using Zinc Oxide Nanoparticles. *Bull. Mater. Sci.* **2006,** *29,* 641–645.
57. Zhang, J.; France, P.; Radomyselskiy, A.; Datta, S.; Zhao, J.; van Ooij, W. Hydrophobic Cotton Fabric Coated by a Thin Nanoparticulate Plasma Film. *J. Appl. Polym. Sci.* **2003,** *88,* 1473–1481.
58. Das, A.; Wang, D. Y.; Leuteritz, A.; Subramaniam, K.; Greenwell, H. C.; Wagenknecht, U., et al. Preparation of Zinc Oxide Free, Transparent Rubber Nanocomposites Using a Layered Double Hydroxide Filler. *J. Mater. Chem.* **2011,** *21,* 7194–7200.
59. Yuan, Z.; Zhou, W.; Hu, T.; Chen, Y.; Li, F.; Xu, Z., et al. Fabrication and Properties of Silicone Rubber/ZnO Nanocomposites via In Situ Surface Hydrosilylation. *Surface Rev. Lett.* **2011,** *18,* 33–38.
60. Mandal, U.; Tripathy, D.; De, S. Dynamic Mechanical Spectroscopic Studies on Plasticization of an Ionic Elastomer Based on Carboxylated Nitrile Rubber by Ammonia. *Polymer* **1996,** *37,* 5739–5742.
61. Ibarra, L.; Marcos-Fernandez, A.; Alzorriz, M. Mechanistic Approach to the Curing of Carboxylated Nitrile Rubber (XNBR) by Zinc Peroxide/zinc Oxide. *Polymer.* **2002,** *43,* 1649–1655.
62. Hamed, G.; Hua, K. C. Effect of ZnO Particle Size on the Curing of Carboxylated NBR and Carboxylated SBR. *Rubber Chem. Technol.* **2004,** *77,* 214–226.
63. Przybyszewska, M.; Zaborski, M. The Effect of Zinc Oxide Nanoparticle Morphology on Activity in Crosslinking of Carboxylated Nitrile Elastomer. *Express Polym. Lett.* **2009,** *3,* 542–552.
64. Sabura Begum, P. M.; Mohammed Yusuff, K. K.; Joseph, R. Preparation and Use of Nano Zinc Oxide in Neoprene Rubber. *Int. J. Polym. Mater. Polym. Biomater.* **2008,** *57,* 1083–1094.

CHAPTER 12

CRYSTALLINE BIOFILM FORMATION ON FOLEY CATHETER: A MACRO PROBLEM WITH MICROBES AND THE NANO WEAPONS TO CONTROL BLOCKAGE

R. MALA* and A. S. RUBY CELSIA

Department of Biotechnology, Mepco Schlenk Engineering College, Sivakasi 626005, Tamil Nadu, India

Corresponding author. E-mail: maalsindia@gmail.com

CONTENTS

ABSTRACT

Catheter-associated urinary tract infection (CAUTI) is a major hospital-acquired infection. Millions of people suffer from this disease annually. *Proteus mirabilis* is a dimorphic organism that is predominantly associated with crystalline biofilm formation and encrustation in urinary catheter. This review addresses the orchestral performance of *P. mirabilis* with various virulence factors in the process of encrustation. The influence of pH on the shape and planes of struvite crystals were elaborated. It explores how silver and other nano weapons are exploited to win over the micro enemy by targeting the different stages of crystalline biofilm formation. The mechanism of silver nanomaterial in reducing the biofilm and its limitations are also discussed. The advantages of functionalizing Foley catheter with an array of bioactive compounds by layer-by-layer assembly was emphasized in nutshell.

12.1 INTRODUCTION

Hospital-acquired infection is responsible for the increased rate of mortality globally. Approximately 40% of hospital-acquired infection is caused by urinary tract infection (UTI). About 80% of UTI is due to the indwelling urinary catheter deployed to drain urine in the healthcare setting.[1,2] Catheter-associated urinary tract infection (CAUTI) is the common hospital-acquired infection. CAUTI rates have continued to rise in almost every healthcare unit.[3,4] In America alone, 4 million people undergo catheterization every year.[5] CAUTI is the root cause of the secondary hospital-acquired bloodstream infection.[6] According to the data by CDC National and State HAI Progress Report 2013, compared to 2009, there is 6% increase in the occurrence of CAUTI in acute care units.[7] Approximately, 13,000 deaths annually are attributed the CAUTI. Approximately, \$340–450 million are spent on the treatment of CAUTI/year.[8] Expenses incurred with every incidence of UTI varies between \$600 and \$3803.[9,10]

In a healthy human being, the flushing action of the urethra will pump out pathogens and hence, drives away infection. Moreover, the slippery surface of the bladder lined by heteropolysaccharide prevents the adhesion of bacteria. If this layer is breached by pathogens, the innate immune system of the host will elicit an immune response and provides protection.

But in patients, the normal physiological process is under stress by the underlying disease and the immune system is not strong enough to rage a war against pathogens.[11] Hence, the indwelling Foley catheter is easily colonized by the invading pathogens leading to CAUTI. CAUTI predisposes the patients to bacteremia, chronic infection, pyelonephritis, encrustation, and blockage of catheter, bladder calculi, and neoplasia of bladder. This increases the hospitalization duration and the expenses associated with the therapy for infection.[12–14] Females are more prone to UTI because of the short urethra than male. On an average, the occurrence of UTI in female is 40% and in male it is 12%. About 25% of women suffer from recurrent UTI within 6–12 months. Prophylaxis with antibiotics is the general strategy followed to circumvent the complications.[15]

12.2 FOLEY CATHETER

A urinary catheter is an inevitable urological device used to relieve urinary incontinence and obstruction. The insertion of Foley catheter into the bladder is shown in Figure 12.1.

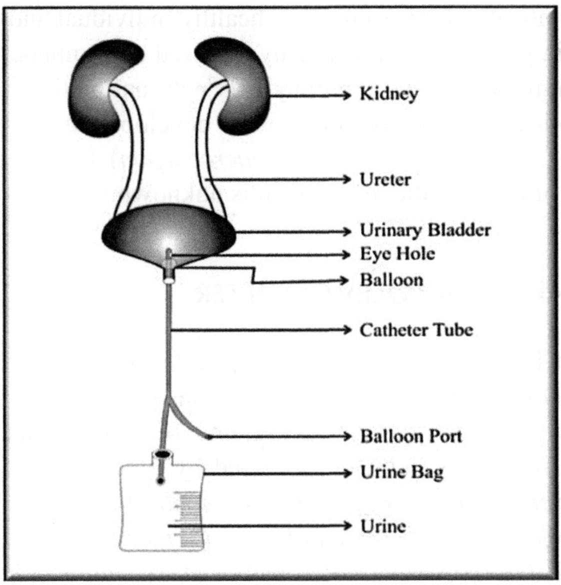

FIGURE 12.1 Catheterized urinary bladder.

To get rid of temporary difficulties in voiding urine, catheters are used for the short term ranging from 1 to 14 days. To manage urinary incontinence and chronic urinary retention, long-term catheterization is adopted. In long-term catheterization, the catheter should be replaced by a new one for every 4–8 weeks.[16] The urinary catheter is otherwise called a Foley catheter. The term Foley catheter is derived from the name of the American urologist, Frederic Foley, who first designed the catheter in the 1930s. C. R. Bard, Inc. of Murray Hill, NJ, first manufactured the urinary catheter by adopting the basic design of Foley and honored him by naming the catheter as Foley catheter.[17] The relative size of a Foley catheter is mentioned in French units (F). Commonly used Foley catheter size ranges from 10 to 28 F where 1 F is equivalent to 0.33 mm. Many modifications were made in the material used to manufacture catheter but the configuration still remains the same. Latex catheters are twined with allergy and urethritis. To minimize these disadvantages, catheters are made from silicon elastomers since 1968. From 2001,the industry has evolved to manufacture antibiotic-coated catheters to prevent infection. Feneley et al.[18] have elaborately narrated the history and evolutionary advancements in the development of functionalized catheters. UTIs are classified into uncomplicated and complicated UTI based on the health status of the host. Uncomplicated UTI occurs in a healthy individual and complicated UTI occurs in persons who are already diseased or immunocompromised. General treatment for UTI is prophylaxis with antibiotics.[19] But therapy with antibiotics is ineffective due to the development of resistance.[20,21] Traditionally, cranberry (*Vaccinium macrocarpon*) fruit is used for the prophylaxis of UTI but the mechanism is unknown.[22,23]

12.3 INVADERS OF FOLEY CATHETER

Predominant infectious agent in uncomplicated UTI is *Escherichia coli* whereas in complicated UTI, it is by a consortium of pathogens most frequently with *P. mirabilis.*[19] Crystalline biofilm formation and the encrustation in urinary catheter is caused by urease positive organisms like *P. mirabilis, Proteus vulgaris, Pseudomonas aeruginosa, Providencia stuartii, Morganella morganii,* and *Klebsiella pneumonia.*

Proteus is the most devastating pathogen causing encrustation in the catheter. Many species of *Proteus* are potent colonizers of urinary catheter. Numerous repositories are available on *P. mirabilis.*[24,25] *Proteus* belongs to

Enterobacteriaceae family. It is an opportunistic pathogen widely distributed in the natural environment and the intestine of human and animals.[26] The name *Proteus* stems from a character named Proteus in Homer's Odyssey. The name represents an inherited attribute to assume different shapes. This portrays the character of *Proteus* as it swiftly switches between a normal cell with limited growth and restricted motility and a highly motile swarmer cell with profuse growth and high mobility as illustrated in Figures 12.2a,b.

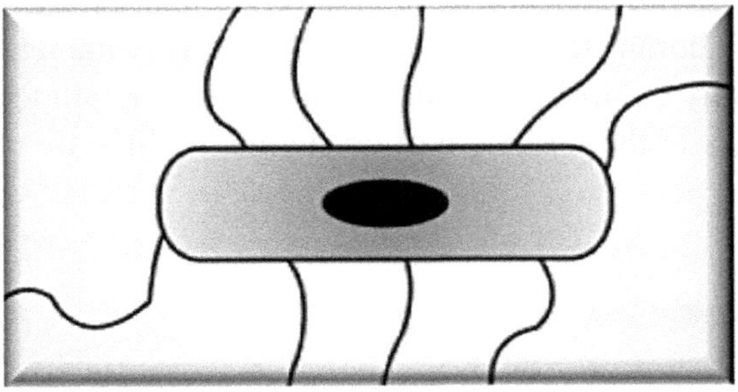

FIGURE 12.2a Normal planktonic cell.

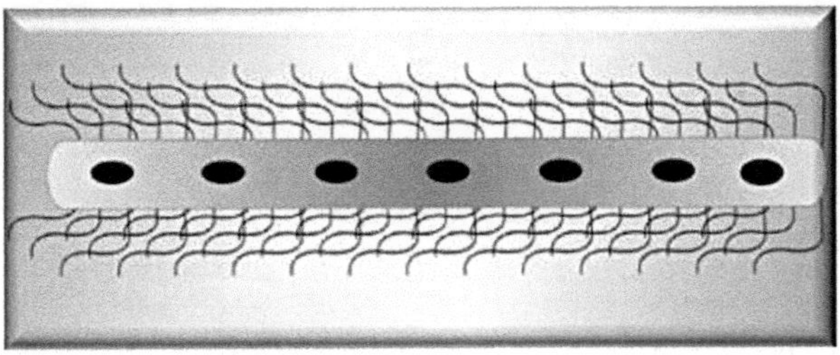

FIGURE 12.2b Highly motile swarmer cell.

It infects an array of medical devices including vascular stents[27] and voice prosthesis.[28] Once it attaches to the catheter, it follows different but overlapping stages of forming a mature biofilm and struvite and apatite crystals. Within 24 h of insertion, a mature biofilm is formed on the catheter. Encrustation in the lumen of the catheter leads to the retention of urine in the bladder itself or it flows out of the catheter. Currently, there is no successful treatment for the prevention of encrustation. The close relationship between UTI and stone formation was recognized by Horton and Smith[29] and Brown.[30]

12.4 STAGES IN CRYSTALLINE BIOFILM FORMATION

Biofilm formation is the underlying root cause of all device-associated infections.[31] Biofilm mode of life permits the pathogens to survive in hostile environment, withstand nutrient limitation, evade from host immune system, and recalcitrant to antimicrobial agents. Crystalline biofilm formation occurs in a series of steps. It starts with the formation of a conditioning film on the surface of the catheter followed by the reversible and irreversible attachment of pathogens, multiplication of bacteria supported by quorum sensing, switching between swimmer and swarmer type, and finally culminates in the formation of crystalline biofilm. The different stages in the formation of crystalline biofilm are represented in Figure 12.3.

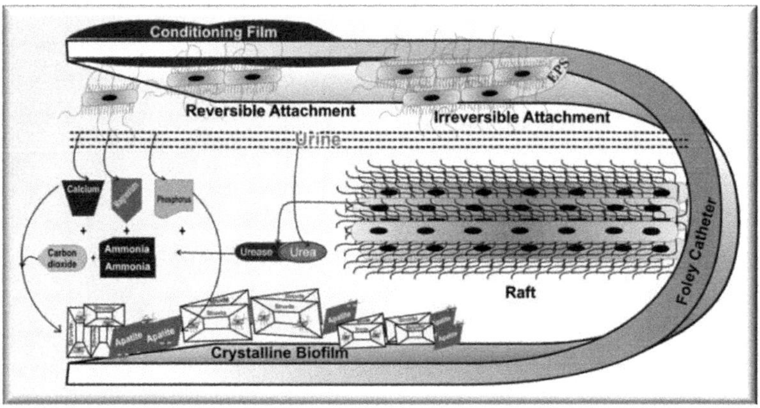

FIGURE 12.3 Stages in the formation of crystalline biofilm.

12.4.1 CONDITIONING FILM

Formation of a conditioning film on the surface of the catheter by hydrophobic, electrostatic, and van der Walls forces of interaction between the urinary catheter and the urine is the first step in the biofilm development. Conditioning film is composed of proteins, electrolytes, and other organic molecules present in the urine and the body fluid in contact with the catheter. The organic and the inorganic constituents attach to the surface of catheter by gravitational force or settle during their flow on the catheter.[32]

12.4.2 REVERSIBLE ATTACHMENT OF BACTERIA ON THE CATHETER

van der Walls force of attraction brings the pathogen close to the conditioning film. Weak and reversible attachment of bacteria on to the conditioning film is the second stage in the formation of biofilm. Initially, the bacterial attachment to the solid surface is weak and reversible. The pathogens attach by their flagella causing a twitching movement. At this stage, the pathogens can be easily eliminated by washing.[33] Within 5–10 nm space between the pathogen and the conditioning film, the turbulent and the shear is less and the reversible physical contact between the pathogen and the conditioning film is mediated by adhesion structures like fimbriae, pili, and flagella.

According to Derjaguin, Landau, Vervey, and Overbeek—DLVO hypothesis, the adhesion of bacteria is the result of balance between the two additive forces: the London van der Walls attractive force and the negative charge repulsion between the double layers of bacterial surface charge and the charge of the solid substratum. When bacteria interact with the negatively charged substratum, there is an interaction between the electrical double layers around the two surfaces. At low concentration of electrolytes, the thickness of the double layer of the counter ions repels the interacting surfaces. But at high concentration of electrolytes, the double layers are compressed, exposing the secondary attraction zone formed by London van der Walls force. This force is responsible for the reversible attachment of bacteria to the solid surface. DLVO theory elaborates the thermodynamic interaction including the osmotic, hydrophilic, and hydrophobic nature of the substratum.[34,35] Vigeant et al.[36] narrate that the DVLO theory is inconsistent and they proposed a three-compartment model for describing the interaction of bacteria with the solid surface. Approaching from the top to bottom, the model describes the presence of three compartments. They are the bulk fluid, near-surface bulk, and near-surface constrained. The distance between the bacteria and the surface and the other physico-chemical factors determine the residence of bacteria in any one of the compartment. According to this hypothesis, "the cells farther than 20 nm from the surface feel no additional forces; those within 10 nm to 20 nm of the surface feel the hydrodynamic effect of the surface, and those closer than 20 nm from the surface feel both the hydrodynamics and the electrostatic influences of the surface." Therefore, the hypothesis suggests that the strategies to prevent the attachment of bacteria should focus on the hydrodynamic properties of the surface environment. The other key factors that govern the attachment of bacteria to the surface are hydrophilicity of the surface, nature of components present in the conditioning film and the constituents of extracellular polymeric substances.[37,38]

12.4.3 IRREVERSIBLE ATTACHMENT

Reversible attachment is followed by the irreversible attachment. Irreversible attachment is governed by properties like temperature, pressure, viscosity, and energy prevailing in the respective microenvironment. Adhesins play a vital role in the attachment of bacteria to solid surfaces and hence, they determine the biofilm formation and virulence.[39] Fimbriae and pili are used by pathogens for the irreversible attachment to a solid substratum in spite of the repulsion between bacteria and other surfaces. Fimbriae and pili are distinguished easily by their length and thickness. Fimbriae are short and thin whereas pili are long and thick. The carbohydrate residues of glycoproteins or glycolipids are the receptors for bacterial pili. Pili protect the pathogen effectively by evading the host immune response. Antibodies generated by the host against pili degrade the pili, but they are immediately replaced by new pili. So the control of biofilm formation by targeting pili is ineffective practically.[40]

12.4.4 MULTIPLICATION AND SWARMING MOVEMENT OF BACTERIA ON CATHETER SURFACE

Once bacteria adhere to the catheter surface, it multiplies and exhibits different phenotype. *Proteus* follows different types of motilities depending upon the surrounding environment and the adhesion structure used for attachment. Twitching represents the movement of single bacteria using type IV pili.[41,42] Gliding represents the movement by the whole body with focal adhesion complexes without flagella or pili.[43] Flagella mediate two different types of motility in liquid and solid surfaces. In liquid, it shows swimming motility. Swimming mentions the movement of single bacteria by flagella. In swimmer type, *Proteus* possess only 6–10 flagella. On solid surfaces, bacteria express swarmer phenotype. In swarmer type, the flagella number is greatly increased to thousand/cell. Swarming motility is defined as a "rapid multi cellular movement of bacteria across a surface, powered by rotating flagella."[44] The term "swarm" represents movement in large numbers. *Proteus* can swiftly change its phenotype to swarmer type after being attached to the catheter. Swarmer cell type is the virulent form causing CAUTI in hospitalized patients. There is a strong correlation between the swarmer cell type, attachment, micro colony formation, movement on the catheter surface, and ascending infection in the urethra. In swarmer phenotype, the size of the cell is 10–20-fold larger (20–80 μm) compared to the normal cell type (2–4 μm). Swarmer cells have certain unique characteristics as listed below: (1) formation of long multinucleate cells, (2) hyper flagellation, (3) the synthesis of surfactants, and (4) the raft like movement[45–47] Within 8–13 h *Proteus* move on the total length of the silicone catheter. Darouchie et al.[48] investigated and documented that *Pseudomonas* can travel the 10 cm length of the catheter in 12 days, *E. coli* takes 20 days, *K. pneumonia* takes 16 days, and *Enterococcus faecalis* takes 27 days. Sabbuba et al.[49] studied the movement of uropathogens on the different urinary catheter. The study recorded *P. mirabilis, P. vulgaris,* and *Serratia marcescens* to be the fastest travelers. Catheters coated with silver also have no exception. From urethra, they migrate to the bladder, ureter, and kidney.[50]

P. mirabilis is dimorphic, that shuttles between swimmer and swarmer types producing a unique Bull's eye pattern of movement on the solid surface. Swarmer phenotype is characterized by the rapid movement of multiple bacteria that lie side by side in a coordinated fashion termed as rafts and sets in biofilm formation in another site with stringent regulation of many genes.[51–53] In in vitro assay of swarming movement, it displays concentric rings as shown in Figure 12.4.

FIGURE 12.4 Swarming mobility of *Proteus mirabilis* in solid medium.

Irreversible attachment is accompanied by increased expression of hemolysin, urease, etc.[54,55] Swarming depends on cell-to-cell contact. Swarming is promoted by organisms in large groups with a high growth rate and hence, high-energy requirements.[56] This energy demand is met by greater force generated by the increased stiffness created by the bundling of all peritrichous flagella.[57] This permits the organisms to move efficiently on a surface with different physical properties. Many swarmer type cells synthesize and secrete surfactants. Surfactants are amphipathic substances containing both hydrophilic and hydrophobic parts. The surfactants reduce the surface tension between the substratum and the cell surface facilitating the movement of bacteria on surfaces. Surfactants interact with lipophilic molecules by their hydrophobic parts and with water-soluble molecules by hydrophilic component. Surfactants can be visualized as a clear layer that precedes the front side of the raft.[58] Few studies have investigated the influence of mutation on the production of surfactant and swarming motility. Interestingly, the results proved that mutation in surfactant production abolishes the swarming mobility. Exogenous supply of surfactant restores the motility. Lipopolysaccharides (LPSs), the constituent of bacteria are also implicated in the swarming motility.

Kearns[51] extensively reviewed the basic requirements and characteristics of swarmer movement in different microbial models. Swarming is directly correlated with the pathogenesis of CAUTI. Mutant swarmer cells lack the ability to migrate in the catheter and cause infection.[49,59] Mutations in some genes essential for the swarmer phenotype impairs raft like movement. Mutants in gene *flaA* encoding flagellin, in wzz region coding for O-antigen chain, genes waaD and waaC coding for the core region of LPS are defective in swarmer motility attesting their role in differentiation into swarmer phenotype.[60] Colony migration factor (Cmf) functions as a surfactant in *Proteus*. It is an extracellular acidic capsular polysaccharide.[61] It is identical to the O-specific part of *Proteus* LPS sero group O6.[62] Belas et al.[63] proposed a hypothesis that the mucous membrane lining the urinary tract traps the swimmers. This induces the swimmers to differentiate into swarmer type that protects itself from the macrophage-mediated first line of defense. It produces a specific IgA protease that degrades the predominant immunoglobulin in the mucus secreted by epithelial surfaces. *P. mirabilis* secretes a metalloprotease ZapA that cleaves the antibodies IgA2 and IgG. It also digests antibodies generated against ZapA and cleaves complement proteins. This weakens the weapons of the innate and acquired immune system of the host.[64,65] They are resistant to treatment by antibiotics.[66,67] This helps the pathogen to ascend freely in the mucus lining without hindrance and cause infection of the urinary tract. There are also controversial reports to this hypothesis where non-motile mutants were isolated from the catheterized urinary tract. Jones et al.[68] reported that non-swarming mutants profusely attach to all types of silicon catheters. Recent studies showed the inverse relationships between the swarmer type and biofilm phenotype.[69–71] Despite a number of reports on the invasiveness of *P. mirabilis*, still there is controversy related to the need for swarmer type to be infectious than the swimmer type. Studies by Allison et al.[72] recorded that only hyper flagellated swarmer type can invade uroepithelial cells. In contrast to this, investigations by Chippendal et al.[73] and Jansen et al.[74] in a mouse model of ascending UTI, showed that swimmer types are invasive than the swarmer types. A different observation was documented by Fujihara et al.[75] The study showed that at acidic pH of urine, the cells differentiate into hyper flagellated and multinucleate forms and at alkaline pH of urine caused by the catalytic activity of urease, the swarmer types dedifferentiates into swimmer type.

12.4.5 VIRULENCE FACTORS AND CRYSTALLINE BIOFILM FORMATION

Swarmer cells produce major virulence factors like flagella, mannose-resistant *Proteus*-like fimbriae, urease, hemolysin, HpmA, sIgA metalloprotease, iron acquisition systems, and ZapA in enormous quantities.[65] There is a strong correlation between the swarmer phenotype and the expression of virulence genes, ability to invade human cells and resistance to antibiotics.[76,77] Inhibition of swarming mobility by many compounds were studied.[78–81]

Flagella are the organelles responsible for adhesion, colonization, and biofilm formation. They are one of the major virulent factors of invasion and infection.[82] *Proteus* is a very sticky bacillus having at least four different adhesins that mediate its attachment to tissue and catheter surfaces. It secretes a hemolysin, an iron-scavenging protein, proteases, and amino acid deaminases, all of which are essential for extracting nutrients from host tissues and fluids. ZapA was identified as a potent virulence factor of *P. mirabilis.*[83] The presence of ZapA operon in *P. vulgaris* and *Proteus penneri* strains were reported by the studies of Kwil et al.[84] *Proteus* uses many other virulence factors like iron acquiring siderophore, phosphate transporter–pst,[65] and zinc transporter system—ZnuABC. Siderophore function mediated by proteobactin and yersiniabactin was supported by the observations of Himpsl et al.[85] These factors assist the pathogen to survive in nutritionally limited regions like the urinary tract.

Microarray analysis by Pearson et al.[86] showed the upregulation of 471 genes and down-regulation of 82 genes in swarmer type cells. Many of the upregulated genes code for virulence factors already mentioned. Flannery et al.[87] identified the pathogenicity island (PAI) and designated as integrative and conjugative elements *P. mirabilis* 1 (ICEPm1) common to *P. mirabilis, P. stuartii, and Morganella morganii.*

P. mirabilis genome has genes coding for fimbriae in a well-structured 17 operons and 13 orphan genes not organized as a complete operon.[88] Six different types of fimbriae and agglutinins were studied. Among the six types of fimbriae, namely, mannose resistant *Proteus*-like fimbriae (MR/P), mannose resistant *Klebsiella* hemagglutinins (MR/K), *P. mirabilis* fimbriae (PMF), non-agglutinating fimbriae (NAF), ambient temperature fimbriae (ATF), and *P. mirabilis* P-like fimbriae, the most important is MR/P.[89,90] MR/P is encoded by 10 genes organized in mrp operon on the chromosome. Expression of MR/P fimbriae is increased under oxygen limitation.[91] These fimbriae contribute to biofilm formation, facilitate colonization of upper urinary tract, and found more often on bacterial strains which cause pyelonephritis. Hola et al.[92] analyzed the relationship between the expression of virulence factors and biofilm forming ability under different media conditions. The study proved the profuse growth of biofilm in brain heart infusion than LB. O'May et al.[93] found that a disruption in a membrane protein, the *pst* phosphate-specific transport system, caused biofilm deficiencies in *P. mirabilis* mutants. sIgAs are a type of antibodies present in mucin secreted by cell linings. They bind with antigen via antigen binding site and with mucin via Fc region. The role of secretory immunoglobulin in protecting against bacterial infection is represented in Figure 12.5. To counteract the effect of antibody, pathogens secrete proteases cleaving sIgA in the hinge region as represented in Figure 12.6. Pathogens are refractory to the attack of complements and they escape all sorts of the first line and second line of defense.[94]

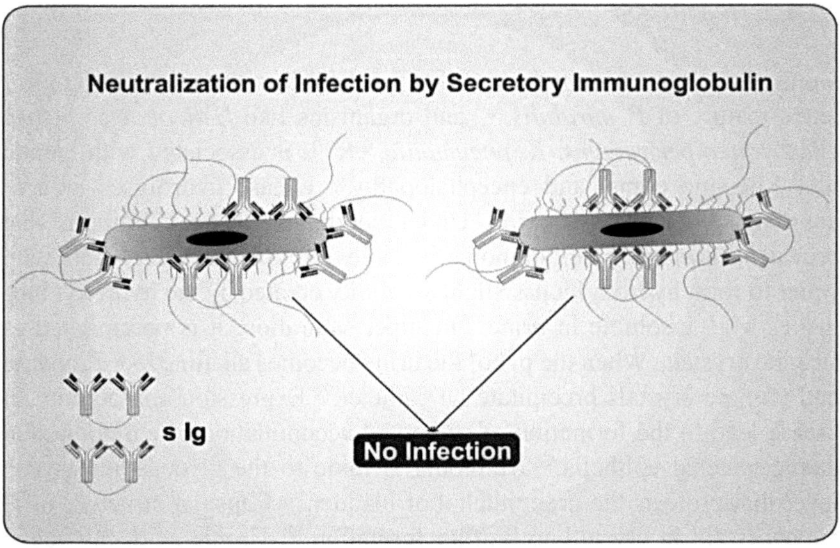

FIGURE 12.5 Role of secretory Ig in neutralizing the bacteria.

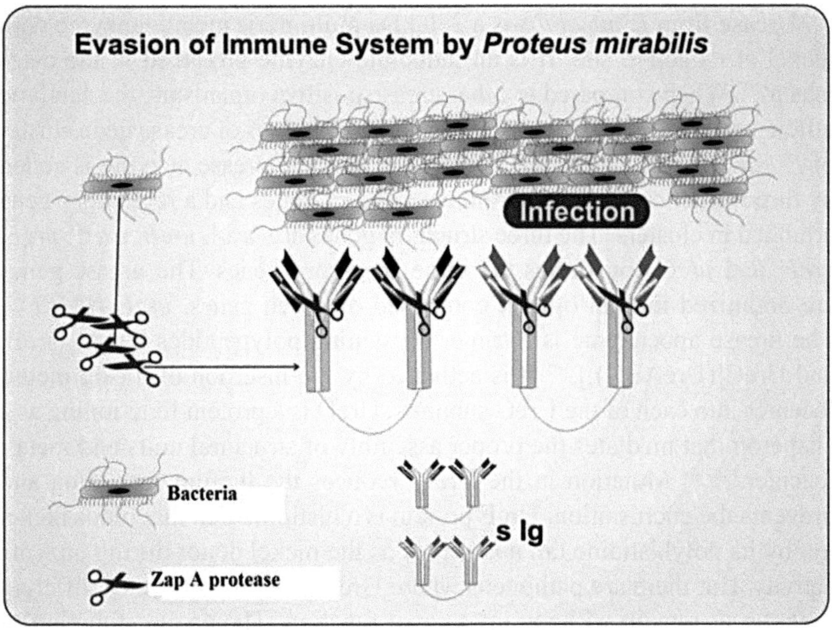

FIGURE 12.6 Digestion of secretory Ig by protease of *Proteus mirabilis.*

12.4.5.1 UREASE

Urease is a urea amidohydrolase (EC 3.5.1.5). It is one of the basic virulence feature of *P. mirabilis*[95,96] and organisms like *Helicobacter pylori, Clostridium perfringens, K. pneumonia,* etc. It is associated with, peptic ulcer, hepatic coma, and encephalopathy.[97] Urease hydrolyses urea to ammonia and carbon dioxide. Under neutral or acidic conditions that prevail in normal urine, ammonia becomes protonated. This reacts with water to form hydroxyl ions. Slight alkalinity created by the hydroxyl ions makes Mg^{2+} insoluble in urine. On super saturation, it is precipitated as struvite crystals. When the pH of the urine becomes alkaline to 8.2, apatite and struvite crystals precipitate on catheter.[98] Expression and activity of urease lead to the formation of stone and accumulation of ammonia that damages renal epithelia.[99] Ammonia is toxic to the glycosaminoglycan layer that protects the uroepithelial of bladder.[100] Capsular structure of *P. mirabilis* aid in crystalline biofilm formation.[101] Uronic acid, the acidic constituent of the capsular polysaccharide binds with the divalent cations like Ca^{2+} and Mg^{2+} present in urine increasing the rate of calculi formation and precipitation.[102]

Urease from *P. mirabilis* is a 250 kDa multimeric metalloenzyme with nickel at its active site. It is an inducible enzyme produced in the cytoplasm.[103] When compared to other urease positive organisms, the catalytic efficiency of urease in *P. mirabilis* is high. The genes of urease gene cluster of *P. mirabilis* has been determined.[104] Complete urease enzyme is coded by three structural genes and three accessory genes and a regulatory gene arranged in clusters. The three structural genes are *ureA*, *ureB*, *ureC*, *ureE*, *ureF*, and *ureG* constitutes the three accessory genes. The urease genes are organized into an operon composed of seven genes, *ureDABCEFG*. The urease apoenzyme is a trimer containing polypeptides UreA, UreB, and UreC[$(UreABC)_3$].[105] It is activated by the insertion of nickel metallocenter into each of the UreC subunits. UreD is a protein functioning as a chaperon that mediates the proper assembly of structural units and metallocenter.[106–108] Mutation in the UreC, reduces the biofilm formation and prevents the encrustation. UreE protein is a histidine rich that binds nickel ion by its polyhistidine tail and serves as the nickel donor during enzyme activity. But there are pathogens where UreE protein performs a different function or remain without any known function. The precise function of UreF protein remains obscure even though it is essential for the functional assembly of urease. UreG peptides contain a nucleotide-binding motif similar to those present in ATP and GTP-binding proteins. Investigation of deletion mutanion on either ureF or ureG revealed the role of them in the complete assembly of Ure D-urease apoprotein complex.[109] In patients infected with *P. mirabilis*, the formation of renal calculi, uriolithiasis, and bladder cancer are more prevalent.[110–112]. Approximately, 62% of CAUTI patients develop bladder calculi. *ureR* is the regulatory gene in the cluster. *UreR* binds to the intergenic region between *ureR* and *ureD* with more affinity in the presence of urea proving the inducible nature.[113,114]

Urease is constitutive in organisms like *Bacillus pasteuri, Sporosarcina ureae,* and in *M. morganii.* But it is inducible in *Proteus* in the presence of urea. It is induced 15–25 times than in the absence of urea[115,116] UreR, mediates the inducible expression and it is a positive regulator of urease.[117] In swarmer type of cells, the expression of urease is manyfold greater than the swimmer type attesting its role in invasion and virulence. It is supported by Donlan[118] who documented the negligible expression of urease in swimming and enormous expression in swarmer type. Urease is active near neutral pH and it is irreversibly denatured under pH 5. Km values of *Proteus* urease are independent of pH whereas *V*max values are pH dependent.[119]

As there is no urease and no nickel containing metallo enzyme in the human system, urease is a potential target to attack crystalline biofilm by *P. mirabilis.*[120] Hydroxamic acids and its derivatives are effective inhibitors of urease.[121,122] They are generally competitive inhibitors. Increase in the biofilm thickness in the presence of N-butanoyl homoserine lactone (BHL) was reported by the observations of Stankowska et al.[123] Investigation by Czerwonka et al.[124] suggested that urease inhibitors were unable to penetrate the biofilm matrix embedded deeper cells and they are ineffective in their ureolytic activity. Many works were initiated with the impact of cranberry material on the expression of different genes, phenotype, and infectivity.[125–129]

12.4.6 CRYSTALLINE PLANES

The struvite and apatite crystals in catheter and kidney are formed by infection by urease positive organisms, mainly *Proteus*. Hence, these renal calculi are termed as infection stones.[130] Crystalline biofilm is an aggregation of crystals, organic matrix, and microbes embedded in it. As mentioned earlier, the urease positive pathogens hydrolyze urea to ammonia and carbon dioxide. At alkaline pH salts from the urine are precipitated as crystals of calcium phosphate (carbonate apatite—$Ca_{10}(PO4)_6CO_3$) and magnesium-ammonium phosphate (struvite—$MgNH_4PO_46H_2O$) as indicated in Figure 12.7.

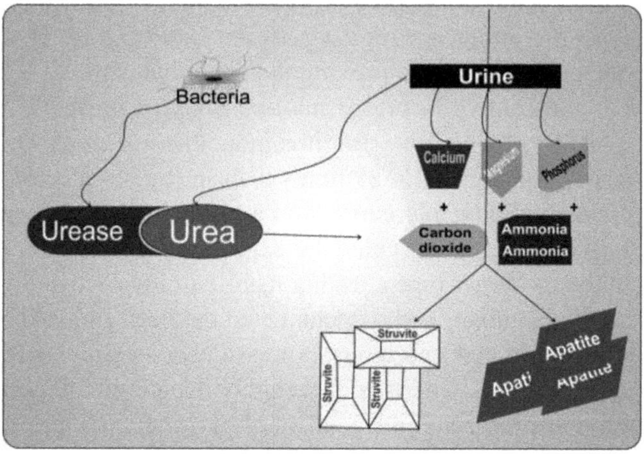

FIGURE 12.7 Precipitation of struvite and apatite crystals from urine.

These crystals complex with the pathogens forming a crystalline biofilm in the catheter ultimately blocking the catheter.[131–133] The number and the type of crystals in the crystalline biofilm are dependent on pH of the urine. From pH 7.5 to 9.0, the crystals are in singles and groups are observed above pH 9.[134] Highly branched dendrite types of crystals are signatures of high pH. LPS of *Proteus* accelerates crystallization.[135]

Pathogens in the crystalline biofilm ascend into the kidney and cause pyelonephritis, septicemia, and endotoxic shock.[136] Hedelin et al.[137] reported that critical pH at which crystallization initiates is 6.8. Nucleation pH is defined as the "pH at which crystals form in urine as it becomes alkaline." Various factors influencing encrustation was investigated by Mathur et al.[138]

Struvite and apatite crystals are the main constituents of infectious urinary calculi. Sadowski et al.[139] extensively studied the structure of struvite crystals in the presence and absence of *P. mirabilis*. In the absence of *Proteus*, they evaluated the crystallization in the presence of ammonia solution. Crystals with well-defined structure, perfect face, and porous structures are formed only in the presence of *P. mirabilis*. The study portrays the significance of pH in the process of crystallization. Crystallization was initiated above pH 7.2. The pH is achieved within a span of 3–4 h in the presence of *Proteus*. At this starting stage of nucleation pH, coffin-like single crystals were formed with perfect crystalline surface.[140–143] With advancement in time, pH raises rapidly leading to the formation of crystals at a faster rate. At this stage, crystals occur in twins. When the growth of *Proteus* is high, then the pH swiftly changes above 9, increasing the rate of crystallization. This change in pH is accompanied by the formation of highly branched crystals called as dendrites. X shaped crystals are fingerprints of rapid change in pH caused by urease activity. The crystals are formed by a central trunk and branching on both sides symmetrically. The influence of pH on the shape of crystals is shown in Figure 12.8.

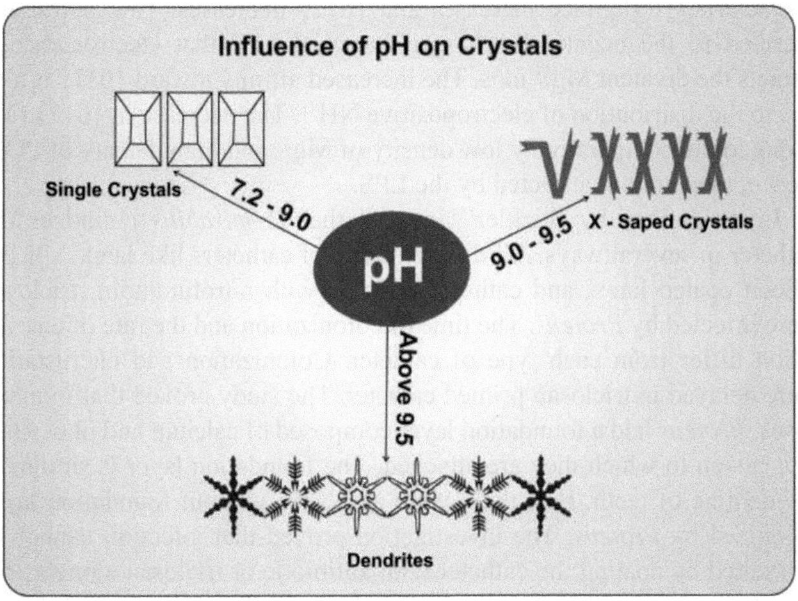

FIGURE 12.8 Influence of urine pH on the shape of crystals.

In the absence of *Proteus* at lower pH, single coffin crystals were formed with truncated ends without {001} face that forms an edge parallel to the *a*-axis. At moderate pH (001) faces were formed with elongation on *a*-axis. At higher pH, crystals were formed with {001} face with elongations on *b*-axis. The absolute sizes of most struvite crystals in the absence of *Proteus* were 19.15–40.72 μm along the *b*-axis. The aspect ratio of the crystals also varied based on the pH. The lowest aspect ratio was 0.60 at lower pH and 2.27 at higher pH. In the presence of *Proteus*, at low pH, struvite crystals were formed without (001) faces with elongation along the *b*-axis. At moderate pH, {001} and {011} faces appeared. With further increase in pH, {011} increases with concomitant decrease in {012}. The lower aspect ratio was 0.70 and the higher aspect ratio in the presence of *Proteus* was 3.2. The absolute sizes of the struvite crystals in the presence of *Proteus* were 27.22–51.46 μm. This indicates the role of *P. mirabilis* in determining the size, shape, and face of crystals precipitated on the catheter. The changes in the aspect ratio in the absence and presence of bacteria are solely dependent on the rate of change of pH. The change in pH is attributed to the catalytic property of urease. So in the presence of bacteria, {001} face increases and {012} decreases. The increase is ascribed to the negatively charged bacterial LPS that electrostatically attracts the divalent Mg^{2+} ions. The increased affinity toward {011} is also due to the distribution of electropositive NH^{4+}. The decrease in {012} face is due to the comparatively low density of Mg^{2+} and high density of PO^{3-}. Hence, they are not attracted by the LPS.

Investigations by Stickler[144] proved that *P. mirabilis* colonizes the catheter in several ways. All different type of catheters like latex, silicon, silicon coated latex, and catheters primed with nitrofurantoin, triclosan were infected by *Proteus*. The time of colonization and the rate of encrustation differ from each type of catheter. Colonization and encrustation were delayed in triclosan primed catheter. The study proved that in many cases, *Proteus* laid a foundation layer composed of calcium and phosphate crystals on to which they are attached. The foundation layer is similar to the dentine of teeth. But, there were also sites without foundation layer colonized by *Proteus*. The investigation proved that infection cannot be prevented by coating the catheter with antibiotic or triclosan alone as the foundation layer safeguards the pathogen from the antimicrobial agents making them ineffective.

Hence, for an effective control of bacterial colonization, the antimicrobial agent must be released at an optimum concentration to kill the pathogens. Another attractive and viable solution to prevent encrustation is to prevent the urine pH rising above the nucleation pH for crystallization.

The pH of the voided urine should be less than the nucleation pH to prevent encrustation. This can be achieved by increased intake of water and citrate. When the urine pH is less than the nucleation pH, catheter blockage is prevented or delayed. Based on this fundamental principle, when measures are taken to prevent the pH above nucleation pH, crystalline biofilm formation was reduced dramatically.[145] So, Broomfield et al.[146] investigated the effect of diluting the urine and the impact of citrate on the formation of crystalline biofilm. The study confirmed that dilution of urine by increased consumption of water prevents encrustation in catheter. Consuming lemon drinks also positively delayed the precipitation of salts on catheter. Citrate in the lemon drinks serves as a chelator of calcium and magnesium thus preventing the nucleation of crystallization. This elevates the nucleation pH of urine above the pH at which *P. mirabilis* crystallizes calcium and ammonium and thus can prevent the formation of crystalline biofilm. Most of the studies focused only on *P. mirabilis* and not on other urease positive infectious agents. The reason is that the nucleation pH of all those pathogens is far less than *P. mirabilis*. If a strategy is developed with the most devastating pathogen, then it will be applicable for all other urease positive organisms which have a less nucleation pH. Both measures increase the nucleation pH to above 8. Therefore, the study throws a light that dietary habits can be fine-tuned to prevent encrustation.

12.5 NANO WEAPONS TO CONTROL CATHETER BLOCKAGE

Nanotechnology opens new avenues to prevent biofilm on medical devices and implants by tailoring–engineering materials at nanoscale. With the outburst of many multi drug-resistant pathogens, measures are in pipeline to use nanotechnology for functionalization of urinary catheters. Coating of catheters with silver nanoparticles and other biocompatible compounds using nano tool become an attractive endeavor to combat crystalline biofilm formation. Biofilm forming and encrustation capacity for no two pathogens are alike. The infectivity depends on the nature of pathogen and the surrounding medium. The time taken for crystallization varies considerably. But once biofilm is formed, it is recalcitrant to therapy with antibiotics irrespective of the infectious agent. So strategies are devised to functionalize catheters to prevent or delay encrustation.

Early approaches were targeted to add value to the catheter by providing them with antibacterial property only. But recently, strategies were focused to hit encrustation from the initial stage itself. So the biomaterials were modified to repel the adhesion of bacteria, or coated with bacteriostatic or bactericidal compounds or with quorum quenching compounds which will block inter cellular communication. The antibacterial compound will battle against the pathogen only on contact with the catheter. The concentration of the antibacterial compound should be sufficient to be bacteriostatic or bactericidal to be effective. The chemical nature of antibacterial compound is natural or synthetic, metallic, or nonmetallic.[147] Metallic antibacterial agents are silver, copper, zinc, and titanium. Perni et al.[148] and Chifiriuc et al.[149] suggested that catheter can be endowed with anti-infectious property either by making them anti-adhesive or by coating them with antimicrobial materials.

Catheters were coated with antibiotics,[150] furanones,[151] and a mixture of protamine sulfate and chlorhexidine.[152] But the antibiotic-coated catheters release the antibacterial agent in a short span of minimum five days[151] paving a way for infection. Kowalczuk et al.[153] investigated the effect of coating catheter with sparfloxacin and heparin against *E. coli, Staphylococcus aureus, and Staphylococcus epidermidis.* The study reported a significant reduction in adhesion of bacteria.

Coating of surface by dipping is effective only for a short term as the antibiotic is easily eluted on contact with body fluids.[154] Uncontrolled burst release at the initial stage is cytotoxic to the host cells. Later the release will be below the MIC leading to the development of resistance by pathogens.[154,155] Urethral stents eluting triclosan was also developed.[156–159] Triclosan leaching device was effective against the attachment of many uropathogenic *E. coli*, *K. pneumoniae*, and *S. aureus.*[160] But triclosan eluting stents were not approved by FDA because of the development of resistance. Local delivery of antibiotics releases antibiotics in sub optimum concentration leading to the emergence of drug resistance.[161] Hence, there is an emergent need to coat the device with nano drug antimicrobial agents.[162]

To circumvent the problem, antibacterial compounds were used along with nano materials which provide slow and sustained release. Moreover, the efficacy of antimicrobial agent in nano formulation is effective at low concentration. Polymeric nano spheres were used to deliver hydrophobic materials. Antibiotics can be physically manipulated as nano antibiotics and they can be used in conjunction with quorum quenching compounds. One will quench the bacterial communication and the other will be bactericidal at very low concentration. Soto et al.[163] reviewed the effect of coating urinary catheter with compounds that quenches quorum sensing. The antibacterial effect of polymer nano sphere encapsulated chlorhexidine was less than free chlorhexidine due to its slow release. But it shows its potency for a long duration compared with free chlorhexidine. It was less effective against *P. aeruginosa* because of its thick LPS layer.[164] Sustained release of bioactive compounds by nano carriers was evaluated by many studies.[165–167] Phuengkham[168] studied the effect of silicone catheter coated with chlorhexidine-loaded PCL nano spheres. They coated catheter by spray coating method and optimized the number of spray coatings and the concentration of polymers required. It showed burst release for a short time and sustained release for two weeks. Spray coating is an inexpensive and simple method to coat medical devices and wound dressings[169–171] and it is scalable at industrial level.[172]

Many studies were focused on the effect of silver-coated implantable medical devices.[173–175] Polyurathane catheter coated with silver nanomaterial offers a greater surface area for antimicrobial activity at very low concentration. Macocinschi et al.[176] coated polyurethane urinary catheter with a nano membrane composed of collagen, elastin, chondroitin sulfate, and hyaluronic acid embedded with silver nanomaterial. Coating was done by solvent casting and electrospinning/electrospraying. Ionic interaction of silver ion with the catheter matrix provided stable antibacterial activity. Høiby et al.[177] showed that silver nanomaterial coating was effective in latex catheters than silicone catheters. Gristina et al.[178] and Tenke et al.[179] addressed the effect of size and shape of nanomaterial on the bactericidal effect and concluded that triangular shaped silver nano particles were toxic to bacteria than spherical shape at a size less than 10 nm.

Catheter coated with silver nanomaterial showed sustained release and offers an additional protection against many different pathogens. The lethal effect of silver on pathogen is attributed to its binding with sulfhydryl groups that interferes with energy transduction and electrolyte transport processes. Coating with silver nanoparticle was investigated by Maki et al.[180] The study showed biphasic release of silver ions from the coated catheter with burst release for the first 3–4 days and slow release from 4 to 10 days. This show that silver at the surface of the catheter was released in first few days and those impregnated in the interior of catheter was released slowly for the next few days. Excretion of silver in the feces also follows the same biphasic pattern.

The negatively charged, bacterial membrane is electrostatically attracted by the positively charged metal nanoparticles. Silver or any metal nanoparticle hits the bacteria in all forms cleaving the protective membrane barrier, arresting the respiratory chain, inactivating the enzymes by binding to the sulfur moieties in amino acids, etc. They generate reactive oxygen species (ROS) and create oxidative stress and toxic effect on pathogens. As the concentration required to elicit the above-mentioned lethal effect is less, it is not capable of destroying the host cell. Chopra et al.[181] studied the antibacterial effect of silver-based agents. Gluggenbichler et al.[182] showed that silver nanoparticle coated surfaces exhibited good protection against infectious agents. The coated catheter on contact with the body fluids releases silver ions which are responsible for the oligodynamic effect. Oligodynamic effect defines a condition in which, the concentration of silver is at an extremely low concentration to be cytotoxic to human beings or higher animals. The effectiveness of silver nanomaterial coated catheter was evaluated against *P. mirabilis*, *Candida Glabrata*, and *Candida tropicalis* by Sahal et al.[183] The results proved that nano layer is capable of restricting biofilm to a greater extent.

Nowadays, many consumer products contain silver. So there is an inherited chance for developing resistance by pathogens against silver. Rupp et al.[184] suggested that, as the mode of action of silver is not single headed, the chances for mutation and developing resistance are remote and rare. One notable disadvantage of silver coating is its ability to precipitate chloride ions. Cook et al.[185] suggested that sliver is potent in bactericidal activity but the dead remnant cells of bacteria act as a site for the attachment of new live cells. Thus, new cells are protected from the silver-coated catheter. So, many silver coated heart valves and urinary catheters are not recommended. Two different controversial reports are available with reference to the coating of urinary catheter with silver nanomaterials.[186–188] Silver or nitrofurazone-coated catheters are effective in vitro, but their in vivo efficiency is questionable in long-term patients to prevent biofilm and encrustation.[189,190] Lai et al.[191] reported that nanostructured titanium coating will give promising results than conventional titanium coating.

Hence, Tran et al.[192] investigated the impact of coating urinary catheter with selenium nanoparticles and its effect on biofilm formation. It showed promising antibacterial activity. The lethal effect was hypothesized to be through the generation of reactive oxygen species. Ron et al.[193] coated urinary catheter with inorganic fullerene-like MoS_2 (IF-MoS_2) nanoparticles. Fullerene-like MoS_2 exhibit smooth surface topology and solid lubrication behavior.[194–196] The study reported the change in the physicochemical properties of coated catheter that offers low surface energy for the attachment of pathogens on the surface of the catheter. The self-assembled nanoarchitecture of the Re:IF-MoS_2 on the catheter offered a tightly packed 2D array of particles which prevented the precipitation of stones on the catheter. The tight packing of nanoparticle provides no contact point on the catheter as a focal point for attachment.[197–199] Moreover, the surface offers low friction and hinders the anchorage of struvite and other crystals on the surface of catheter.[195,196]

To strengthen the anti-adhesive and antibacterial property, chitosan, and heparin were coated layer by layer onto aminolyzed poly(ethylene terephthalate) (PET) films by Fu et al.[200] Antibacterial layer-by-layer coatings were fabricated on silicone that efficiently prevents *P. aeruginosa* biofilm formation for extended time than currently available catheters. The coatings were composed of intact, highly antibacterial polycationic nano spheres processed from aminated cellulose and bacteria-degrading glycosaminoglycan hyaluronic acid. Biofilms are 1000 times recalcitrant to antibiotic therapy.[201] Salicylic acid-eluting device was investigated by Nowatzki.[202] By an unknown mechanism, it inhibits quorum sensing by bacteria which is a key factor in mass multiplication of bacteria.

Layer-by-layer assembly of polyelectrolytes is an easy and cost-effective method for the impregnation of bioactive compounds. The principle behind this layer-by-layer assembly is the alternative adsorption of oppositely charged electrolytes. Biomolecules like polysaccharides, polypeptides,[203,204] enzymes,[205] drugs,[206] and nanoparticles[207] can be coated without causing structural changes. Ivanova et al.[208] employed this method to functionalize catheter with acylase that quenches quorumsensing signals and inhibits the intercellular communication and growth. Francesko et al.[209] coated urinary catheter with polycationic amino cellulose conjugated nano spheres to render the surface resistant to bacteria. Membrane disruption and lethal effect on bacteria by nano sphere is more than the free molecule in bulk.

Surface modification with hydrophilic polymers to provide a smooth surface inhibiting the attachment.[210] The hydrophilic surface reduces the attachment of protein. This renders the surface less favorable for the formation of conditioning film formation which is the first stage in crystalline biofilm formation. But PEGylation did not inhibit the attachment completely. Catheter can be coated with anti-adhesive materials with engineered nano particles[211,212] and antibiotics[213–215] The surface offers a slippery floor to the microbes and can be easily washed away.

Catheter coated with polymers like chitosan, PVP[216–218], and PVP based polymers also have high antibacterial activity.[219] Polymer coating will provide a smooth surface where pathogens cannot stick to and from a crystalline biofilm. Further, it provides a lubricating function also. They also confer an additional advantage of stability and high drug loading.[220] But a greater limitation with a single mode of action of any compound the possibility of pathogens to develop resistance.[221] The stability and durability of the coated catheter are not sufficient to maintain the bactericidal activity for a long term.[222,223]

Other methods to restrict microbial colonization are PEGylation[224] coating with agarose,[225] heparin[226], or zwitterionic residues.[227,228] Blanco et al.[229] coated catheter with gallic acid by laccase-mediated free radical polymerization. This increased the hydrophilicity of the urinary catheter and hence, adhesion of bacteria on the catheter. Chitosan is a potent antimicrobial agent and can be used to functionalize catheter. Moreover, it is acidic and can keep the pH lower than the nucleation pH for crystallization and encrustation in catheter.[230] Coating of catheter with hyaluronic acid[231] offers a biocompatible surface resembling the smooth surface of the urinary tract.

An ideal catheter coating technology should be cheap, safe, long lasting, and active against a wide spectrum of pathogens.[232] Generally, the methods used to coat the surface of medical devices are dipping, thin film coating, matrix loading, plasma coating, PVD, CVD, RF magnetron sputtering, etc. The choice of the technique depends on the physicochemical nature of the medical device and the coating materials. Dipping is a surface adsorption phenomenon. But in case of catheter material, the adsorbed silver nanoparticles are loaded into the silicon or latex matrix providing an antibacterial activity. In this method, the coating is heterogeneous and not uniform. In other methods like CVD, PVD, and sputtering, the coating is of uniform thickness and it is reproducible. Placement of silver wires at the cutaneous extremity of the catheters and release of silver by iontophoresis[233] and catheter impregnation with silver nanoparticles using supercritical carbon dioxide[234] were also investigated.

12.6 FUTURE SCOPE

With advancement in the healthcare unit, urinary catheter becomes inevitable. Nanotechnology offers a variety of biomaterials and the tools to functionalize the catheter to prevent encrustation. But, pathogens are evolving at a faster pace than the technology to evade from all measures of control. Catheter should be armed with a cocktail of compounds each targeting different stage of encrustation. They should be impregnated as the nerve and fiber of the catheter with nano tools for the burst, slow, and sustainable release on contact with pathogens. The attractive properties of biopolymers can be exploited to render the surface smooth that prevents adhesion of pathogen. Biopolymers are a versatile scaffold to hold the drugs and release for a long term in a sustainable manner. Maintaining good clinical practices will minimize the opportunities to the micro enemy to become an opportunistic pathogen.

KEYWORDS

- **CAUTI**
- **bacteria**
- **biofilm**
- **flagella**
- **crystals**

REFERENCES

1. Hartstein, A. I.; Garber, S. B.; Ward, T. T.; Jones, S. R.; Morthland, V. H. Nosocomial Urinary Tract Infection: A Prospective Evaluation of 108 Catheterized Patients. *Infect. Control.* **1981,** *2,* 380–386.
2. Carr, H. A Short History of the Foley Catheter: From Handmade Instrument to Infection-prevention Device. *J. Endourol.* **2000,** *14,* 5–8.
3. Zarb, P.; Coignard, B.; Griskevicienne, J.; Muller, A.; Vankerckho ven Weist, K.; Goossens, M. M.; Vaerenberg, S.; Hopkins, S.; Catry, B.; Monnet, D. L.; Goosens, H.; Suetens, C. The European Centre for Disease Prevention and Control (ECDC) Pilot Point Prevalence Survey of Healthcare-associated Infections and Antimicrobial Use. *Euro Surveill.* **2012,** *17* (46), 20316.

4. Magill, S. S.; Edwards, J. R.; Bamberg, W.; Beldavs, Z. G.; Dumyati, G.; Kainer, M. A.; Lynfield, R.; Maloney, M.; McAllister-Hollod, L.; Nadle, J.; Ray, S. M.; Thompson, D. L.; Wilson, L. E.; Fridkin, S. K.; Emerging Infections Program Healthcare-Associated Infections and Antimicrobial Use Prevalence Survey Team. Multistate Point-prevalence Survey of Health Care-associated Infections. *N. Eng. J. Med.* **2014,** *370,* 1198–1208.
5. Doyle, B.; Mawji, Z.; Horgan, M.; Stillman, P.; Rinehart, A.; Bailey, J.; Mullin, E. Decreasing Nosocomial Urinary Tract Infection in a Large Academic Community Hospital. *Lippincotts Case Manag.* **2001,** *6,* 127–136.
6. Gould, C. V.; Umschied, C. A.; Agarwal, K. A.; Kuntz, G.; Pegues, D. A.; Guidelines for Prevention of Catheter Associated Urinary Tract Infection 2009. *Infect. Control. Hosp. Epidemiol.* **2009,** *31* (4), 319–326. Available from: https://www.cdc.gov/infectioncontrol/guidelines/cauti/
7. Canterbury District Health Board. 2013. Catheter Care Guidelines 2013. https://www.cdhb.health.nz/Hospitals-Services/Health-Professionals/CDHB-Policies/Nursing-Policies-Procedures/Documents/Catheter-Care-Guidelines.pdf
8. UTI and CAUTI. https://www.vdh.virginia.gov/epidemiology/surveillance/hai/uti.htm (accessed May 23, 2016).
9. Saint, S. Clinical and Economic Consequences of Nosocomial Catheter-related Bacteriuria. *Am. J. Infect. Control.* **2000,** *28,* 68–75.
10. McConnell, E. New Catheters Decrease Nosocomial Infections. *Nurs. Manag.* **2000,** *31,* 52–55.
11. Zhang, D.; Zhang, G.; Hayden, M. S.; Greenblatt, M. B.; Bussey, C.; Flavell, R. A.; Ghosh, H. A Toll-like Receptor That Prevents Infection by Uropathogenic Bacteria. *Science* **2014,** *303,* 1522–1526.
12. Saint, S. Catheter-associated Urinary Tract Infection and the Medicare Rule Changes. *Ann. Intern. Med.* **2009,** *150,* 877–884.
13. Khan, A.; Housami, I.; Melloti, R.; Timoney, A.; Stickler, D. Strategy to Control Catheter Encrustation with Citrated Drinks: A Randomized Crossover Study. *J. Urol.* **2010,** *183,* 1390–1394.
14. Stickler, D. Clinical Complications of Urinary Catheters Caused by Crystalline Biofilms: Something Needs to be Done. *J. Intern. Med.* **2014,** *276,* 120–129.
15. O'Hanley, P. Molecular Pathogenesis and Clinical Management. In *Urinary Tract Infections;* Mobley, H. L. T., Warren, J. W., Eds.; ASM Press: Washington, DC, 1996; pp 405–425.
16. Long Term Catheter Urinary Catheter. In *Infection: Prevention and Control of Healthcare-Associated Infections in Primary and Community Care*; Partial Update of NICE Clinical Guideline 2. National Clinical Guideline Centre (UK). Royal College of Physicians (UK); 2012; Chapter 10.
17. Foley, F. E. A Hemostatic Bag Catheter: One Piece Latex Rubber Structure for Control of Bleeding and Constant Drainage Following Prostatic Resection. *J. Urol.* **1937,** *38,* 134–139.
18. Feneley, R.; Hopley, I.; Wells, P. Urinary Catheters: History, Current Status, Adverse Events and Research Agenda. *J. Med. Eng. Technol.* **2015,** *39,* 459–470.
19. Nielubowicz, G.; Mobley, H. Host–Pathogen Interactions in Urinary Tract Infection. *Nat. Rev. Urol.* **2010,** *7,* 430–441.

20. Prais, D.; Straussberg, R.; Avitzur, Y.; Nussinovitch, M.; Harel, L.; Amir, J. Bacterial Susceptibility to Oral Antibiotics in Community Acquired Urinary Tract Infection. *Arch. Dis. Child.* **2003,** *88* (3), 215–218.
21. Foxman, B. The Epidemiology of Urinary Tract Infection. *Nat. Rev. Urol.* **2010,** *7,* 653–660.
22. Howell, A. Bioactive Compounds in Cranberries and Their Role in Prevention of Urinary Tract Infections. *Mol. Nutr. Food Res.* **2007,** *51,* 732–737.
23. Pinzón-Arango, P.; Liu, Y.; Camesano, T. Role of Cranberry on Bacterial Adhesion Forces and Implications for *Escherichia coli*–Uroepithelial Cell Attachment. *J. Med. Food* **2009,** *12,* 259–270.
24. Sosa, V. *Proteus mirabilis* Isolates of Different Origins Do Not Show Correlation with Virulence Attributes and Can Colonize the Urinary Tract of Mice. *Microbiology* **2006,** *152,* 2149–2157.
25. Nzakizwanayo, J.; Hanin, A.; Alves, D. R.; McCutcheon, B.; Dedi, C.; Salvage, J.; Knox, K.; Stewart, B.; Metcalfe, A.; Clark, J.; Gilmore, B. F.; Gahan, C. G.; Jenkins, A. T.; Jones, B. V. Bacteriophage Can Prevent Encrustation and Blockage of Urinary Catheters by *Proteus mirabilis*. *Antimicrob. Agents Chemother.* **2015,** *60* (3), 1530–1536.
26. Janda, J. M.; Abbott, S. L. The Genera Klebsiella and Raoultella. *The Enterobacteria,* 2nd ed.; ASM Press: Washington, DC, 2006; pp 115–129.
27. Reed, W. P.; Moody, M. R.; Newman, K. A; Light, P. D; Costerton, J. W. Bacterial Colonization of Hemasite Access Devices. *Surgery* **1986,** *99,* 308–317.
28. Tićac, B.; Tićac, R.; Rukavina, T.; Kesovija, P.; Pedisić, D.; Maljevac, B.; Starčević, R. Microbial Colonization of Tracheoesophageal Voice Prostheses (Provox2) Following Total Laryngectomy. *Eur. Arch. Otorhinolaryngol.* **2010,** *267,* 1579–1586.
29. Horton-Smith, P. On Bacillus *Proteus* Urinæ: A New Variety of the *Proteus* Group, Discovered in the Urine of a Patient Suffering from Cystitis. *J. Pathol.* **1897,** *4,* 210–215.
30. Brown, T. R. On the Relation between the Variety of Microorganisms and the Composition of Stone in Calculous Pyelonephritis. *JAMA* **1901,** *36,* 1395–1397.
31. Coenye, T.; Nelis, H. In Vitro and In Vivo Model Systems to Study Microbial Biofilm Formation. *J. Microbiol. Methods.* **2010,** *83,* 89–105.
32. Tenke, P.; Köves, B.; Nagy, K.; Hultgren, S.; Mendling, W.; Wullt, B.; Grabe, M.; Wagenlehner, F.; Cek, M.; Pickard, R.; Botto, H.; Naber, K. G.; Bjerklund-Johansen, T. E. Update on Biofilm Infections in the Urinary Tract. *World J. Urol.* **2011,** *30,* 51–57.
33. Trautner, B. W.; Darouiche, R. O. Role of Biofilm in Catheter-Associated Urinary Tract Infection. *Am. J. Infect. Control.* **2004,** *32* (3), 177–183.
34. Chang, Y.; Chang, P. The Role of Hydration Force on the Stability of the Suspension of *Saccharomyces cerevisiae*–Application of the Extended DLVO Theory. *Colloids Surf. A Physicochem. Eng. Asp.* **2002,** *211,* 67–77.
35. Gallardo-Moreno, A.; González-Martín, M.; Pérez-Giraldo, C.; Bruque, J.; Gómez-García, A. The Measurement Temperature: An Important Factor Relating Physicochemical and Adhesive Properties of Yeast Cells to Biomaterials. *J. Colloid Interface Sci.* **2004,** *271,* 351–358.

36. Vigeant, M.; Ford, R.; Wagner, M.; Tamm, L. Reversible and Irreversible Adhesion of Motile *Escherichia coli* Cells Analyzed by Total Internal Reflection Aqueous Fluorescence Microscopy. *Appl. Environ. Microbiol.* **2002,** *68,* 2794–2801.
37. Marshall, K. C.; Stout, R.; Mitchell, R. Mechanism of the Initial Events in the Sorption of Marine Bacteria to Surfaces. *J. Gen. Microbiol.* **1971,** *68,* 337–348.
38. Denstedt, J.; Wollin, T.; Reid, G. Biomaterials Used in Urology: Current Issues of Biocompatibility, Infection, and Encrustation. *J. Endourol.* **1998,** *12,* 493–500.
39. Sauer, F. G.; Mulvey, M. A.; Schilling, J. D.; Martinez, J. J; Hultgren, S. J. Bacterial Pili: Molecular Mechanisms of Pathogenesis. *Curr. Opin. Microbiol.* **2000,** *3* (1), 65–72.
40. Giltner, C.; Nguyen, Y.; Burrows, L. Type IV Pilin Proteins: Versatile Molecular Modules. *Microbiol. Mol. Biol. Rev.* **2012,** *76,* 740–772.
41. Burrows, L. *Pseudomonas aeruginosa* Twitching Motility: Type IV Pili in Action. *Annu. Rev. Microbiol.* **2012,** *66,* 493–520.
42. Ni, L.; Yang, S.; Zhang, R.; Jin, Z.; Chen, H.; Conrad, J.; Jin, F. Bacteria Differently Deploy Type-IV Pili on Surfaces to Adapt to Nutrient Availability. *NPJ Biofilms Microbiomes* **2016,** *2,* 15029.
43. Mignot, T. The Elusive Engine in *Myxococcus xanthus* Gliding Motility. *Cell. Mol. Life Sci.* **2007,** *64,* 2733–2745.
44. Henrichsen, J. Bacterial Surface Translocation: A Survey and a Classification. *Bacteriol. Rev.* **1972,** *36,* 478–503.
45. Tuson, H.; Copeland, M.; Carey, S.; Sacotte, R.; Weibel, D. Flagellum Density Regulates *Proteusmirabilis* Swarmer Cell Motility in Viscous Environments. *J. Bacteriol.* **2012,** *195,* 368–377.
46. Stickler, D.; Hughes, G. Ability of *Proteus mirabilis* to Swarm over Urethral Catheters. *Eur. J. Clin. Microbiol. Infect. Diseases* **1999,** *18,* 206–208.
47. Morgenstein, R.; Clemmer, K.; Rather, P. Loss of the Waal O-Antigen Ligase Prevents Surface Activation of the Flagellar Gene Cascade in *Proteus mirabilis*. *J. Bacteriol.* **2010,** *192,* 3213–3221.
48. Darouiche, R. O.; Wall, M. J.; Itani, K. M.; Otterson, M. F.; Webb, A. L.; Carrick, M. M.; Miller, H. J.; Awad, S. S.; Crosby, C. T.; Mosier, M. C.; Alsharif, A.; Berger, D. H. Chlorhexidine–Alcohol versus Povidone–Iodine for Surgical-Site Antisepsis. *N. Engl. J. Med.* **2010,** *362,* 18–26.
49. Sabbuba, N.; Hughes, G.; Stickler, D. The Migration of *Proteus mirabilis* and Other Urinary Tract Pathogens over Foley Catheters. *BJU Int.* **2008,** *89,* 55–60.
50. Armbruster, C.; Mobley, H. Merging Mythology and Morphology: The Multifaceted Lifestyle of *Proteus mirabilis*. *Nat. Rev. Microbiol.* **2012,** *10,* 743–754.
51. Kearns, D. A Field Guide to Bacterial Swarming Motility. *Nat. Rev. Microbiol.* **2010,** *8,* 634–644.
52. Copeland, M.; Weibel, D. Bacterial Swarming: A Model System for Studying Dynamic Self-Assembly. *Soft Matter.* **2009,** *5,* 1174.
53. Gibbs, K.; Wenren, L.; Greenberg, E. Identity Gene Expression in *Proteus mirabilis*. *J. Bacteriol.* **2011,** *193,* 3286–3292.
54. Ariison, C.; Lai, H.; Hughes, C. Co-Ordinate Expression of Virulence Genes during Swarm-cell Differentiation and Population Migration of *Proteus mirabilis*. *Mol. Microbiol.* **1992,** *6,* 1583–1591.

55. Allison, C.; Emody, L.; Coleman, N.; Hughes, C. The Role of Swarm Cell Differentiation and Multicellular Migration in the Uropathogenicity of *Proteus mirabilis*. *J. Infect. Dis.* **1994,** *169,* 1155–1158.
56. Patrick, J.; Kearns, D. Swarming Motility and the Control of Master Regulators of Flagellar Biosynthesis. *Mol. Microbiol.* **2011,** *83,* 14–23.
57. Atsumi, T; Maekawa, Y; Yamada, T; Kawagishi, I; Imae, Y; Homma, M. Effect of Viscosity on Swimming by the Lateral and Polar Flagella of *Vibrio alginolyticus*. *J. Bacteriol.* **1996,** *178,* 5024–5026.
58. Julkowska, D.; Obuchowski, M.; Holland, I.; Seror, S. Comparative Analysis of the Development of Swarming Communities of *Bacillus subtilis* 168 and a Natural Wild Type: Critical Effects of Surfactin and the Composition of the Medium. *J. Bacteriol.* **2004,** *187,* 65–76.
59. Jones, B.; Young, R.; Mahenthiralingam, E.; Stickler, D. Ultrastructure of *Proteus mirabilis* Swarmer Cell Rafts and Role of Swarming in Catheter-associated Urinary Tract Infection. *Infect. Immun.* **2004,** *72,* 3941–3950.
60. Morgenstein, R.; Szostek, B.; Rather, P. Regulation of Gene Expression during Swarmer Cell Differentiation in *Proteus mirabilis*. *FEMS Microbiol. Rev.* **2010,** *34,* 753–763.
61. Gygi, D.; Rahman, M.; Lai, H.; Carlson, R.; Guard-Petter, J.; Hughes, C. A Cell-surface Polysaccharide that Facilitates Rapid Population Migration by Differentiated Swarm Cells of *Proteus* mirabilis. *Mol. Microbiol.* **1995,** *17,* 1167–1175.
62. Knirel, Y. A.; Perepelov, A. V.; Kondakova, A.; Senchekova, S. N.; Sidorczyk, Z.; Różalski, A.; Kaca, W. Structure and Serology of O Antigens as the Basis for Classification of *Proteus* Strains. *Inn. Immun.* **2011,** *17,* 70–96.
63. Belas, R.; Manos, J.; Suvanasuthi, R. *Proteus mirabilis* ZapA Metalloprotease Degrades a Broad Spectrum of Substrates, Including Antimicrobial Peptides. *Infect. Immun.* **2004,** *72,* 5159–5167.
64. Cusick, K.; Lee, Y.; Youchak, B.; Belas, R. Perturbation of FliL Interferes with *Proteus mirabilis* Swarmer Cell Gene Expression and Differentiation. *J. Bacteriol.* **2011,** *194,* 437–447.
65. Jacobsen, S.; Stickler, D.; Mobley, H.; Shirtliff, M. Complicated Catheter-associated Urinary Tract Infections Due to *Escherichia coli* and *Proteus mirabilis*. *Clin. Microbiol. Rev.* **2008,** *21,* 26–59.
66. Lai, S.; Tremblay, J.; Déziel, E. Swarming Motility: A Multicellular Behaviour Conferring Antimicrobial Resistance. *Environ. Microbiol.* **2009,** *11,* 126–136.
67. Kim, W.; Killam, T.; Sood, V.; Surette, M. Swarm-cell Differentiation in *Salmonella enterica* Serovar *Typhimurium* Results in Elevated Resistance to Multiple Antibiotics. *J. Bacteriol.* **2003,** *185,* 311–3117.
68. Jones, B. Role of Swarming in the Formation of Crystalline *Proteus mirabilis* Biofilms on Urinary Catheters. *J. Med. Microbiol.* **2005,** *54,* 807–813.
69. Pearson, M. M.; Sebaihia, M.; Churcher, C.; Quail, M. A.; Seshasayee, A. S.; Luscombe, N. M.; Abdellah, Z.; Arrosmith, C.; Atkin, B.; Chillingworth, T.; Hauser, H.; Jagels, K.; Moule, S.; Mungall, K.; Norbertczak, H.; Rabbinowitsch, E.; Walker, D.; Whithead, S.; Thomson, N. R.; Rather, P. N.; Parkhill, J.; Mobley, H. L. Complete Genome Sequence of Uropathogenic *Proteus* mirabilis, a Master of Both Adherence and Motility. *J. Bacteriol.* **2008,** *190,* 4027–4037.

70. Verstraeten, N.; Braeken, K.; Debkumari, B.; Fauvart, M.; Fransaer, J.; Vermant, J.; Michiels, J. Living on a Surface: Swarming and Biofilm Formation. *Trends Microbiol.* **2008,** *16,* 496–506.
71. Jacobsen, S.; Shirtliff, M. *Proteus mirabilis* Biofilms and Catheter-associated Urinary Tract Infections. *Virulence* **2011,** *2,* 460–465.
72. Allison, C.; Lai, H.; Gygi, D.; Hughes, C. Cell Differentiation of *Proteus mirabilis* is Initiated by Glutamine, a Specific Chemo Attractant for Swarming Cells. *Mol. Microbiol.* **1993,** *8,* 53–60.
73. Chippendale, G. R.; Warren, J. W.; Trifillis, A. L.; Mobley, H. L. Internalization of *Proteus mirabilis* by Human Renal Epithelial Cells. *Infect. Immun.* **1994,** *62,* 3115–3121.
74. Jansen, A.; Lockatell, C.; Johnson, D.; Mobley, H. Visualization of *Proteus mirabilis* Morphotypes in the Urinary Tract: The Elongated Swarmer Cell is Rarely Observed in Ascending Urinary Tract Infection. *Infect. Immun.* **2003,** *71,* 3607–3613.
75. Fujihara, M.; Obara, H.; Watanabe, Y.; Ono, H.; Sasaki, J.; Goryo, M.; Harasawa, R. Acidic Environments Induce Differentiation of *Proteus mirabilis* into Swarmer Morphotypes. *Microbiol. Immunol.* **2011,** *55,* 489–493.
76. Pearson, M.; Rasko, D.; Smith, S.; Mobley, H. Transcriptome of Swarming *Proteus mirabilis*. *Infect. Immun.* **2010,** *78,* 2834–2845.
77. Tremblay, J.; Déziel, E. Gene Expression in *Pseudomonas aeruginosa* Swarming Motility. *BMC Genomics* **2010,** *11,* 587.
78. Kopp, R.; Müller, J.; Lemme, R. Inhibition of Swarming of *Proteus* by Sodium Tetradecyl Sulfate, Beta-Phenethyl Alcohol, and *p*-Nitrophenylglycerine. *Appl. Microbiol.* **1966,** *14* (6), 873–878.
79. Smith, D. G. Inhibition of Swarming in *Proteus* spp. by Tannic Acid. *J. Appl. Bacteriol.* **1975,** *38,* 29–32.
80. Cole, S.; Smith, D. Inhibition of Swarming in *Proteus mirabilis* by Calcium Ions. *FEMS Microbiol. Lett.* **1981,** *10,* 17–20.
81. Liaw, S.; Ho, S.; Wang, W.; Lai, H.; Luh, K. Inhibition of Virulence Factor Expression and Swarming Differentiation in *Proteus mirabilis* by P-Nitrophenylglycerol. *J. Med. Microbiol.* **2000,** *49,* 725–731.
82. Kirov, S. Bacteria that Express Lateral Flagella Enable Dissection of the Multifunctional Roles of Flagella in Pathogenesis. *FEMS Microbiol. Lett.* **2003,** *224,* 151–159.
83. Phan, V.; Belas, R.; Gilmore, B.; Ceri, H.; Zapa, A. Virulence Factor in a Rat Model of *Proteus mirabilis*-Induced Acute and Chronic Prostatitis. *Infect. Immun.* **2008,** *76,* 4859–4864.
84. Kwil, I.; Babicka, D.; Stączek, P.; Różalski, A. In *Applying the PCR Method for Searching Genes Encoding Main Structural Fimbrial Proteins,* 5th Conference on Molecular Biology in Diagnostics of Infectious Disease and Biotechnology, SGGW Publisher: Warsaw, Poland, 2002; pp 88–91.
85. Himpsl, S.; Pearson, M.; Arewång, C.; Nusca, T.; Sherman, D.; Mobley, H. Proteobactin and a Yersiniabactin-Related Siderophore Mediate Iron Acquisition in *Proteus mirabilis*. *Mol. Microbiol.* **2010,** *78,* 138–157.
86. Pearson, M.; Mobley, H. Repression of Motility during Fimbrial Expression: Identification of 14 mrpJ Gene Paralogues in *Proteus mirabilis*. *Mol. Microbiol.* **2008,** *69,* 548–558.

87. Flannery, E.; Mody, L.; Mobley, H. Identification of a Modular Pathogenicity Island that is Widespread among Urease-producing Uropathogens and Shares Features with a Diverse Group of Mobile Elements. *Infect. Immun.* **2009,** *77,* 4887–4894.
88. Jansen, A.; Lockatell, V.; Johnson, D.; Mobley, H. Mannose-resistant Proteus-like Fimbriae are Produced by Most *Proteus mirabilis* Strains Infecting the Urinary Tract, Dictate the In Vivo Localization of Bacteria, and Contribute to Biofilm Formation. *Infect. Immun.* **2004,** *72,* 7294–7305.
89. Coker, C.; Poore, C.; Li, X.; Mobley, H. Pathogenesis of *Proteus mirabilis* Urinary Tract Infection. *Microb. Infect.* **2000,** *2,* 1497–1505.
90. Rocha, S.; Pelayo, J.; Elias, W. Fimbriae of Uropathogenic *Proteus mirabilis*. *FEMS Immunol. Med. Microbiol.* **2007,** *51,* 1–7.
91. Lane, M.; Li, X.; Pearson, M.; Simms, A.; Mobley, H. Oxygen-limiting Conditions Enrich for Fimbriate Cells of Uropathogenic *Proteus mirabilis* and *Escherichia coli. J. Bacteriol.* **2008,** *191,* 1382–1392.
92. Hola, V.; Peroutkova, T.; Ruzicka, F. Virulence Factors in *Proteus*bacteria from Biofilm Communities of Catheter-associated Urinary Tract Infections. *FEMS Immunol. Med. Microbiol.* **2012,** *65,* 343–349.
93. O'May, G.; Jacobsen, S.; Longwell, M.; Stoodley, P.; Mobley, H.; Shirtliff, M. The High-affinity Phosphate Transporter Pst in *Proteus mirabilis* HI4320 and Its Importance in Biofilm Formation. *Microbiology* **2009,** *155,* 1523–1535.
94. Pastorello, I.; Rossi Paccani, S.; Rosini, R.; Mattera, R.; Ferrer Navarro, M.; Urosev, D.; Nesta, B.; Lo Surdo, P.; Del Vecchio, M.; Rippa, V.; Bertoldi, I.; Moriel, D. G.; Laarman, A. J.; Van Strijp, J. A. G.; Daura, X.; Pizza, M.; Serino, L.; Soriania, M. EsiB, a Novel Pathogenic *Escherichia coli* Secretory Immunoglobulin A-Binding Protein Impairing Neutrophil Activation. *MBio* **2013,** *4* (4), e00206–e00213.
95. Stickler, D.; Feneley, R. The Encrustation and Blockage of Long-Term Indwelling Bladder Catheters: A Way Forward in Prevention and Control. *Spinal Cord.* **2010,** *48,* 784–790.
96. Schaffer, J. N.; Norsworthy, A. N.; Sun, T. T.; Pearson, M. M. *Proteus mirabilis* Fimbriae- and Urease-dependent Clusters Assemble in an Extracellular Niche to Initiate Bladder Stone Formation. *Proc. Natl. Acad. Sci.* **2016,** *113,* 4494–4499.
97. Mora, D.; Arioli, S. Microbial Urease in Health and Disease. *PLoS Pathog.* **2014,** *10,* e1004472.
98. Stickler, D.; Morgan, S. Observations on the Development of the Crystalline Bacterial Biofilms that Encrust and Block Foley Catheters. *J. Hosp. Infect.* **2008,** *69,* 350–360.
99. Musher, D.; Griffith, D.; Yawn, D.; Rossen, R. Role of Urease in Pyelonephritis Resulting from Urinary Tract Infection with *Proteus*. *J. Infect. Dis.* **1975,** *131,* 177–181.
100. Griffith, D. Urease Stones. *Urol. Res.* **1979,** *7* (3), 215–221.
101. Clapham, L.; McLean, R.; Nickel, J.; Downey, J.; Costerton, J. The Influence of Bacteria on Struvite Crystal Habit and Its Importance in Urinary Stone Formation. *J. Cryst. Growth* **1990,** *104,* 475–484.
102. Rozalski, A.; Sidorczyk, Z.; Kotelko, K. Potential Virulence Factors of *Proteus bacilli*. *Microbiol. Mol. Biol. Rev.* **1997,** *61,* 65–89.
103. Coker, C.; Poore, C. A.; Li, X.; Mobley, H. L. Pathogenesis of *Proteus mirabilis* Urinary Tract Infection. *Microb. Infect.* **2000,** *2,* 1497–1505.

104. Hashmi, S.; Kelly, E.; Rogers, S.; Gates, J. Urinary Tract Infection in Surgical Patients. *Am. J. Surg.* **2003,** *186,* 53–56.
105. Mobley, H. L.; Island, M. D.; Hausinger, R. P. Molecular Biology of Microbial Ureases. *Microbiol. Rev.* **1995,** *59,* 451–480.
106. Mobley, H.; Belas, R. Swarming and Pathogenicity of *Proteus mirabilis* in the Urinary Tract. *Trends Microbiol.* **1995,** *3,* 280–284.
107. Maki, D. Prevention of Catheter-associated Urinary Tract Infection. *JAMA.* **1972,** *221,* 1270.
108. Marre, R.; Hacker, J.; Henkel, W.; Goebel, W. Contribution of Cloned Virulence Factors from Uropathogenic *Escherichia coli* Strains to Nephro Pathogenicity in an Experimental Rat Pyelonephritis Model. *Infect. Immun.* **1986,** *54,* 761–767.
109. Marcus, R. Narrow-Bore Suprapubic Bladder Drainage in Uganda. *Lancet.* **1967,** *289,* 748–750.
110. Fraser, G.; Furness, R.; Hughes, C.; Claret, L.; Gupta, S. Swarming-Coupled Expression of the *Proteus mirabilis hpmBA* Haemolysin Operon A. *Microbiology* **2002,** *148,* 2191–2201.
111. Fudala, R.; Kondakova, A.; Bednarska, K.; Senchenkova, S.; Shashkov, A.; Knirel, Y.; Zähringer, U.; Kaca, W. Structure and Serological Characterization of the O-Antigen of *Proteus mirabilis* O18 with a Phosphocholine-containing Oligosaccharide Phosphate Repeating Unit. *Carbohydr. Res.* **2003,** *338,* 1835–1842.
112. D'Orazio, S.; Thomas, V.; Collins, C. Activation of Transcription at Divergent Urea-dependent Promoters by the Urease Gene Regulator Urer. *Mol. Microbiol.* **1996,** *21,* 643–655.
113. Thomas, V.; Collins, C. Identification of UreR Binding Sites in the *Enterobacteriaceae* Plasmid-Encoded and *Proteus mirabilis* Urease Gene Operons. *Mol. Microbiol.* **1999,** *31,* 1417–1428.
114. McLean, R.; Nickel, J.; Cheng, K.; Costerton, J.; Banwell, J. The Ecology and Pathogenicity of Urease-producing Bacteria in the Urinary Tract. *Crit. Rev. Microbiol.* **1988,** *16,* 37–79.
115. Haraoka, M.; Hang, L.; Frendéus, B.; Godaly, G.; Burdick, M.; Strieter, R.; Svanborg, C. Neutrophil Recruitment and Resistance to Urinary Tract Infection. *J. Infect. Dis.* **1999,** *180,* 1220–1229.
116. Koukalova, D.; Hajek, V.; Kodusek, R. Development of a Vaccine for Treatment of Urinary Tract Infections. *Bratisl. Lek. Listy* **1999,** *100,* 92–95.
117. Li, X.; Zhao, H.; Geymonat, L.; Bahrani, F.; Johnson, D. E.; Mobley, H. L. *Proteus mirabilis* Mannose-resistant, *Proteus*-like Fimbriae: Mrpg Islocated at the Fimbrial Tip and is Required for Fimbrial Assembly. *Infect. Immun.* **1997,** *65,* 1327–1334.
118. Donlan, R. Biofilms and Device-associated Infections. *Emerg. Infect. Dis*. **2001,** *7,* 277–281.
119. Bibby, J. M.; Cox A. J.; Hukins D. W. L. Feasibility of Preventing Encrustation on Urinary Catheters. *Cell. Mater*. **1995,** *2,* 183–195.
120. Rutherford, J. The Emerging Role of Urease as a General Microbial Virulence Factor. *PLoS Pathog.* **2014,** *10,* e1004062.
121. Gaonkar, T.; Sampath, L.; Modak, S. Evaluation of the Antimicrobial Efficacy of Urinary Catheters Impregnated with Antiseptics in an *In Vitro* Urinary Tract Model. *Infect. Control Hosp. Epidemiol.* **2003,** *24,* 506–513.

122. Li, X.; Lockatell, C.; Johnson, D.; Mobley, H. Identification of Mrpl as the Sole Recombinase that Regulates the Phase Variation of MR/P Fimbria, a Bladder Colonization Factor of Uropathogenic *Proteus mirabilis*. *Mol. Microbiol.* **2002,** *45,* 865–874.
123. Stankowska, D.; Czerwonka, G.; Rozalska, S.; Grosicka, M.; Dziadek, J.; Kaca, W. Influence of Quorum Sensing Signal Molecules on Biofilm Formation in *Proteus mirabilis* O18. *Folia Microbiol.* **2011,** *57,* 53–60.
124. Czerwonka, G.; Arabski, M.; Wąsik, S.; Jabłońska-Wawrzycka, A.; Rogala, P.; Kaca, W. Morphological Changes in *Proteus mirabilis* O18 Biofilm under the Influence of a Urease Inhibitor and a Homoserine Lactone Derivative. *Arch. Microbiol.* **2014,** *196,* 169–177.
125. Harmidy, K.; Tufenkji, N.; Gruenheid, S. Perturbation of Host Cell Cytoskeleton by Cranberry Proanthocyanidins and Their Effect on Enteric Infections. *PLoS One* **2011,** *6* (11), e27267.
126. Hidalgo, G.; Chan, M.; Tufenkji, N. Cranberry Materials Inhibit *Escherichia coli* CFT073 *fliC* Expression and Motility. *Appl. Environ. Microbiol.* **2011,** *77* (19), 6852–6857.
127. Hidalgo, G.; Ponton, A.; Fatisson, J.; O'May, C.; Asadishad, B.; Schinner, T.; Tufenkji, N. Induction of a State of Iron Limitation in Uropathogenic *Escherichia coli* CFT073 by Cranberry-derived Proanthocyanidins as Revealed by Microarray Analysis. *Appl. Environ. Microbiol.* **2011,** *77* (4), 1532–1535.
128. O'May, C.; Tufenkji, N. The Swarming Motility of *Pseudomonas aeruginosa* is Blocked by Cranberry Proanthocyanidins and Other Tannin-containing Materials. *Appl. Environ. Microbiol.* **2011,** *77,* 3061–3067.
129. O'May, C.; Ciobanu, A.; Lam, H.; Tufenkji, N. Tannin Derived Materials Can Block Swarming Motility and Enhance Biofilm Formation in *Pseudomonas aeruginosa*. *Biofouling.* **2012,** *28,* 1063–1076.
130. Clapham, L.; McLean, R.; Nickel, J.; Downey, J.; Costerton, J. The Influence of Bacteria on Struvite Crystal Habit and Its Importance in Urinary Stone Formation. *J. Cryst. Growth.* **1990,** *104,* 475–484.
131. Holling, N.; Dedi, C.; Jones, C.; Hawthorne, J.; Hanlon, G.; Salvage, J.; Patel, B.; Barnes, L.; Jones, B. Evaluation of Environmental Scanning Electron Microscopy for Analysis of *Proteus mirabilis* Crystalline Biofilms In Situ on Urinary Catheters. *FEMS Microbiol. Lett.* **2014,** *355,* 20–27.
132. Burr, R. G.; Nuseibeh, I. M. Urinary Catheter Blockage Depends on Urine pH, Calcium and Rate of Flow. *Spinal Cord.* **1997,** *35* (8), 521–525.
133. Wilks, S. A.; Fader, M. J.; William Keevil, C. Novel Insights into the *Proteus mirabilis* Crystalline Biofilm Using Real-Time Imaging. *PLoS One.* **2015,** *10,* e0141711.
134. Prayer, J.; Torzewska, A. Bacterially Induced Struvite Growth from Synthetic Urine: Experimental and Theoretical Characterization of Crystal Morphology. *Cryst. Growth Des.* **2009,** *9,* 3538–3543.
135. Dumanski, A. J.; Hedelin, H.; Edin-Liljegren, A.; Beauchemin, D.; McLean, R. J. Unique Ability of the *Proteus mirabilis* Capsule to Enhance Mineral Growth in Infectious Urinary Calculi. *Infect. Immun.* **1994,** *62,* 2998–3003.
136. Kunin, C.; McCormack, R. Prevention of Catheter-induced Urinary-tract Infections by Sterile Closed Drainage. *N. Engl. J. Med.* **1966,** *274,* 1155–1161.

137. Hedelin, H.; Eddeland, A.; Larsson, L.; Petersson, S.; Öhman, S. The Composition of Catheter Encrustations, Including the Effects of Allopurinol Treatment. *Br. J. Urol.* **1984,** *56,* 250–254.
138. Mathur, S.; Suller, M.; Stickler, D.; Feneley, R. Factors Affecting Crystal Precipitation from Urine in Individuals with Long-term Urinary Catheters Colonized with Urease-positive Bacterial Species. *Urol. Res.* **2006,** *34,* 173–177.
139. Sadowski, R. R.; Prywer, J.; Torzewska, A. Morphology of Struvite Crystals as an Evidence of Bacteria Mediated Growth. *Cryst. Res. Technol.* **2014,** *49,* 478–489.
140. Weil, M. The Struvite-type Compounds $[Mg(H_2O)_6](XO_4)$, Where M = Rb, Tl and X = P. *Cryst. Res. Technol.* **2008,** *43,* 1286–1291.
141. 141. Wierzbicki, A.; Sallis, J.; Stevens, E.; Smith, M.; Sikes, C. Crystal Growth and Molecular Modeling Studies of Inhibition of Struvite by Phosphocitrate. *Calcif. Tissue Int.* **1997,** *61,* 216–222.
142. Rakovan, J. Hemimorphism. *Rocks Minerals* **2007,** *82,* 329–337.
143. Bazin, D.; André, G.; Weil, R.; Matzen, G.; Emmanuel, V.; Carpentier, X.; Daudon, M. Absence of Bacterial Imprints on Struvite-containing Kidney Stones: A Structural Investigation at the Mesoscopic and Atomic Scale. *Urology.* **2012,** *79,* 786–790.
144. Stickler, D. Modulation of Crystalline *Proteus mirabilis* Biofilm Development on Urinary Catheters. *J. Med. Microbiol.* **2006,** *55,* 489–494.
145. Stickler, D.; Morris, N.; Moreno, M.; Sabbuba, N. Studies on the Formation of Crystalline Bacterial Biofilms on Urethral Catheters. *Eur. J. Clin. Microbiol. Infect. Dis.* **1998,** *17,* 649–652.
146. Broomfield, R.; Morgan, S.; Khan, A.; Stickler, D. Crystalline Bacterial Biofilm Formation on Urinary Catheters by Urease-producing Urinary Tract Pathogens: A Simple Method of Control. *J. Med. Microbiol.* **2009,** *58,* 1367–1375.
147. Zhao, L.; Wang, H.; Huo, K.; Cui, L.; Zhang, W.; Ni, H.; Zhang, Y.; Wu, Z.; Chu, P. Antibacterial Nano-structured Titania Coating Incorporated with Silver Nanoparticles. *Biomaterials* **2011,** *32,* 5706–5716.
148. Perni, S.; Pratten, J.; Wilson, M.; Piccirillo, C.; Parkin, I.; Prokopovich, P. Antimicrobial Properties of Light-activated Polyurethane Containing Indocyanine Green. *J. Biomater. Appl.* **2009,** *25,* 387–400.
149. Chifiriuc, C.; Grumezescu, V.; Grumezescu, A.; Saviuc, C.; Lazăr, V.; Andronescu, E. Hybrid Magnetite Nanoparticles/*Rosmarinus officinalis* Essential Oil Nanobiosystem with Antibiofilm Activity. *Nanoscale Res. Lett.* **2012,** *7,* 209.
150. Reid, G.; Sharma, S.; Advikolanu, K.; Tieszer, C.; Martin, R. A.; Bruce, A. W. Effects of Ciprofloxacin, Norfloxacin, and Ofloxacin on *In Vitro* Adhesion and Survival of *Pseudomonas aeruginosa* AK1 on Urinary Catheters. *Antimicrob. Agents Chemother.* **1994,** *38,* 1490–1495.
151. Baveja, J.; Willcox, M.; Hume, E.; Kumar, N.; Odell, R.; Poole-Warren, L. Furanones as Potential Anti-bacterial Coatings on Biomaterials. *Biomaterials* **2004,** *25,* 5003–5012.
152. Darouiche, R.; Mansouri, M.; Gawande, P.; Madhyastha, S. Efficacy of Combination of Chlorhexidine and Protamine Sulphate against Device-associated Pathogens. *J. Antimicrob. Chemother.* **2008,** *61,* 651–657.
153. Kowalczuk, D.; Ginalska, G.; Golus, J. Characterization of the Developed Antimicrobial Urological Catheters. *Int. J. Pharm.* **2010,** *402* (1–2), 175–183.

154. Noimark, S.; Dunnill, C.; Wilson, M.; Parkin, I. The Role of Surfaces in Catheter-associated Infections. *Chem. Soc. Rev.* **2009,** *38,* 3435.
155. Walder, B.; Pittet, D.; Tramer, M. R. Prevention of Bloodstream Infections with Central Venous Catheters Treated with Anti-infective Agents Depends on Catheter Type and Insertion Time: Evidence from a Meta-analysis. *Infect. Control Hosp. Epidemiol.* **2002,** *23,* 748–756.
156. Minardi, D.; Cirioni, O.; Ghiselli, R.; Silvestri, C.; Mocchegiani, F.; Gabrielli, E.; d'Anzeo, G.; Conti, A.; Orlando, F.; Rimini, M.; Brescini, L.; Guerrieri, M.; Giacometti, A.; Muzzonigro, G. Efficacy of Tigecycline and Rifampin Alone and in Combination against *Enterococcus faecalis* Biofilm Infection in a Rat Model of Ureteral Stent. *J. Surg. Res.* **2012,** *176,* 1–6.
157. Stickler, D. J.; Jones, G. L.; Russell, A. D. Control of Encrustation and Blockage of Foley Catheters. *Lancet.* **2003,** *361,* 1435–1437.
158. Chew, B. H.; Cadieux, P. A.; Reid, G.; Denstedt, J. D. In-Vitro Activity of Triclosan-Eluting Ureteral Stents against Common Bacterial Uropathogens. *J. Endourol.* **2006,** *20,* 949–958.
159. Cadieux, P.; Chew, B.; Nott, L.; Seney, S.; Elwood, C.; Wignall, G.; Goneau, L.; Denstedt, J. Use of Triclosan-eluting Ureteral Stents in Patients with Long-term Stents. *J. Endourol.* **2009,** *23,* 1187–1194.
160. Lange, D.; Elwood, C. N.; Choi, K.; Hendlin, K.; Monga, M.; Chew, B. H. Uropathogen Interaction with the Surface of Urological Stents Using Different Surface Properties. *J. Urol.* **2009,** *182,* 1194–1200.
161. Krishnasami, Z.; Carlton, D.; Bimbo, L.; Taylor, M.; Balkovetz, D.; Barker, J.; Allon, M. Management of Hemodialysis Catheter-Related Bacteremia with an Adjunctive Antibiotic Lock Solution. *Kidney Int.* **2002,** *61,* 1136–1142.
162. Yahav, D.; Rozen-Zvi, B.; Gafter-Gvili, A.; Leibovici, L.; Gafter, U.; Paul, M. Antimicrobial Lock Solutions for the Prevention of Infections Associated with Intravascular Catheters in Patients Undergoing Hemodialysis: Systematic Review and Meta-analysis of Randomized, Controlled Trials. *Clin. Infect. Dis.* **2008,** *47,* 83–93.
163. Soto, S.M. Importance of Biofilms in Urinary Tract Infections: New Therapeutic Approaches. *Adv. Biol.* **2014,** *2014,* 1–13.
164. McDonnell, G.; Russell, A. D. Antiseptics and Disinfectants: Activity, Action, and Resistance. *Clin. Microbiol. Rev.* **1999,** *12,* 147–179.
165. Vasilev, K.; Cook, J.; Griesser, H. Antibacterial Surfaces for Biomedical Devices. *Exp. Rev. Med. Devices* **2009,** *6,* 553–567.
166. Theerasilp, M.; Nasongkla, N. Comparative Studies of Poly(E-Caprolactone) and Poly(D,L-Lactide) as Core Materials of Polymeric Micelles. *J. Microencapsul.* **2012,** *30,* 390–397.
167. Nasongkla, N.; Bey, E.; Ren, J.; Ai, H.; Khemtong, C.; Guthi, J.; Chin, S.; Sherry, A.; Boothman, D.; Gao, J. Multifunctional Polymeric Micelles as Cancer-targeted, MRI-Ultrasensitive Drug Delivery Systems. *Nano Lett.* **2006,** *6,* 2427–2430.
168. Phuengkham, H.; Nasongkla, N. Development of Antibacterial Coating on Silicone Surface via Chlorhexidine-loaded Nanospheres. *J. Mater. Sci. Mater. Med.* **2015,** *26* (2), 78.

169. Chen, M.; Liang, H.; Chiu, Y.; Chang, Y.; Wei, H.; Sung, H. A Novel Drug-eluting Stent Spray-Coated with Multi-Layers of Collagen and Sirolimus. *J. Control. Release* **2005,** *108,* 178–189.
170. Shukla, A.; Fang, J.; Puranam, S.; Hammond, P. Release of Vancomycin from Multilayer Coated Absorbent Gelatin Sponges. *J. Control. Release* **2012,** *157,* 64–71.
171. Sanpo, N.; Ang, S.; Cheang, P.; Khor, K. Antibacterial Property of Cold Sprayed Chitosan-Cu/Al Coating. *J. Therm. Spray Technol.* **2009,** *18,* 600–608.
172. Dierendonck, M.; De Koker, S.; De Rycke, R.; De Geest, B. Just Spray It–LbL Assembly Enters a New Age. *Soft Matter* **2014,** *10,* 804–807.
173. White, R. Silver in Healthcare: Its Antimicrobial Efficacy and Safety in Use Alan B.G. Lansdown Price: £121.99 RSC Publishing Http://Www.Rsc.Org. *J. Wound Care.* **2011,** *20,* 26.
174. Sousa, C.; Henriques, M.; Oliveira, R. Mini Review: Antimicrobial Central Venous Catheters–Recent Advances and Strategies. *Biofouling* **2011,** *27,* 609–620.
175. Cao, Z.; Sun, X.; Yao, J.; Sun, Y. Silver Sulfadiazine-Immobilized Celluloses as Biocompatible Polymeric Biocides. *J. Bioact. Compat. Polym.* **2013,** *28,* 398–410.
176. Macocinschi, D.; Filip, D.; Paslaru, E.; Munteanu, B.; Dumitriu, R.; Pricope, G.; Aflori, M.; Dobromir, M.; Nica, V.; Vasile, C. Polyurethane-extracellular Matrix/Silver Bionanocomposites for Urinary Catheters. *J. Bioact. Compat. Polym.* **2014,** *30,* 99–113.
177. Høiby, N.; Bjarnsholt, T.; Givskov, M.; Molin, S.; Ciofu, O. Antibiotic Resistance of Bacterial Biofilms. *Int. J. Antimicrob. Agents* **2010,** *35,* 322–332.
178. Gristina, A. G.; Giridhar, G.; Gabriel, B. L.; Naylor, P. T.; Myrvik, Q. N. Cell Biology and Molecular Mechanisms in Artificial Device Infections. *Int. J. Artif. Organs* **1993,** *16,* 755–763.
179. Tenke, P.; Riedl, C. R.; Jones, G. L.; Williams, G. J.; Stickler, D.; Nagy, E. Bacterial Biofilmformation on Urologic Devices and Heparin Coating as Preventive Strategy. *Int. J. Antimicrob. Agents* **2004,** *23,* S67–S74.
180. Maki, D. Engineering out the Risk of Infection with Urinary Catheters. *Emerg. Infect. Dis.* **2001,** *7,* 342–347.
181. Chopra, I. The Increasing Use of Silver-based Products as Antimicrobial Agents: A Useful Development or a Cause for Concern? *J. Antimicrob. Chemother.* **2007,** *59,* 587–590.
182. Guggenbichler, J. P; Böswald, M.; Lugauer, S.; Krall, T. A New Technology of Microdisperse Silver in Polyurethane Induces Antimicrobial Activity in Central Venous Catheters. *Infection* **1999,** *27,* S16–S23.
183. Sahal, G.; Nasseri, B.; Bilkay, I.; Piskin, E. Anti-Biofilm Effect of Nanometer Scale Silver (NmsAg) Coatings on Glass and Polystyrene Surfaces against *P. mirabilis, C. Glabrata* and *C. Tropicalis* Strains. *J. Appl. Biomater. Funct. Mater.* **2015,** *13,* e351–355.
184. Rupp, T. M. E; Fitzgerald, T; Marion, N. Effect of Silver-coated Urinary Catheters: Efficacy, Cost-effectiveness, and Antimicrobial Resistance. *Am. J. Infect. Control.* **2004,** *32* (8), 445–450.
185. Cook, G.; Costerton, J.; Darouiche, R. Direct Confocal Microscopy Studies of the Bacterial Colonization *In Vitro* of a Silver-coated Heart Valve Sewing Cuff. *Int. J. Antimicrob. Agents* **2000,** *13,* 169–173.

186. Logghe, C. Evaluation of Chlorhexidine and Silver-sulfadiazine Impregnated Central Venous Catheters for the Prevention of Bloodstream Infection in Leukaemic Patients: A Randomized Controlled Trial. *J. Hosp. Infect.* **1997,** *37,* 145–156.
187. Bach, A.; Eberhardt, H.; Frick, A.; Schmidt, H.; Bottiger, B.; Martin, E. Efficacy of Silver-coating Central Venous Catheters in Reducing Bacterial Colonization. *Crit. Care Med.* **1999,** *27,* 515–521.
188. Johnson, J. Systematic Review: Antimicrobial Urinary Catheters to Prevent Catheter-associated Urinary Tract Infection in Hospitalized Patients. *Ann. Intern. Med.* **2006,** *144,* 116.
189. Beyth, N.; Houri-Haddad, Y.; Baraness-Hadar, L.; Yudovin-Farber, I.; Domb, A.; Weiss, E. Surface Antimicrobial Activity and Biocompatibility of Incorporated Polyethylenimine Nanoparticles. *Biomaterials.* **2008,** *29,* 4157–4163.
190. Schierholz, J.; Yücel, N.; Rump, A.; Beuth, J.; Pulverer, G. Antiinfective and Encrustation-inhibiting Materials—Myth and Facts. *Int. J. Antimicrob. Agents* **2002,** *19,* 511–516.
191. Lai, L.; Tyson, D.; Clayman, R.; Earthman, J. Encrustation of Nanostructured Ti in a Simulated Urinary Tract Environment. *Mater. Sci. Eng.* C. **2008,** *28,* 460–464.
192. Tran, P.; Hammond, A.; Mosley, T.; Cortez, J.; Gray, T.; Colmer-Hamood, J.; Shashtri, M.; Spallholz, J.; Hamood, A.; Reid, T. Organo Selenium Coating on Cellulose Inhibits the Formation of Biofilms by *Pseudomonas aeruginosa* and *Staphylococcus aureus*. *Appl. Environ. Microbiol.* **2009,** *75,* 3586–3592.
193. Ron, R.; Zbaida, D.; Kafka, I.; Rosentsveig, R.; Leibovitch, I.; Tenne, R. Attenuation of Encrustation by Self-assembled Inorganic Fullerene-like Nanoparticles. *Nanoscale* **2014,** *6,* 5251.
194. Rapoport, L.; Bilik, Y.; Feldman, Y.; Homyonfer, M.; Cohen, S. R; Tenne, R. Hollownanoparticles of WS_2 as Potential Solid-state Lubricants. *Nature* **1997,** *387,* 791–793.
195. Chhowalla, M.; Unalan, H. Thin Films of Hard Cubic Zr_3N_4 Stabilized by Stress. *Nat. Mater.* **2005,** *4,* 317–322.
196. Goldbart, O.; Elianov, O.; Shumalinskym, D.; Lobik, L.; Cytron, S.; Rosentsveig, R.; Wagner, H. D.; Tenne, R. *Nanoscale* **2013,** *5,* 8526–8532.
197. Marmur, A. The Lotus Effect: Superhydrophobicity and Metastability. *Langmuir* **2004,** *20,* 3517–3519.
198. Wong, T.; Kang, S.; Tang, S.; Smythe, E.; Hatton, B.; Grinthal, A.; Aizenberg, J. Bioinspired Self-repairing Slippery Surfaces with Pressure-stable Omniphobicity. *Nature* **2011,** *477,* 443–447.
199. Feng, L.; Zhang, Y.; Xi, J.; Zhu, Y.; Wang, N.; Xia, F; Jiang, L. Petal Effect: A Superhydrophobic State with High Adhesive Force. *Langmuir* **2008,** *24,* 4114–4119.
200. Fu, J.; Ji, J.; Yuan, W.; Shen, J. Construction of Anti-Adhesive and Antibacterial Multilayer Films via Layer-By-Layer Assembly of Heparin and Chitosan. *Biomaterials* **2005,** *26,* 6684–6692.
201. Hoiby, N.; Ciofu, O.; Johansen, H. K.; Song, Z. J.; Moser, C.; Jensen, P. O.; Molin, S.; Givskov, M.; Tolker-Nielsen, T.; Bjarnsholt, T. The Clinical Impact of Bacterial Biofilms. *Int. J. Oral Sci.* **2011,** *3,* 55–65.
202. Nowatzki, P. J.; Koepsel, R. R.; Stoodley, P.; Min, K.; Harper, A.; Murata, H.; Donfack, J.; Hortelano, E. R.; Ehrlich, G. D.; Russell, A. J. Salicylic Acid-Releasing

Polyurethane Acrylate Polymers as Anti-biofilm Urological Catheter Coatings. *Acta Biomater.* **2012,** *8,* 1869–1880.

203. Crouzier, T.; Boudou, T.; Picart, C. Polysaccharide-based Polyelectrolyte Multilayers. *Curr. Opin. Colloid Interface Sci.* **2010,** *15,* 417–426.
204. Haynie, D.; Zhang, L.; Rudra, J.; Zhao, W.; Zhong, Y.; Palath, N. Polypeptide Multilayer Films. *Biomacromolecules* **2005,** *6,* 2895–2913.
205. Caruso, F.; Schüler, C. Enzyme Multilayers on Colloid Particles: Assembly, Stability, and Enzymatic Activity. *Langmuir* **2000,** *16,* 9595–9603.
206. Berg, M.; Zhai, L.; Cohen, R.; Rubner, M. Controlled Drug Release from Porous Polyelectrolyte Multilayers. *Biomacromolecules* **2006,** *7,* 357–364.
207. Kolasinska, M.; Gutberlet, T.; Krastev, R. Ordering of Fe_3O_4 Nanoparticles in Polyelectrolyte Multilayer Films. *Langmuir* **2009,** *25,* 10292–10297.
208. Ivanova, K.; Fernandes, M.; Mendoza, E.; Tzanov, T. Enzyme Multilayer Coatings Inhibit *Pseudomonas aeruginosa* Biofilm Formation on Urinary Catheters. *Appl. Microbiol. Biotechnol.* **2015,** *99,* 4373–4385.
209. Francesko, A.; Fernandes, M. M.; Ivanova, K.; Amorim, S.; Reis, R. L.; Pashkuleva, I.; Mendoza, E.; Pfeifer, A.; Heinze, T.; Tzanov, T. Bacteria-responsive Multilayer Coatings Comprising Polycationic Nanospheres for Bacteria Biofilm Prevention on Urinary Catheters. *Acta Biomater.* **2016,** *33,* 203–212.
210. Desai, D.; Liao, K.; Cevallos, M.; Trautner, B. Silver or Nitrofurazone Impregnation of Urinary Catheters has a Minimal Effect on Uropathogen Adherence. *J. Urol.* **2010,** *184,* 2565–2571.
211. Dong, Y.; Li, X.; Bell, T.; Sammons, R.; Dong, H. Surface Microstructure and Antibacterial Property of an Active-screen Plasma Alloyed Austenitic Stainless Steel Surface with Cu and N. *Biomed. Mater.* **2010,** *5,* 054105.
212. Fadeeva, E.; Truong, V.; Stiesch, M.; Chichkov, B.; Crawford, R.; Wang, J.; Ivanova, E. Bacterial Retention on Superhydrophobic Titanium Surfaces Fabricated by Femtosecond Laser Ablation. *Langmuir* **2011,** *27,* 3012–3019.
213. Wang, X.; Venkatraman, S.; Boey, F.; Loo, J.; Tan, L. Controlled Release of Sirolimus from a Multilayered PLGA Stent Matrix. *Biomaterials* **2006,** *27,* 5588–5595.
214. Xu, Q.; Czernuszka, J. Controlled Release of Amoxicillin from Hydroxyapatite-coated Poly(Lactic-Co-Glycolic Acid) Microspheres. *J. Control. Release* **2008,** *127,* 146–153.
215. Hickok, N.; Shapiro, I. Immobilized Antibiotics to Prevent Orthopaedic Implant Infections. *Adv. Drug Deliv. Rev.* **2012,** *64,* 1165–1176.
216. Gao, G.; Lange, D.; Hilpert, K.; Kindrachuk, J.; Zou, Y.; Cheng, J. T.; Kazemzadeh-Narbat, M.; Yu, K.; Wang, R.; Straus, S. K.; Brooks, D. E.; Chew, B. H.; Hancock, R. E.; Kizhakkedathu, J. N. The Biocompatibility and Biofilm Resistance of Implant Coatings Based on Hydrophilic Polymer Brushes Conjugated with Antimicrobial Peptides. *Biomaterials* **2011,** *32,* 3899–3909.
217. Francois, P. Physical and Biological Effects of a Surface Coating Procedure on Polyurethane Catheters. *Biomaterials* **1996,** *17,* 667–678.
218. Tunney, M.; Gorman, S. Evaluation of a Poly(Vinyl Pyrollidone)-Coated Biomaterial for Urological Use. *Biomaterials* **2002,** *23,* 4601–4608.
219. Buffet-Bataillon, S.; Tattevin, P.; Bonnaure-Mallet, M.; Jolivet-Gougeon, A. Emergence of Resistance to Antibacterial Agents: The Role of Quaternary Ammonium Compounds—A Critical Review. *Int. J. Antimicrob. Agents* **2012,** *39,* 381–389.

220. Soppimath, K.; Aminabhavi, T.; Kulkarni, A.; Rudzinski, W. Biodegradable Polymeric Nanoparticles as Drug Delivery Devices. *J. Control. Release* **2001,** *70,* 1–20.
221. Bridier, A.; Briandet, R.; Thomas, V.; Dubois-Brissonnet, F. Resistance of Bacterial Biofilms to Disinfectants: A Review. *Biofouling* **2011,** *27,* 1017–1032.
222. Eke, G.; Kuzmina, A.; Goreva, A.; Shishatskaya, E.; Hasirci, N.; Hasirci, V. In Vitro and Transdermal Penetration of PHBV Micro/Nanoparticles. *J. Mater. Sci. Mater. Med.* **2014,** *25,* 1471–1481.
223. Durá, N.; De Oliveira, A.; De Azevedo, M. In Vitro Studies on the Release of Isoniazid Incorporated in Poly(E-Caprolactone). *J. Chemother.* **2006,** *18,* 473–479.
224. Harris, C. A; Resau, J. H.; Hudson, E. A.; West, R. A.; Moon, C.; Black, A. D.; McAllister, J. P. Reduction of Protein Adsorption and Macrophage and Astrocyte Adhesion on Ventricular Catheters by Polyethylene Glycol and N-Acetyl-L-Cysteine. *J. Biomed. Mater. Res. A.* **2011,** *98,* 425–433.
225. Li, M.; Neoh, K.; Kang, E.; Lau, T.; Chiong, E. Surface Modification of Silicone with Covalently Immobilized and Crosslinked Agarose for Potential Application in the Inhibition of Infection and Omental Wrapping. *Adv. Funct. Mater.* **2013,** *24,* 1631–1643.
226. Follmann, H.; Martins, A.; Gerola, A.; Burgo, T.; Nakamura, C.; Rubira, A.; Muniz, E. Antiadhesive and Antibacterial Multilayer Films via Layer-By-Layer Assembly of TMC/Heparin Complexes. *Biomacromolecules* **2012,** *13,* 3711–3722.
227. Kuo, W. H.; Wang, M. J.; Chien, H. W.; Wei, T. C.; Lee, C.; Tsai, W. B. Surface Modification with Poly(Sulfobetainemethacrylateco-Acrylic Acid) to Reduce Fibrinogen Adsorption, Platelet Adhesion, and Plasma Coagulation. *Biomacromolecules* **2011,** *12,* 4348−4356.
228. Zhou, J.; Yuan, J.; Zang, X.; Shen, J.; Lin, S. Platelet Adhesion and Protein Adsorption on Silicone Rubber Surface by Ozone-induced Grafted Polymerization with Carboxybetaine Monomer. *Colloids Surf.* B. **2005,** *41,* 55−62.
229. Blanco, C. D; Ortner, A.; Dimitrov, R.; Navarro, A; Mendoza, E; Tzanov, T. Building an Antifouling Zwitterionic Coating on Urinary Catheters Using an Enzymatically Triggered Bottom-Up Approach. *ACS Appl. Mater. Interfaces* **2014,** *6,* 11385−11393.
230. Tan, H.; Peng, Z.; Li, Q.; Xu, X.; Guo, S.; Tang, T. The Use of Quaternised Chitosan-Loaded PMMA to Inhibit Biofilm Formation and Downregulate the Virulence-associated Gene Expression of Antibiotic-resistant *Staphylococcus. Biomaterials* **2012,** *33,* 365–377.
231. Choong, S. K.; Wood, S.; Whitfield, H. N. A Model to Quantify Encrustation on Ureteric Stents, Urethral Catheters and Polymers Intended for Urological Use. *BJU Int.* **2000,** *86,* 414–421.
232. Roe, D.; Karandikar, B.; Bonn-Savage, N.; Gibbins, B.; Roullet, J. B. Antimicrobial Surface Functionalization of Plastic Catheters by Silver Nanoparticles. *J. Antimicrob. Chemother.* **2008,** *61,* 869–876.
233. Raad, I. Intravascular-Catheter Related Infections. *Lancet.* **1998,** *351,* 893–898.
234. Furno, F.; Morley, K. S.; Wong, B.; Sharp, B. L.; Arnold, P. L.; Howdle, S. M.; Bayston, R.; Brown, P. D.; Winship, P. D.; Reid, H. J. Silver Nanoparticles and Polymeric Medical Devices: A New Approach to Prevention of Infection? *J. Antimicrob. Chemother.* **2004,** *54,* 1019–1024.

CHAPTER 13

NANOSTRUCTURE DEFORMABLE ELASTIC VESICLES AS NANOCARRIERS FOR THE DELIVERY OF DRUGS INTO THE SKIN

SUDHAKAR C. K.[1,2,*], NITISH UPADHYAY[2], SANJAY JAIN[3], and R. NARAYANA CHARYULU[4]

[1]*Department of Pharmaceutics, School of Pharmaceutical Sciences, Lovely Professional University, Jalandhar-Delhi G. T. Road, Phagwara 144411, Punjab, India*

[2]*Smriti College of Pharmaceutical Education, Indore, India*

[3]*Department of Pharmacognosy, Indore Institute of Pharmacy, Indore, India*

[4]*Department of Pharmaceutics, NGSMIPS, Mangalore, Karnataka, India*

[*]*Corresponding author. E-mail: ckbhaipharma@gmail.com*

CONTENTS

ABSTRACT

Dressing of active pharmaceutical ingredient (API) should be in such a manner that it looks simple and descent with more beneficial way. Phospholipids (phosphatidylcholine) with accessories like ethanol and water are dressing material for many drugs which overcome many problems related to poor bioavailability, poor solubility, etc. A plethora of nanomedicine has been highlighted for various purposes. Human immunodeficiency virus (HIV) is not curable but it can be controlled by nanomedicine or drug delivery system which boosts the immune system of the body. Ethosomes (deformable vesicles) are the perfect dress for antiretroviral drugs and it is nano-drug delivery system which enhances the permeation of drugs through the skin and it is able to heighten the immune system of body during HIV infection. Different formulations of ethosomes are prepared and characterized for size, zeta potential, and entrapment efficacy in in vitro and in vivo studies. Result reveals that ratio of phospholipids and ethanol should be optimal to achieve superior entrapment, permeation of model drug into skin and to have sustained release pattern. The entrapment efficacies are ET-1 (69.78 ± 1.24), ET-2 (78.95 ± 1.54), ET-3 (79.54 ± 2.10), and LP (59.87 ± 2.12). The EL-2 and El-3 have superior release profile than all other formulation with 84.52 ± 2.63 and 86.65 ± 2.39%, respectively, for 12 h. Plasma concentration profile divulges the sustained release of model drug and penetration of model drugs into deeper layers of skin due to the presence of ethanol.

13.1 INTRODUCTION

From the primeval time, humans used to apply substances on the skin for healing or therapeutic effects and in the current modern era, numerous transdermal and topical formulations have been developed to treat systemic and local infections.[46,52] Skin is the biggest interface between human body and external environment. The impermeable nature of skin (stratum corneum, SC) acts as a barrier for delivering of drugs to the skin as systemic or local effect.[45,54] Topical delivery to the various regions of the skin and underlying tissues, transdermal drug delivery and dermal exposure to environmental chemicals are essential areas of research.[4,18,57] The skin is the largest surface area of human body (1.8–2.0 m^2). It is composed of three main layers; the epidermis, dermis, and subcutaneous layer and rational target for drug delivery, the main obstacle is the SC which limits the entry of drugs to the skin.[5,39,43,44,47,59]. During last decades, research on transdermal drug delivery and vesicular drug delivery system has been a hot topic for delivering the drugs through the skin to systemic and local effect.[43,54] Vesicular carrier or lipid vesicles drug delivery has been extensively studied for delivering drug to and through the skin. Liposomes, sheath liposomes[34,35] niosomes[31] transfersomes,[12] flexosomes,[3] ethosomes,[37] transethosomes,[6] invasomes,[16] etc. are some vesicular carrier developed for transdermal drug delivery system.

Nanomedicine is an engineered nanostructure which acts as a driver for the delivery of drugs to specific site. Nano-drug delivery because of its very small size is able to intermingle with excessively small systems; and biological systems are today the concourse for nano-drug delivery applications having desire to address unmet needs in biology. Controlling and operating things at the nanometer scale allows exploring and interacting at the cellular level with unique fashion.[58] Dressing of active pharmaceutical ingredient (API) should be in such a manner that it looks simple and descent with more beneficial way.[49] Selection of the dressing material (polymers) should fulfill all desire of the formulation to be achieved. Lipids/polymers which are known to be best clothes for API can be transformed into vesicular formulation which put forward elegant solutions to a number of problems. Phospholipids (phosphatidylcholine) with accessories like ethanol and water are dressing material for many drugs which overcome many problems related to poor bioavailability, poor solubility, etc. A plethora of nanomedicine has been highlighted for various purposes. Ethosomal systems are soft malleable nanovesicles consisting essentially of phospholipids, water, and a high quantity of ethanol (Fig. 13.1).[7,52] Ethosome along with surfactant can be used to enhance the permeation of drugs through the skin by a dual mechanism. Ethosomes presents an opening to outwit many challenges linked with antiretroviral drug therapy.[37,49,50,52] Human immunodeficiency virus (HIV) is not curable but it can be controlled by nanomedicine or drug delivery system which boost the immune system of the body. Lamivudine (3TC) is 2′,3′-dideoxy-3′-thiacytidine and nucleoside analog reverse transcriptase inhibitor (NRTI) class water-soluble antiretroviral drug for treatment of HIV. It is an antiretroviral drug that decreases the amount of HIV in the body. Lamivudine maintains viral suppression and decreases disease progression and helps in the building up of the immune system.[2,10,42,48] Lamivudine has short biological half-life; so it requires frequent dosing as antiretroviral drugs are used lifelong in acquired immunodeficiency syndrome (AIDS).[10,42,53] Transdermal application has numerous advantages such as less frequent dosing; maintain constant plasma concentration of drug in blood for a longer time.[25] Ethosomes which are second generation of the vesicular carrier are used for transdermal delivery of lamivudine. Ethosomes are soft malleable, phospholipid nanovesicles bubble like which, due to their structure, are able to overcome the natural dermal barrier SC, delivering drugs through the skin layers.[9,17,30,52] Ethanol may provide the vesicles with soft flexible characteristics which allow them to more easily penetrate into deeper layers of the skin.[52] Transethosomes are form of the ethosomes which contains edge surfactant and have combination properties of transfersomes and ethosomes. The physicochemical features of vesicles such as the fluidity, lamellarity, elasticity, and size are significant elements for their dermal drug delivery properties. Drug penetration can be influenced by modifying the surface charge of the vesicles.[36,41] The skin may act as negatively charged membrane due to the presences of negative charged lipids in the SC.[20,60] Researcher have stressed that drug penetration can be influenced by modifying the surfaces of liposomes, because this parameter can affect the transcutaneous diffusion of drugs.

Negatively charged vesicles may experience a weak electrostatic repulsion in the intercellular domain of the SC. Negative-charged vesicles usually result in higher flux and greater diffusion coefficient than positive-charged vesicles due to the lipid region/corneocytes of skin is negative-charged or neutral vesicles.[15,21,28,32] Positive charged vesicles formulation could bind to the negative charge of corneocytes of the SC and drug remains as a depot in the outer layer of the skin.[23] Negative-charged vesicles mixed with lipids of SC induce ultrastructural changes due to an impaired barrier function in the deeper layer of the SC providing enhancer effect of the drug to deeper skin layer or systemic circulation or lymph capillaries.[22] Ethosomes with additional permeation enhancer boost the penetration of the lamivudine to deeper skin tissue and show systemic effect. The objective of the study was to deliver the antiretroviral drugs to deeper skin and targets the lymph capillaries to lymph nodes (HIV reservoir) by negative-charged vesicular system. A negative-charged vesicle directs toward to lymph system helping the reduction of virus in the blood.[23] Ethosomes (deformable vesicles) are promising approach for delivering antiretroviral drugs and being nano-size vesicles, which will enhance the penetration of drugs through the skin and is able to increase the immune system of body during HIV infection.

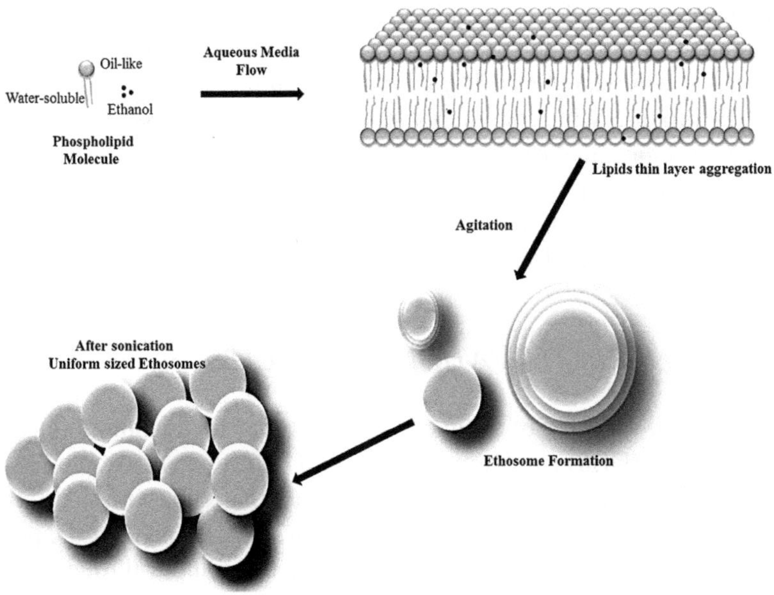

FIGURE 13.1 Schematic of formation of ethosomes from phospholipid, ethanol, and aqueous media.

13.2 MATERIAL AND METHODS

Lamivudine is procured from Mylan, Hyderabad; Lecithin (soybean-derived phosphatidylcholine) was obtained from Lipoid, Germany; dicetyl phosphate, polycarbonate membrane, 6-carboxy fluorescein, propylene glycol, ethanol, and cholesterol were purchased from Sigma, India and other solvents are of HPLC grade from HiMedia.

13.2.1 PREPARATION OF DEFORMABLE VESICLES

Deformable vesicles bearing lamivudine were prepared according to previous methods[26,27,55] with slight modification. The composition of deformable vesicles system consists of phospholipids, ethanol, lamivudine/6-carboxy fluorescein (CF) and aqueous phase. In this method phospholipid (soybean-derived phosphatidylcholine) and lamivudine/6-carboxy fluorescein is dissolved in ethanol with continued stirring in a vessel maintained to temperature 60°C. To ethanolic mixture dicetylphosphate (DCP) was added as a negative inducer, and stirred it for another 15 min. In a separate vessel, aqueous phase/phosphate buffer system 7.4 was heated at 60°C and then ethanolic mixture was added to the PBS mixture with continuous stirring at 700 rpm for 30 min. After mixing, the ethosomal preparation was sonicated and passed through polycarbonate membrane for the desired size of vesicles. Total volume of the preparation was 30 mL and formula was given in Table 13.1. In a similar manner the liposomal formulation containing lamivudine was prepared.

13.2.2 CHARACTERIZATION OF VESICLES

Zeta potential and size of vesicles were analyzed by dynamic light scattering method (DLS) (Malvern Zetamaster, UK). It consisted of a He–Ne laser (5 mW) and a small volume sample-holding cell, with a stirrer so that the sample, diluted with distilled water, was stirred all over the period of measurement so as to provide stirring of the sample throughout the measurement period. Measurements were performed using a 45 mm focus objective and a beam length of 2.4 mm.[24]

TABLE 13.1 Composition of Different Ethosomal and Liposomal Formulation.

Ingredient	Formulation			
	ET-1	ET-2	ET-3	LP
Phospholipid (%wt/wt)	2.0	2.0	2.0	2.0
Dicetyl phosphate (mg)	13 mg	15 mg	17 mg	–
Drug (%wt/wt)	0.4	0.4	0.4	0.4
Ethanol (%wt/wt)	20	30	40	–
Cholesterol (%wt/wt)	–	–	–	0.10

13.2.3 ENTRAPMENT EFFICIENCY DETERMINATION

Entrapment efficiency of vesicles will influence the release kinetics of the drug. The entrapment efficiency of drug in ethosome and liposome was determined to calculate the unentrapped content with the use of HPLC and ultracentrifugation method. Add 10 mL of ethosomal solution in centrifuge tube and vesicular suspension to centrifuge in a cooling centrifugation (Remi C852, India) at 20,000 rpm at 4°C for 1 h, where upon the sedimentation of vesicles and the supernatant containing free drug were obtained. The sediment was washed again with distilled water to remove any unentrapped drug by centrifugation. The combined supernatant layer was diluted with solvent suitably and lamivudine concentration was determined at 271 nm in all samples. Elastic vesicles prepared without drug, in a similar manner, served as blank for the above studies conducted in triplicate. The entrapment efficiency of the vesicles was calculated from the difference between the initial drug added and the free drug detected in the supernatant.[38,40] Same procedure is followed for all ethosomal formulation and liposomal formulation.

Permeation studies of ethosomal formulation: All animal trials were performed according to the protocol standard by the Institutional Animal Ethics Committee (IAEC) of SCOPE, Indore. (IAEC Reg No.1227/ac/08/CPCSEA/SCOPE/14-15/105) The in vitro skin penetration of lamivudine from different ethosomal and liposomal formulation was studied using Franz diffusion cell with diffusional area of 3.14 cm^2. The temperature and stirring were monitored and maintained in receptor cell at 37 ± 0.5°C and 120 rpm, respectively. The hairs on abdominal of rat skin were removed with an electric clipper cautiously. Fats and extra tissues were removed from the rat skin. Prepared dorsal rat skin was placed between the donor and the receiver compartments. In receiver compartment, fill fresh buffer phosphate buffer sulfate (PBS) 7.4 was filled and assembly was placed on the magnetic stirrer. Freshly prepared ethosomal and liposomal formulation on the upper surface of skin and the content of diffusion cell was kept under constant stirring. The donor compartment of the Franz diffusion cell was shielded with parafilm to avoid any evaporation process. At regular interval 1 mL of sample was withdrawn from diffusion cell and replaced with fresh buffer PBS 7.4 to maintain sink condition. The withdrawn samples were examined by spectroscopic method at 271 nm. Triplicate experiments were conducted for skin penetration study.

13.2.4 IN VIVO STUDIES

In male Wistar rats, formulation was applied on trimmed dorsal surface of rat on the dorsal surface of skin. At regular intervals, blood was withdrawn from the retro-orbital vein and centrifuged in cooling centrifuge to separate plasma and blood components. Drug in plasma was analyzed by HPLC using a PDA detector at 271 nm.

13.2.5 HPLC ASSAY

The calibration curve of drug done by HPLC using mobile phase methanol:water:acetonitrile was (70:20:10 v/v) at a flow rate of 1 mL/min (Waters HPLC). The drug penetrated in the receptor compartment of Franz diffusion cell during in vivo skin permeation experiments were examined by HPLC. The samples were investigated using PDA detector at 271 nm. Similarly, drug content in plasma of in vivo studies was analyzed.

13.2.6 FLUORESCENCE STUDY

Fluorescence study was performed to endorse the skin permeation ability of ethosomal and liposomal formulations. The fluorescent labeling was carried out by formulating the optimized ethosomal formulations with fluorescence marker 6-carboxy fluorescein indicator. Then 6-carboxy fluorescein loaded formulation was applied topically to dorsal surface of trimmed rats. After 3 h of application, the rats were sacrificed and skin was removed, cut into small pieces, fixed into fixative solution for 3 h. The paraffin blocks were made by embedding the tissue in hard paraffin, matured at 60°C. Using microtome the sections were cut into 2–5 μm and observed under a fluorescence microscope.[27,49]

13.2.7 STABILITY OF VESICLES

All ethosomal formulations were kept for stability study and stored in refrigerated condition, 4–8°C at a temperature of 25 ± 2°C and at body temperature (37 ± 2°C). Percentage of drug entrapment and size of vesicles were determined at regular time intervals (0, 1, 30, and 45 days).

13.3 RESULT AND DISCUSSION

Sometimes an API needs more than a simple formulation to become a successful drug.[51] The selection of a vehicle can considerably impinge on drug delivery and subsequently the efficacy of topical/transdermal preparations. Selecting the correct delivery medium is also of paramount importance. DCP is used as negative charge inducer which makes the surface of ethosome negatively charged and helps in targeting the lymph.[28,49,50,52,55]

Ethosome is ethanolic flexible liposome and it has the ability to permeate drug into the skin. If the ethosomes were combined with surfactant DCP, it will boost the permeation of drug into deeper skin tissue and show systemic action. Ethanol is an effective permeation enhancer and plays a vital role in ethosomal systems by giving the vesicles unique characteristics in terms of size, zeta potential, stability, entrapment efficacy, and enhanced skin permeability.[1,6,11] Ethanol plays a significant role in controlling the size of vesicles, increase in ethanol concentration, and decrease the size of vesicles (Fig. 13.2). A presence of ethanol also provides extra negative charge to ethosomes and improves the stability of vesicles. The entrapment efficiency of the vesicles are shown in Figure 13.3 and entrapment depends on many parameters like concentration of phospholipids, ethanol concentration, and zeta potential. The entrapment efficiency of the ethosome formulation is ET-1 (69.78 ± 1.24), ET-2 (78.95 ± 1.54), ET-3 (79.54 ± 2.10) and liposomes formulation is LP (59.87 ± 2.12). Due to the presence of ethanol the flexibility of vesicles is increased and entrapped drug do not leak out of the vesicle membrane easily. Increasing ethanol in ethosomes does not significantly increase the entrapment efficiency as results shown in Figure 13.3. Absences of ethanol in liposomes result in lower entrapment efficiency. The size of vesicles also plays a vital role in permeation of drug in the skin. The formulation of ethosomes and liposomes are sonicated and passed through polycarbonate membrane, and the size of ethosomes ranges from 267 ± 8.34 to 287 ± 12.43 nm, whereas the size of liposomes is 487.7 ± 21.34 nm.

Ethanol and DCP are responsible for the size reduction of vesicles. The presence of DCP in ethosome formulation increases the zeta potential up to ET-1 (−10.76 ± 0.23 ζmv), ET-2 (−17.56 ± 1.02 ζmv), ET-3 (−26.74 ± 0.98 ζmv), and LP (−2.73 ± 0.09 ζmv). The entrapment efficiency increases with higher value of zeta potential. Progresses in the current understanding of the role of the lymphatics in pathological change and immunity have driven the recognition that lymph-targeted delivery has the potential to transform disease treatment and vaccination.[29,56] Lymph node and lymph capillaries are garbage of the viruses. There has been increased interest in developing charged liposomes as carriers for transdermal drug to target lymph. While modifying the basic composition of the vesicles will enhance the permeation of drugs in the vesicles through the skin. Charged vesicles offer several benefits compared with uncharged vesicles system.[13,14,19,23,33] Ethosomes were surface-engineered by DCP

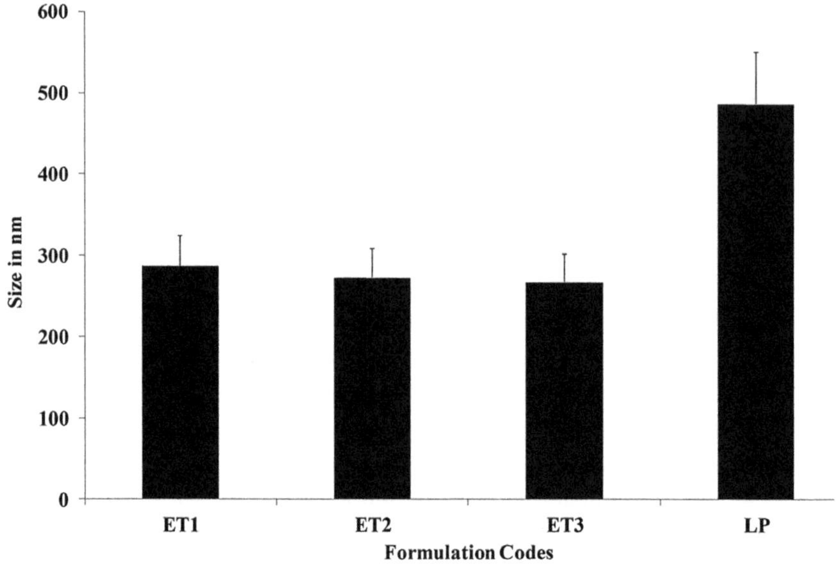

FIGURE 13.2 Vesicles size of different formulations of vesicles.

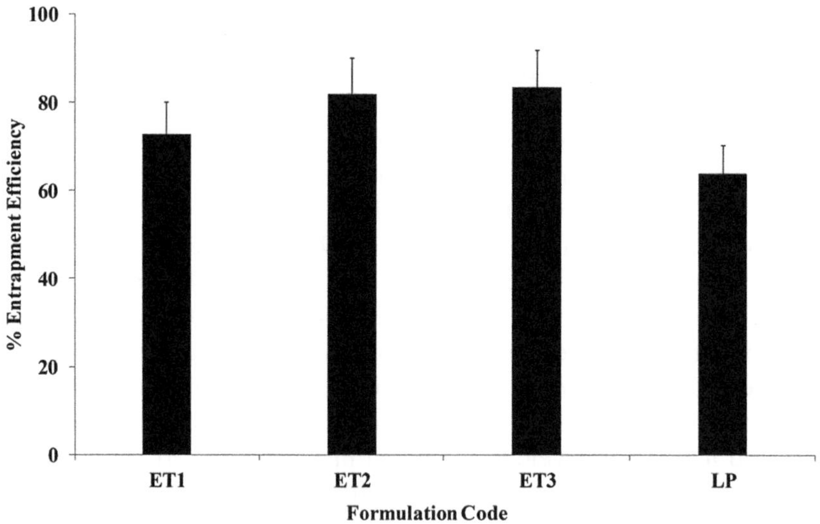

FIGURE 13.3 Entrapment efficiency of different formulations of vesicular system.

which enhances lymphatic targeting of lamivudine. Comprehension of the zeta potential of a vesicles preparation can help to forecast the fate of the vesicles in vivo. The fluorescence photograph of ethosomes and liposomes formulation in Figure 13.4 shows that ethosomes have maintained the size in nano range compared to liposomes. In Figures 13.5 and 13.6, fluorescence photograph of ET-2 and ET-3 has shown permeation of CF into the deeper tissue layer of skin (dermis) and ET-1 also shown permeation of CF into the dermis, but LP has been restricted toward to the SC to epidermis only as shown in Figure 13.6. The fluorescence's study of histological section of skin reveals that DCP and ethanol help in penetrating the CF to deeper skin layer and lymph capillaries. The negative-charged vesicle

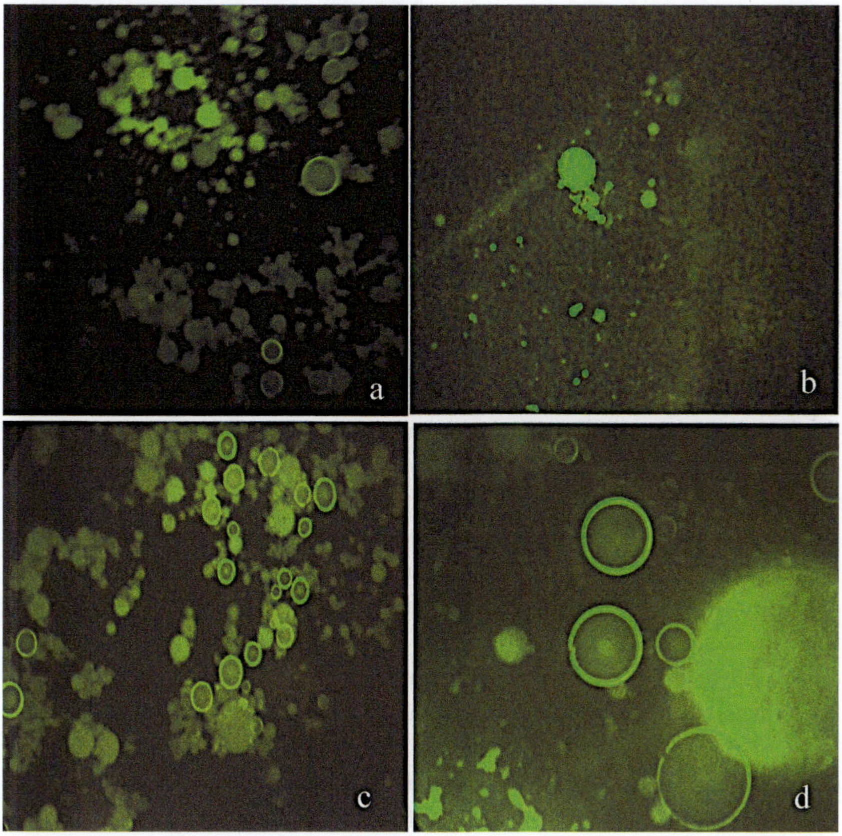

FIGURE 13.4 Fluorescence photograph of ethosome and liposome formulation: (a) ethosome formulation-I (ET-1), (b) ethosome formulation-II (ET-2), (c) ethosome III (ET-3), and (d) liposome formulation (LP).

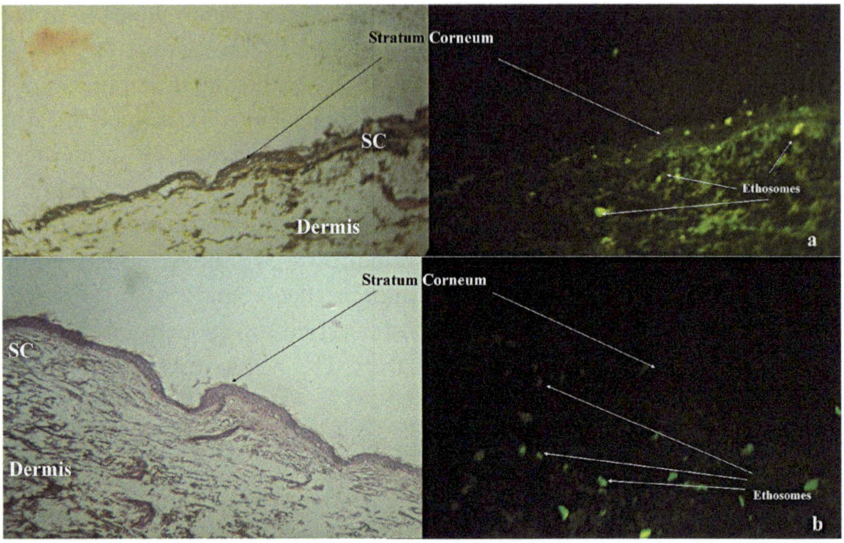

FIGURE 13.5 Fluorescence photograph of ethosome penetration in the skin after 3 h: (a) ethosome formulation-III (ET-3) and (b) ethosome formulation-II (ET-2) (SC—stratum corneum).

containing CF will move toward lymph capillaries due to affinity toward the lymph.[29] DCP perturbs the SC, the phospholipid packaging characteristics and thus, fluidize the vesicle bilayer, which further penetrates to the deeper skin. Histological studies publicized that the negatively charged flexible liposomes diffused to the dermis by electrostatic forces and the lower portion of hair follicles through the SC and the follicles much faster than the neutral vesicles after application of formulation. Thus, the rapid permeation of negatively charged flexible liposomes would contribute to the increased permeation of drugs through the skin.[8,13,41] Permeation studies of ethosome and liposome formulations show that ethosome formulation permeates faster than liposomes (Fig. 13.7a) where lag time is less compared to liposomes. All ethosome formulation has shown more than 75% release in 12 h, whereas liposomes have shown less than 50% release in 12 h (Fig. 13.7). The EL-2 and El-3 have superior release profile than all other formulation with 84.52 ± 2.63 and 86.65 ± 2.39%, respectively. LP also had shown 48.3 ± 2.36% for 12 h. The ethosomal formulation ET-3, liposomal, and drug solution were taken for in vivo study. The plasma concentration profile shows that the ethosomes ET-3 with higher C_{max} had longer duration compared to liposome and drug solution (Fig. 13.8).

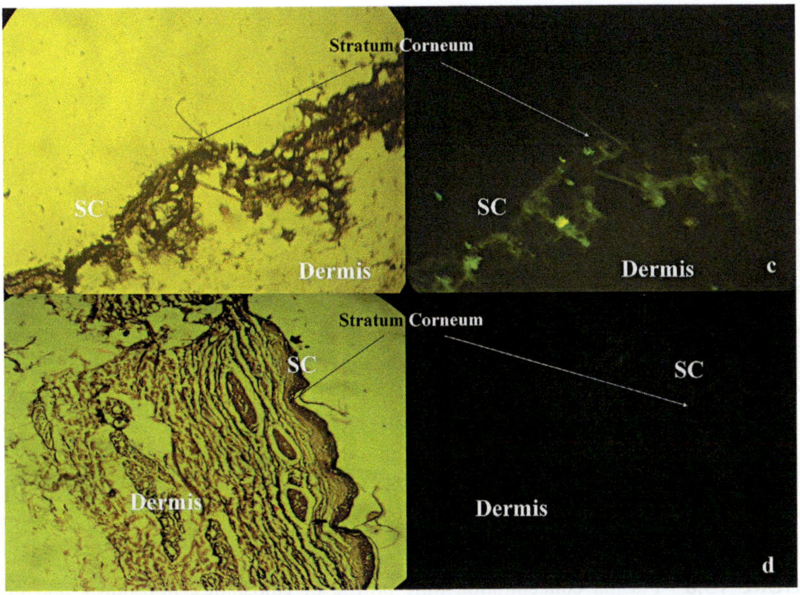

FIGURE 13.6 Fluorescence photograph of ethosome and liposome penetration in the skin after 3 h: (c) ethosome formulation-I (ET-1) and (d) liposome formulation (SC—stratum corneum).

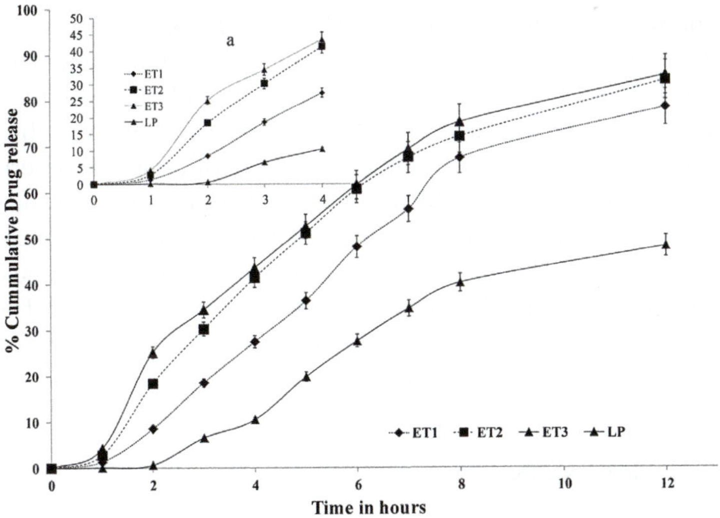

FIGURE 13.7 In vitro release study of ethosomes and liposome bearing lamivudine ET-1, ET-2, ET-3, and LP (a) lag time of all ethosome and liposome formulation.

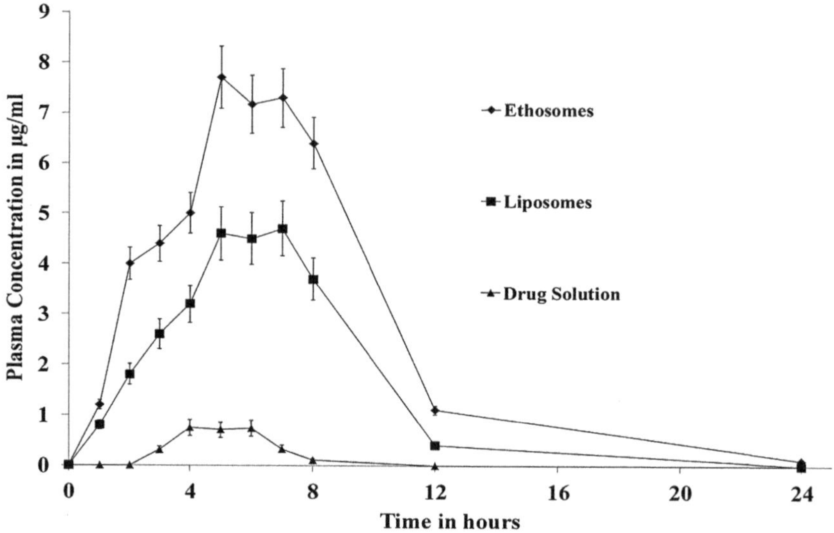

FIGURE 13.8 Plasma concentration time profile of ethosomes, liposomes, and drug solution.

Plasma concentration profile divulges the sustained release of lamivudine drug and penetration of lamivudine into deeper layers of skin due to the presence of ethanol and DCP. The vesicles formulation were kept at three different temperatures and when comparing the entrapment efficiency at 4–8 and 25 ± 2°C, there was no significant leakage of drug from vesicle was witnessed, and at 37 ± 2°C minimum leakage was observed. From liposomal formulation significant leakage of drug was observed at 37 ± 2 and 25 ± 2°C (Fig. 13.9). Ethosomal formulation kept in three temperatures was concluded that they are stable in term of entrapment efficiency for 45 days. Owing to the high negative zeta potential, ethosomes has higher colloidal stability than its liposomal counterparts.[13,55] The size of vesicles remain stable during the storage at 25 ± 2 and 4–8°C when compared to 37 ± 2°C temperature.

13.4 CONCLUSIONS

Nanomedicine field is going to provide multiple new solutions and products that will solve healthcare challenges in the coming decades. The use of nanotechnology to advance nanodelivery systems with more precision and targeting toward the unhealthy or diseased tissues reduces the toxicity of drugs to healthy tissues. Deformable vesicles enhance penetration of drugs with its ethanol effect and show lipid perturbation. The negative charge on ethosomes also enhances the permeation of drug in the skin and help in targeting the lymph capillaries. Ethosomes are ethanol-modified liposomes that act as reservoir systems and offer incessant delivery of medication to the desired site.

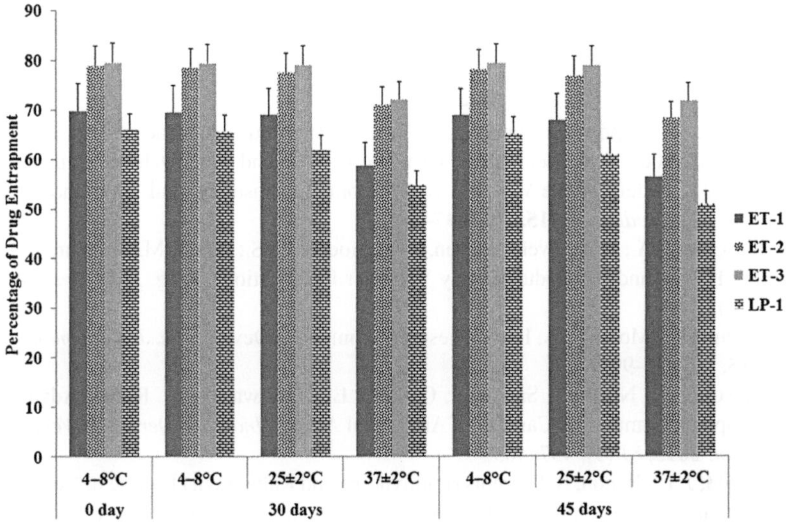

FIGURE 13.9 Stability study of ethosomes and liposomes.

KEYWORDS

- **ethosomes**
- **ethanol**
- **permeation**
- **antiretroviral drugs**
- **deformable vesicles**
- **nanomedicine**

REFERENCES

1. Abdulbaqi, I. M.; Darwis, Y.; Khan, N. A.; Assi, R. A.; Khan, A. A. Ethosomal Nanocarriers: The Impact of Constituents and Formulation Techniques on Ethosomal Properties, In Vivo Studies, and Clinical Trials. *Int. J. Nanomedicine* **2016,** *11,* 2279–2304.
2. Aliasagar, S.; Misra, A. Studies in Topical Application of Niosomally Entrapped Nimesulide. *J. Pharm. Pharm. Sci.* **2005,** *5* (3), 220–225.
3. Alomrani, A. H.; Badran, M. M. Flexosomes for Transdermal Delivery of Meloxicam: Characterization and Anti-inflammatory Activity. *Artif. Cells Nanomed. Biotechnol.* **2016,** *45,* 305–312.
4. Anissimov, Y. G.; Jepps, O. G.; Dancik, Y.; Roberts, M. S. Mathematical and Pharmacokinetic Modelling of Epidermal and Dermal Transport Processes. *Adv. Drug Deliv. Rev.* **2013,** *65* (2), 169–90.
5. Arda, O.; Goksugur, N.; Tuzun, Y. Basic Histological Structure and Functions of Facial Skin. *Clin. Dermatol.* **2014,** *32* (1), 3–13.
6. Ascenso, A.; Raposo, S.; Batista, C.; Cardoso, P.; Mendes, T.; Praça, F. G.; Bentley, M. V.; Simões, S. Development, Characterization, and Skin Delivery Studies of Related Ultradeformable Vesicles: Transfersomes, Ethosomes, and Transethosomes. *Int. J. Nanomedicine* **2015,** *10,* 5837–5851.
7. Bouwstra, J. A.; Honeywell-Nguyen, P. L.; Gooris, G. S.; Ponec, M. Structure of the Skin Barrier and Its Modulation by Vesicular Formulations. *Prog. Lipid Res.* **2003,** *42* (1), 1–36.
8. Bozzuto, G.; Molinari, A. Liposomes as Nanomedical Devices. *Int. J. Nanomedicine.* **2015,** *10,* 975–999.
9. Bseiso, E. A.; Nasr, M.; Sammour, O.; Abd, E. L.; Gawad, N. A. Recent Advances in Topical Formulation Carriers of Antifungal Agents. *Indian J. Dermatol. Venereol. Leprol.* **2015,** *81* (5), 457–463.
10. Casado, J. L.; Banon, S. Dutrebis (Lamivudine and Raltegravir) for Use in Combination with Other Antiretroviral Products for the Treatment of HIV-1 Infection. *Expert. Rev. Clin. Pharmacol.* **2015,** *8* (6), 709–718.
11. Charles, M. H. Ethanol and Other Alcohols: Old Enhancers, Alternative Perspectives. In *Percutaneous Penetration Enhancers Chemical Methods in Penetration Enhancement;* Dragicevic, N., Maibach, H. I., Eds.; Springer: Berlin, Germany, 2016; pp 151–172.
12. Chaudhary, H.; Kohli, K.; Kumar, V. Nano-transfersomes as a Novel Carrier for Transdermal Delivery. *Int. J. Pharm.* **2013,** *454* (1), 367–380.
13. Chen, J.; Lu, W. L.; Gu, W.; Lu, S. S.; Chen, Z. P.; Cai, B. C. Skin Permeation Behaviour of Elastic Liposomes: Role of Formulation Ingredients. *Expert Opin. Drug Deliv.* **2013,** *10* (6), 845–856.
14. Chen, M.; Kumar, S.; Anselmo, A. C.; Gupta, V.; Slee, D. H.; Muraski, J. A.; Mitragotri, S. Topical Delivery of Cyclosporine A into the Skin Using SPACE-peptide. *J. Control Release* **2015,** *199,* 190–197.
15. Cui, Z.; Mumper, R. J. Chitosan-based Nanoparticles for Topical Genetic Immunization. *J. Control. Release* **2001,** *75,* 409–419.
16. Duangjita, S.; Nimcharoenwan, T.; Chomya, N.; Locharoenrat, N.; Ngawhirunpat, T. Design and Development of Optimal Invasomes for Transdermal Drug Delivery Using Computer Program. *Asian J. Pharm. Sci.* **2016,** *11* (1), 52–53.

17. Faisal, W.; Soliman, G. M.; Hamdan, A. M. Enhanced Skin Deposition and Delivery of Voriconazole Using Ethosomal Preparations. *J. Liposome Res.* **2016,** 1–25. http://www.tandfonline.com/doi/abs/10.1080/08982104.2016.1239636?journalCode=ilpr20
18. Firooz, A.; Nafisi, S.; Maibach, H. I. Novel Drug Delivery Strategies for Improving Econazole Antifungal Action. *Int. J. Pharm.* **2015,** *495* (1), 599–607.
19. Gajbhiye, V.; Ganesh, N.; Barve, J.; Jain, N. K. Synthesis, Characterization and Targeting Potential of Zidovudine Loaded Sialic Acid Conjugated-mannosylated Poly (Propyleneimine) Dendrimers. *Eur. J. Pharm. Sci.* **2013,** *48* (4–5), 668–679.
20. Gelfuso, G. M.; Gratieri, T.; Souza, J. G.; Thomazine, J. A.; Lopez, R. F. The Influence of Positive or Negative Charges in the Passive and Iontophoretic Skin Penetration of Porphyrins Used in Photodynamic Therapy. *Eur. J. Pharm. Biopharm.* **2011,** *77* (2), 249–256.
21. Gillet, A.; Compère, P.; Lecomte, F.; Hubert, P.; Ducat, E.; Evrard, B.; Piel, G. Liposome Surface Charge Influence on Skin Penetration Behaviour. *Int. J. Pharm.* **2011** *411,* 223–231.
22. Gonzalez-Rodríguez, M. L.; Cózar-Bernal, M. J.; Fini, A.; Rabasco, M. A. Surface-Charged Vesicles for Penetration Enhancement. In *Percutaneous Penetration Enhancers Chemical Methods in Penetration Enhancement: Nanocarriers;* Dragicevic, N., Maibach, H. I., Eds.; Springer: Berlin, Germany, 2016; pp 121–136.
23. Gonzalez-Rodríguez, M. L.; Rabasco, A. M. Charged Liposomes as Carriers to Enhance the Permeation through the Skin. *Expert Opin Drug Deliv.* **2011,** *8* (7), 857–871.
24. Guinedi, A. S.; Mortada, N. D.; Mansour, S.; Hathout, R. M. Preparation and Evaluation of Reverse Phase Evaporation and Multilamellar Niosomes as Ophthalmic Carrier of Acetazolamide. *Int. J. Pharm.* **2005,** *306,* 71–82.
25. Ham, S. A.; Buckheit, R. W. Current and Emerging Formulation Strategies for the Effective Transdermal Delivery of HIV Inhibitors. *Ther. Deliv.* **2015,** *6* (2), 217–229.
26. Jain, S.; Sapre, R.; Tiwary, A. K.; Jain, N. K. Proultraflexible Lipid Vesicles for Effective Transdermal Delivery of Levonorgestrel: Development, Characterization and Performance Evaluation. *AAPS Pharm. Sci. Tech.* **2005,** *6* (3), E513–E522.
27. Jain, S.; Tiwary, A. K.; Sapra, B.; Jain, N. K. Formulation and Evaluation of Ethosomes for Transdermal Delivery of Lamivudine. *AAPS Pharm. Sci. Tech.* **2007,** *8* (4), E111.
28. Khaizan, A. N.; Wong, T. W. Pectin and Its Roles in Transdermal Drug Delivery. In *Handbook of Sustainable Polymers: Processing and Applications;* Thakur, V. K., Thakur, M. K., Eds.; Pan Stanford Publication: Boca Raton, FL, 2015; pp 453–472.
29. Khan, A. A.; Mudassir, J.; Mohtar, N.; Darwis, Y. Advanced Drug Delivery to the Lymphatic System: Lipid-based Nanoformulations. *Int. J. Nanomedicine* **2013,** *8,* 2733–2744.
30. Khan, N. R.; Wong, T. W. Microwave-aided Skin Drug Penetration and Retention of 5-Fluorouracil-loaded Ethosomes. *Expert Opin Drug Deliv.* **2016,** *13* (9),1209–1219.
31. Khan, R.; Irchhaiya, R. Niosomes: A Potential Tool for Novel Drug Delivery. *J. Pharm. Invest.* **2016,** *46* (3), 195–204.
32. Kohli, A. K.; Alpar, H. O. Potential Use of Nanoparticles for Transcutaneous Vaccine Delivery: Effect of Particle Size and Charge. *Int. J. Pharm.* **2004,** *275,* 13–17.

33. Kong, M.; Hou, L.; Wang, J.; Feng, C.; Liu, Y.; Cheng, X.; Chen, X. Enhanced Transdermal Lymphatic Drug Delivery of Hyaluronic Acid Modified Transfersomes for Tumor Metastasis Therapy. *Chem. Commun. (Camb).* **2015,** *51* (8), 1453–1456.
34. Li, Z.; Kang, H.; Che, N.; Liu, Z.; Li, P.; Li, W.; Zhang, C.; Cao, C.; Liu, R.; Huang, Y. Controlled Release of Liposome-encapsulated Naproxen from Core-sheath Electrospun Nanofibers. *Carbohydr. Polym.* **2014a,** *111,* 18–24.
35. Li, Z.; Kang, H.; Li, Q.; Che, N.; Liu, Z.; Li, P.; Zhang, C.; Liu, R.; Huang, Y. Ultrathin Core-sheath Fibers for Liposome Stabilization. *Colloids Surf. B Biointerfaces* **2014b,** *122,* 630–637.
36. Manosroi, A.; Podjanasoonthon, K.; Manosroi, J. Development of Novel Topical Tranexamic Acid Liposome Formulations. *Int. J. Pharm.* **2002,** *235* (1–2), 61–70.
37. Marto, J.; Vitor, C.; Guerreiro, A.; Severino, C.; Eleutério, C.; Ascenso, A.; Simões, S. Ethosomes for Enhanced Skin Delivery of Griseofulvin. *Colloids Surf. B Biointerfaces* **2016,** *146,* 616–623.
38. Maurya, S. D.; Prajapati, S.; Gupta, A. K.; Saxena, G. K.; Dhakar, R. K. Formulation Development and Evaluation of Ethosome of Stavudine. *Indian J. Pharm. Educ. Res.* **2010,** *44* (1), 102–108.
39. McLafferty, E.; Hendry, C.; Alistair, F. The Integumentary System: Anatomy, Physiology and Function of Skin. *Nurs. Stand.* **2012,** *27* (3), 35–42.
40. Murthy, R. S. R.; Subramanian, N.; Yajnik, A. Artificial Neural Network as an Alternative to Multiple Regression Analysis in Optimizing Formulation Parameters of Cytarabine Liposomes. *AAPS Pharm. Sci. Tech.* **2004,** *5* (1), 1–9.
41. Ogiso, T.; Yamaguchi, T.; Iwaki, M.; Tanino, T.; Miyake, Y. Effect of Positively and Negatively Charged Liposomes on Skin Permeation of Drugs. *J. Drug Target.* **2001,** *9* (1), 49–59.
42. Panwar, P.; Pandey, B.; Lakhera, P. C.; Singh, K. P. Preparation, Characterization, and In Vitro Release Study of Albendazole-encapsulated Nanosize Liposomes. *Int. J. Nanomedicine* **2010,** *5,* 101–108.
43. Perumal, O.; Murthy, S. N.; Kalia, Y. N. Turning Theory into Practice: The Development of Modern Transdermal Drug Delivery Systems and Future Trends. *Skin Pharmacol. Physiol.* **2013,** *26* (4–6), 331–342.
44. Planz, V.; Lehr, C. M.; Windbergs, M. In Vitro Models for Evaluating Safety and Efficacy of Novel Technologies for Skin Drug Delivery. *J. Control Release* **2016,** *42,* 89–104.
45. Prausnitz, M. R.; Elias, P. M.; Franz, T. J.; Schmuth, M.; Tsai, J. C.; Menon, G. K.; Holleran, W. M.; Feingold, K. R. *Skin Barrier and Transdermal Drug Delivery, Dermatology,* 3rd ed.; Bolognia, J., Jorizzo, J., Schaffer, J., Eds.; Elsevier Health Sciences: Amsterdam, the Netherlands, 2012; Vol. 19, pp 2065–2073.
46. Prausnitz, M. R.; Langer, R. Transdermal Drug Delivery. *Nat. Biotechnol.* **2008,** *26* (11), 1261–1268.
47. Redlarski, G.; Palkowski, A.; Krawczuk, M. *Body Surface Area Formulae: An Alarming Ambiguity;* Scientific Reports, 2016; 6, 27966.
48. Sivasubramanian, G.; Frempong, M. E.; MacArthur, R. D. Abacavir/Lamivudine Combination in the Treatment of HIV: A Review. *Ther. Clin. Risk Manag.* **2010,** *6* (1), 83–94.

49. Sudhakar, C. K.; Charyulu, R. N.; Jain, S. Ethosomes: A New Corridor for Transdermal Drug Delivery System for Hydrophobic and Hydrophilic Moieties. PP-07. AJP Abstract Book-Special Online Issue. *Asian J. Pharm.* **2015,** *9,* S11–25.
50. Sudhakar, C. K.; Charyulu, R. N. Influence of Permeation Enhancer on Ethosomes Bearing Lamivudine for Transdermal Drug Delivery. *Res. J. Recent. Sci.* **2014,** *3* (IVC–2014), 155–160.
51. Sudhakar, C. K.; Jain, S.; Charyulu, R. N. A Comparison Study of Liposomes, Transfersomes and Ethosomes Bearing Lamivudine. *Int. J. Pharm. Sci. Res.* **2016,** *7* (10), 4214–4221.
52. Sudhakar, C. K.; Upadhyay, N.; Jain, S.; Charyulu, R. N. Ethosomes as Non-invasive Loom for Transdermal Drug Delivery. In *Nanomedicine and Drug Delivery;* Sebastian, M., Ninan, N., Haghi, A. K., Eds.; Apple Academic Press: San Diego, CL, 2012; Vol. 1, pp 1–16.
53. Suresh, S. N. V.; Veerreddy, K. Formulation and In-Vitro Evaluation of Controlled Release Floating Tablets of Lamivudine. *Int. J. Pharm. Sci. Res.* **2014,** *5* (3), 900–906.
54. Tiwary, A. K.; Sapra, B.; Jain, S. Innovations in Transdermal Drug Delivery: Formulations and Techniques. *Recent Pat. Drug Deliv. Formul.* **2007,** *1* (1), 23–36.
55. Touitou, E.; Dayan, N.; Bergelson, L.; Godin, B.; Eliaz, M. Ethosomes-novel Vesicular Carrier for Enhanced Delivery Characterization and Skin Penetration Properties. *J. Cont. Release* **2000,** *65,* 403–418.
56. Trevaskis, N. L.; Kaminskas, L. M.; Porter, C. J. From Sewer to Saviour-targeting the Lymphatic System to Promote Drug Exposure and Activity. *Nat. Rev. Drug Discov.* **2015,** *14* (11), 781–803.
57. Vincent, N.; Ramya, D. D.; Vedha, H. B. Progress in Psoriasis Therapy via Novel Drug Delivery Systems. *Dermatol. Rep.* **2014,** *6* (1), 5451.
58. Virginie, S. The "Big" Revolution: Nanotechnology to Nanomedicine. https://polaris.mysciencework.com/news/show/the-big-revolution-nanotechnology-to-nanomedicine-3 (accessed on April 15, 2016).
59. Uchechi, O.; Ogbonna, J. D. N.; Attama, A. A. Nanoparticles for Dermal and Transdermal Drug Delivery. In *Applications of Nanotechnology in Drug Delivery*; Sezer, A. D., Ed.; InTech, 2014; 193–235.
60. Yoo, D. et al. Theranostic Magnetic Nanoparticles. *Acc. Chem. Res.* **2011**, *44* (10), 863–74.

INDEX

D

E

R

S